Singlet Oxygen

Singlet Oxygen

Reactions with Organic Compounds and Polymers

Edited by

B. Rånby and J. F. Rabek

Department of Polymer Technology
The Royal Institute of Technology
Stockholm, Sweden

A Wiley–Interscience Publication

JOHN WILEY & SONS

Chichester · New York · Brisbane · Toronto

Library of Congress Cataloging in Publication Data:

Main entry under title:

Singlet oxygen reactions with organic compounds and polymers.

Based on papers presented at the EUCHEM Conference on Singlet Oxygen Reactions with Polymers held at Södergarn, Sweden, Sept. 2–4, 1976.

'A Wiley–Interscience publication.'

1. Oxidation—Congresses. 2. Polymers and polymerization—Congresses. 3. Organic compounds—Congresses. I. Rånby, Bengt, G. II. Rabek, J. F. III. Euchem Conference on Singlet Oxygen Reactions with Polymers, Södergarn, 1976.

QD281.09S53 547'.23 77-2793

ISBN O 471 99535 5

Made and printed in Great Britain by
William Clowes & Sons Limited
London, Beccles and Colchester

Contents

Preface ix

1. The History of Singlet Oxygen—An Introduction . *B. Rånby* 1

2. The Nature of Singlet Oxygen . . . . . *E. A. Ogryzlo* 4

3. Lifetime of the $^1\Sigma_g^+$ Oxygen State Produced in a Corona Discharge and other Measurements on $^1\Sigma_g^+$ and $^1\Delta_g$ States in an Electron-beam Controlled Discharge
F. Bastien, M. Lecuiller, G. Fournier, D. Pigache, and D. Proust 12

4. Physical Quenching of Singlet Oxygen . . . *E. A. Ogryzlo* 17

5. Physical Properties of Singlet Oxygen in Fluid Solvents
F. Wilkinson 27

6. The Mechanism of Quenching of Singlet Oxygen
R. H. Young and D. R. Brewer 36

7. The Use of β-Carotene as a Probe for Singlet Oxygen Yields and Lifetime . . . . . . *A. Garner and F. Wilkinson* 48

8. Kinetics of Singlet Oxygen Peroxidation . . . . *B. Stevens* 54

9. Quenchers of Singlet Oxygen—A Critical Review . *D. Belluš* 61

10. Mechanism and Kinetics of Chemical Reactions of Singlet Oxygen with Organic Compounds . . . . . *K. Gollnick* 111

11. Mechanisms of Photo-oxidation . . . . *C. S. Foote* 135

12. Photo-stimulated Oxidation and Peroxidation in Crystalline Anthracene and some Related Derivatives
S. E. Morsi and J. O. Williams 147

13. Oxidation Reactions of Substrates Containing Vinyl Halide Structures
K. Griesbaum, H. Keul, M. P. Hayes, and M. Haji Javad 159

14. **Singlet Oxygen Reactions with Polyisoprene Model Molecules 4-Methyl-4-Octene and 4,8-Dimethyl-4,8-Dodecadiene**
J. Chaineaux and C. Tanielian 164

15. **Reactions of Ketenes with Molecular Oxygen. Formation of Peroxylacetones and a Polymeric Peroxide. Catalysed Thermal Generation of Singlet Oxygen** . *N. J. Turro, Ming-Fea Chow, and Y. Ito* 174

16. **Formation of Dioxetanes and Related Species**
C. W. Jefford, A. F. Boschung, and C. G. Rimbault 182

17. **Reaction of Singlet Oxygen in Solids with Electron-rich Aromatics. Peroxidic Intermediates in Indole Oxidation**
I. Saito and T. Matsuura 186

18. **Reaction of Singlet Oxygen with α-Ketocarboxylic Acids of Biological Interest** . . *C. W. Jefford, A. F. Boschung, and P. A. Cadby* 192

19. **Mechanism of the Dye-sensitized Photodynamic Inactivation of Lysozyme** . *P. Rosenkranz, Ahmed Al-Ibrahim, and H. Schmidt* 195

20. **Chemical Behaviour of Singlet Oxygen Produced by Energy Transfer from Dye Adsorbed on a Micelle** . . . . *Y. Usui* 203

21. **Singlet Oxygen Reactions with Synthetic Polymers**
B. Rånby and J. F. Rabek 211

22. **Theoretical Investigations on the Degradation of Polymers by Singlet Oxygen** . . . . . *J. J. Lindberg and P. Pyykkö* 224

23. **The Role of Singlet Oxygen in the Photo-oxidation of Polymers—Some Practical Considerations** . . . . . *G. Scott* 230

24. **The Degradation of Anthracene-doped Polymers by Singlet Oxygen**
M. Kryszewski and B. Nadolski 244

25. **The Use of β-Carotene for the Evaluation of the Role of Singlet Oxygen in Longwave UV Photo-oxidation of Polystyrene**
M. Nowakowska 254

26. **Photosensitized Singlet Oxygen Oxidation of Polydienes**
J. F. Rabek, Y. J. Shur, and B. Rånby 264

27. **Photodegradation of Polyisoprene Catalysed by Singlet Oxygen**
H. C. Ng and J. E. Guillet 278

28. Improving the Self-adhesion of EPDM by Means of a Singlet Oxygen Reaction
E. Th. M. Wolters, C. A. van Gunst, and H. J. G. Paulen 282

29. Photosensitized Singlet Oxygen Oxidation of Polysulphides
H. S. Laver and J. R. MacCallum 288

30. Singlet Oxygen and Chemiluminescence in the Oxidation of Polymers Resembling Humic Acids . . *S. Sławińska and T. Michalska* 294

31. Singlet Oxygen Oxidation of Lignin Structures
G. Gellerstedt, K. Kringstad, and E. L. Lindfors 302

32. The Use of Polymer-based Singlet Oxygen Sensitizers in the Study of Light-induced Yellowing of Lignin
G. Brunow, I. Forsskåhl, A. C. Grönlund, G. Lindström, and K. Nyberg 311

33. Deactivation of Singlet Oxygen by Polyolefin Stabilizers
H. Furue and K. E. Russell 316

34. Stabilization of Polymers Against Singlet Oxygen *D. M. Wiles* 320

Index 329

Preface

Excited forms of molecular oxygen of high chemical activity were discovered in Germany in the 1930s and classified spectroscopically in Canada as two singlet forms. The remarkable properties of singlet oxygen in organic reactions were re-discovered in the USA in 1964. Since then, we have witnessed a rapid increase in activity in studies of singlet oxygen. Organic chemists have developed singlet oxygen as a specific oxidative reagent. Singlet oxygen has been interpreted to be an active photochemical intermediate in polluted air. More recently, efforts have been made to relate photo-oxidative reactions of polymers to singlet oxygen.

This book describes the structure and properties of singlet oxygen and its reactions with organic compounds and polymers. This material is the basis for important practical and technical considerations of the role of singlet oxygen in environmental pollution and health hazards. The chapters of the book are based on papers presented at the EUCHEM Conference on Singlet Oxygen Reactions with Polymers held at Södergarn on Lidingö, Stockholm, Sweden, on September 2–4, 1976. Several chapters describe the formation of singlet oxygen, the kinetics of physical deactivation, and the kinetics and mechanisms of chemical oxidation reactions. Approximately half of the chapters deal with singlet oxygen reactions with synthetic and natural polymers and the problems of stabilization of polymers against singlet oxygen oxidation.

It is our pleasure to thank the participants of the EUCHEM Conference for their excellent contributions and cooperation in the preparation of this volume. We also acknowledge financial support for the conference from the Swedish Government through the Department of Education and a special grant from Kjell och Märta Beijers Stiftelse.

Stockholm, March, 1977 BENGT RÅNBY AND JAN F. RABEK

1

The History of Singlet Oxygen—An Introduction

B. Rånby

Department of Polymer Technology, The Royal Institute of Technology, Stockholm, Sweden

The discovery of oxygen by Scheele and Priestly independently about 200 years ago was a turning-point in the development of chemical concepts (e.g. ref. 1). Until then the theory of phlogiston (with negative weight) was accepted and used for the interpretation of oxidation of metals and burning of organic matter. The crucial role of oxygen in life processes became well established during the 1800s. Oxygen had unique properties in various reactions. Already Faraday had discovered that oxygen was paramagnetic and in this way differed from other known permanent gases. This property, based on a proposed electron structure of the diatomic oxygen molecule, was interpreted by Mulliken[2] in 1928 as being due to its content of two outer electrons with parallel spins. This uncoupled electron pair classified oxygen as a triplet in its lowest energy state ($^3\Sigma_g^-$). A few years later, two higher energy states of oxygen were found spectroscopically by Childe and Mecke[3] (the $^1\Sigma_g^+$ in 1931) and by Herzberg[4] (the $^1\Delta_g$ in 1934). These two forms of the oxygen molecule had an excess energy of a fraction of a chemical bond (37·5 and 22·5 kcal mol^{-1}, respectively). They were classified as spectroscopic singlets, and this classification was interpreted as being the result of their outer electron pair having antiparallel spins.

These physical discoveries of oxygen in triplet and singlet forms attracted the attention of a chemist, Dr. Hans Kautsky, a 'Docent' (Assistant Professor) at Heidelberg University in Germany. He was studying the photoluminescence of organic dyes in solution and adsorbed, e.g. chlorophyll, trypaflavin, malachite green, and porphyrins. When illuminated, the dyes absorbed quanta and were excited to higher energy states. They could return to the original low-energy state rapidly (fluorescence) or slowly (phosphorescence) by emission of quanta. Kautsky found that the presence of oxygen could largely extinguish the fluorescence of some irradiated dyes, e.g. the red fluorescence from chlorophyll.[5] He made the correct interpretation that the excess energy of the excited dye was transferred to oxygen molecules. The energy required for the excitation of oxygen was relatively small as even the red fluorescence from chlorophyll was sufficient. Kautsky showed further that the excited oxygen molecules (he called them 'active' oxygen) could oxidize dyes which did not react with molecular oxygen in its normal energy state.[6] Trypaflavine (I) and leuco-malachite green (II) were adsorbed on silica gel. Illumination excited I but not II. With oxygen present with I and II mixed under illumination, the leuco-form II was oxidized to a green colour. Kautsky concluded correctly that the excitation energy of the illuminated dye I was transferred to oxygen and produced 'metastable active oxygen molecules', possibly in the $^1\Sigma$ state, which then oxidized the dye II. In other experiments, Kautsky[7] demonstrated that the 'active oxygen' was deactivated ('quenched') by collision with other oxygen molecules and that the lifetime of 'active oxygen' was longer (10 min or more) with decreasing oxygen pressure (from 0·4 to 0·0004 mm).

Kautsky's results were extensive and his conclusions basically correct from what is known today. However, Kautsky's views were not accepted at the time and his work remained unnoticed for about 25 years. The reason was that the excited singlet oxygen, if it existed, under practical conditions (atmospheric pressure, room temperature) was considered to be a very short-lived intermediate. Instead, the photochemical reactions were interpreted as being due to atomic oxygen and ozone and this alternative was widely accepted.[8] There was some discussion of singlet oxygen in air pollution as a possible reactive intermediate in about 1960.[9] The re-discovery of singlet oxygen was made in 1964 by photo-oxidation experiments by Foote and Wexler[10] and Corey and Taylor.[11] The photoinduced addition of oxygen to double bonds with an allyl hydrogen atom was well interpreted as a singlet oxygen reaction. Since then several groups have been working on the mechanism of singlet oxygen reactions in organic synthesis.

Studies also continued on the photochemical reactions in the atmosphere. Singlet oxygen was considered as a possible intermediate in 'photochemical smog' in 1966 but the mechanism of formation was unknown.[12] Direct absorption of light quanta by oxygen molecules ('a spectroscopically forbidden transition') could occur at atmospheric pressure due to the perturbed state of $^3O_2$. The evidence for $^1O_2$ as an intermediate was obtained by Pitts *et al.*[13] in 1969 in studies of the conversion of NO into NO_2 both in the atmosphere and under laboratory conditions. Among the reaction mechanisms possible, the best established involve olefins, ketonic radicals and ketoperoxides, and probably also ozone with singlet oxygen ($^1\Delta_g$ or $^1\Sigma_g^+$) as critical intermediates.

Singlet oxygen reactions with polymers have been studied in the last few years.[14] Such mechanisms were first proposed by Trozzolo and Winslow[15] in 1968 without much experimental evidence. They studied the photo-oxidation of commercial polyethylene containing carbonyl groups as sensitizers. Exposure of polyethylene surfaces to a stream of singlet oxygen was later shown to result in the formation of hydroperoxide groups.[16] More extensive evidence for singlet oxygen reactions was obtained for polydienes (*cis*- and *trans*-1,4- and -1,2-vinyl).[17,18] It has also been proposed by Rabek and Rånby[19] that singlet oxygen attacks phenyl groups in polystyrene by a ring-opening reaction, giving dialdehyde side-groups containing two conjugated double bonds. Of particular interest are the initiation reactions with singlet oxygen, formed by energy transfer from excited trace amounts of impurities in polymer samples. Such sensitizers may be deposited materials from polluted air or water or extraneous groups from the production or processing of the polymeric material.[20]

There have been recent attempts to use singlet oxygen from a chemical process or a physical generator as a reagent for specific oxidation of chemical compounds, as a bleaching agent for paper and textile fibres and for surface treatment of plastics and rubber.[21]

REFERENCES

1. J. R. Partington, *A History of Chemistry*, Vol. 1, 1970, Vols. 2–4, Macmillan, London, 1961–64.
2. R. S. Mulliken, *Nature, Lond.*, **122**, 505 (1928).
3. W. H. J. Childe and R. Mecke, *Z. Physik*, **68**, 344 (1931).
4. G. Herzberg, *Nature, Lond.*, **133**, 759 (1934).
5. H. Kautsky and H. de Bruijn, *Naturwissenschaften*, **19**, 1043 (1931).
6. H. Kautsky, H. de Bruijn, R. Neuwirth, and W. Baumeister, *Chem. Ber.*, **66B**, 1588 (1933).
7. H. Kautsky, *Trans. Faraday Soc.*, **35**, 216 (1939).
8. G. O. Schenk, *Naturwissenschaften*, **35**, 28 (1948).
9. P. A. Leighton, *Photochemistry of Air Pollution*, Academic Press, New York, 1961.
10. C. S. Foote and S. Wexler, *J. Amer. Chem. Soc.*, **86**, 3879 (1964).
11. E. J. Corey and W. C. Taylor, *J. Amer. Chem. Soc.*, **86**, 3881 (1964).
12. A. M. Winer and K. O. Bayes, *J. Phys. Chem.*, **70**, 302 (1966); R. H. Kummelen, M. H. Bortner, and T. Baurer, *Environ. Sci. Technol.*, **3**, 248 (1969).
13. J. N. Pitts, Jr., A. U. Khan, E. B. Smith, and R. P. Wayne, *Environ. Sci. Technol.*, **3**, 241 (1969).
14. B. Rånby and J. F. Rabek, *Photodegradation, Photo-oxidation and Photostabilization of Polymers*, Wiley, London, 1975, Ch. 5.
15. A. M. Trozzolo and F. H. Winslow, *Macromolecules*, **1**, 98 (1968).
16. M. L. Kaplan and P. G. Kelleher, *J. Polym. Sci. B*, **9**, 565 (1971).
17. M. L. Kaplan and P. G. Kelleher, *J. Polym. Sci. A1*, **8**, 3163 (1970); *Rubb. Chem. Technol.*, **45**, 423 (1972).
18. J. F. Rabek and B. Rånby, *J. Polym. Sci. A1*, **14**, 1463 (1976).
19. J. F. Rabek and B. Rånby, *J. Polym. Sci. A1*, **12**, 273 (1974).
20. B. Rånby and J. F. Rabek, *Amer. Chem. Soc. Symp. Ser.*, No. 25, 391 (1976).
21. Chapters 10, 11, 15–18, 31, 32 of this volume.

2

The Nature of Singlet Oxygen

E. A. OGRYZLO

Department of Chemistry, University of British Columbia, Vancouver, Canada

Although the oxygen atom accounts for less than 0·1% of the atoms in the universe, the forces which formed the earth some 5×10^9 years ago have made oxygen the most abundant element in the earth's crust, and it accounts for more than half the atoms now present. However, it is clear from the geological record that molecular oxygen was not an important species in the early planetary atmosphere.[1] On the contrary, the small amounts formed by the direct photolysis of inorganic oxides probably acted as a poison, destroying amino acids formed in the planets' 'Haldane soup'.[2]

The appearance of significant amounts of molecular oxygen in the atmosphere had to await the development of photosynthetic organisms capable of driving the process

$$m\mathrm{CO_2} + n\mathrm{H_2O} \xrightarrow{h\nu} \mathrm{C}_m(\mathrm{H_2O})_n + m\mathrm{O_2}$$

This appears to have occurred in the sheltered depths of warm pools of water, when the earth was about half its present age.[3] It is likely that the process was autocatalytic as protection from lethal ultraviolet radiation afforded by O_2 and O_3 undoubtedly led to the rapid spread of life out of the deep waters in which they originated. Exposure to oxygen at these higher levels would have caused organisms which based their metabolism on anaerobic photoreduction and fermentation to change to photosynthesis and respiration.[3]

The explosion of life forms associated with an increase of oxygen to something approaching 1% of our present level has been identified with the Cambrian period some 600 000 000 years ago.[3] A build-up of the O_2 to about 10% of the present concentration together with the associated O_2 layer would have reduced the UV radiation to non-lethal levels outside water, and have made possible the movement of life forms to dry land. Whether the O_3 level simply rose to something like the present level, or overshot, causing an ice-age or two, is still open to debate. However, it is evident that present levels of O_2 are the result of a dynamic equilibrium maintained principally by the autotropic and heterotropic organisms. The autotropes consume carbon dioxide and *produce* O_2 in the photosynthetic process while the heterotropes *consume* O_2 (and organic matter) in respiratory processes. The balance between these organisms keeps the molar fraction of oxygen in the atmosphere at a remarkably constant value of 0·2091. It has been estimated that all of the oxygen in the atmosphere passes through this photosynthetic cycle in 2000 years.[3]

Thus, molecular oxygen currently permeates the entire chemosphere and biosphere, playing a central role in the process by which radiation from the sun is used to sustain life on the surface of the earth.

Credit for the discovery of oxygen is usually given to both Scheele and Priestley, who independently isolated the species in about 1772, although Priestley still believed he had 'dephlogisticated air' as late as 1800. In 1811, Avogadro recognized that oxygen is a diatomic molecule and in 1848 Faraday demonstrated that it is unique among diatomic molecules with an even number of electrons: it is paramagnetic in its ground state. However, a satisfactory explanation of this paramagnetism was presented only in 1928 by Mulliken[4] in what is considered a major triumph of molecular orbital theory. We will give the details of this description in a later section. For our present purposes it is sufficient to note that although O_2 has a 'double bond', its outermost pair of electrons are in different orbitals and have their spins parallel, making the ground state a 'triplet' ($^3\Sigma_g^-$). Furthermore, two other arrangements of these two electrons are possible and should result in the presence of two low-lying singlet states ($^1\Delta_g$ and $^1\Sigma_g^+$). Three years after Mulliken's classic paper, Childe and Mecke[5] were able to verify these predictions with the spectroscopic observation of the $^1\Sigma_g^+$ state 37.51 kcal above the ground state, and in 1934 Herzberg reported the observation of the $^1\Delta_g$ state.[6] It is this singlet delta state ($O_2\ ^1\Delta_g$) which has come to be called singlet oxygen. It lies only 22·54 kcal above the ground state and has a remarkable lifetime of 1 h in the absence of collisions with other molecules.

The role of singlet oxygen in a variety of photo-oxygenation processes has been demonstrated by many workers in the last 14 years but the literature reveals that in the period between 1931 and 1964 singlet oxygen had only one champion: Hans Kautsky,[7–9] who in 1931 correctly proposed that this species is responsible for the photo-oxygenation of a variety of unsaturated molecules. With hindsight, it seems remarkable that Kautsky's beautiful experiments and careful reasoning did not convince his contemporaries. Although his papers continued to point to the importance of singlet oxygen throughout the 1930s, a moloxide mechanism for photo-oxygenation proposed in 1935 by Schönberg[10] appeared more attractive to chemists. This alternative theory received the support of workers who dominated the field[11] and hence singlet oxygen was considered unimportant in the oxygenation of organic compounds until 1964 when the experiments of both Foote and Wexler[12] and Corey and Taylor[13] provided new evidence for a singlet oxygen mechanism in several photo-oxygenations. Since then a great deal has been learned about the properties of the molecule and it is now becoming possible to assess its importance in a number of more complex systems such as polymer degradation.

The most universal and probably the more important mechanism by which singlet oxygen is formed in the earth's biosphere is the following. The absorption of visible and ultraviolet light (from the sun) usually leads to the electronic excitation of the molecules. Since oxygen permeates most organic matter, O_2 is a common quencher of this electronic excitation. In this process (especially when the excited state is a triplet), singlet oxygen is formed. Whether the species plays a significant role in such a system is determined by the rate of reaction with the molecules present relative to the rate of relaxation to the ground state. To understand these two processes we must consider the electronic structure of singlet oxygen and how it is affected by interactions with other molecules.

THE QUANTUM MECHANICAL DESCRIPTION OF SINGLET OXYGEN

The potential energy curve for singlet oxygen $O_2(^1\Delta_g)$ is given by the solid curve in Figure 1 together with the curves for the other two double-bonded states which arise from the LCAO–MO configuration:

$$\mathrm{KK}(2\sigma_g)^2(2\sigma_u)^2(3\sigma_g)^2(1\pi_u)^4(1\pi_g)^2$$

Because the σ bond in O_2 remains intact in the reactions which interest us, we can ignore all electrons except the six which are found in π orbitals. Four of these electrons fill the degenerate bonding π_x and π_y orbitals and two electrons half-fill the degenerate antibonding π^*_x and π^*_y orbitals. It is the arrangement of these two electrons in the antibonding orbitals that determine which of the three low-lying states shown in Figure 1 we have. The ground state is well represented by the following conventional diagram:

$$\overset{\uparrow}{\pi^*_x} \quad \overset{\uparrow}{\pi^*_y} \quad \cdots \quad {}^3\Sigma_g^-$$

since the two electrons must be in different orbitals with their spins parallel to yield a $^3\Sigma_g^-$ state.

The singlet oxygen state is more difficult to represent in this manner.[14] It is two-fold degenerate. One component, $^1\Delta_g^-$, has the two electrons in different orbitals and can therefore be represented by the diagram

$$\overset{\uparrow}{\pi^*_x} \quad \overset{\uparrow}{\pi^*_y} \quad \cdots \quad {}^1\Delta_g^-$$

However, this cannot be the component responsible for the interesting reactivity of singlet oxygen. The other component, Δ_g^+, can be represented by

$$\overset{\uparrow\downarrow}{\pi^*_x} \quad \pi^*_y \quad \cdots \quad {}^1\Delta_g^+$$

since the wave function indicates that both electrons are in the same orbital.[14] Inspection of the orbital wave function for the $^1\Sigma_g^+$ state indicates that it also has the electrons paired. Therefore, we can represent it with the diagram

$$\pi^*_x \quad \overset{\uparrow\downarrow}{\pi^*_y} \quad \cdots \quad {}^1\Sigma_g^+$$

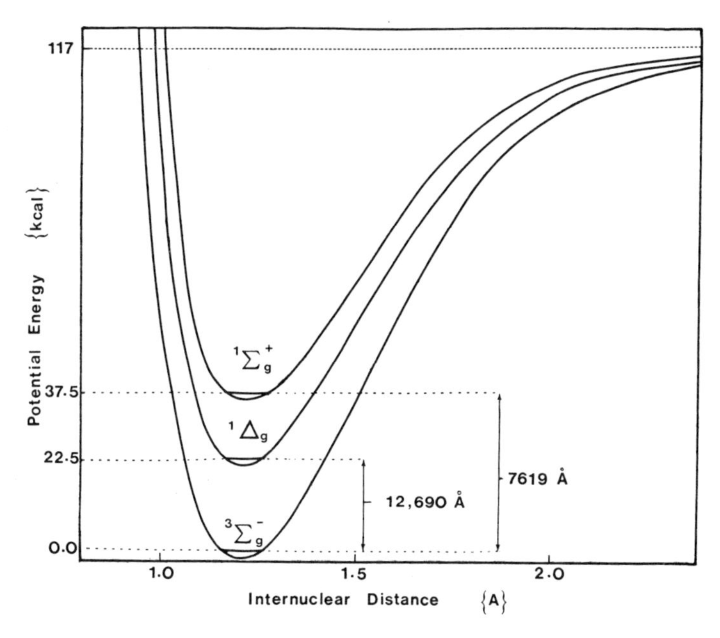

Figure 1 Potential energy curves for the three low-lying electronic states of molecular oxygen

The above representations of the $^1\Delta_g^+$ and $^1\Sigma_g^+$ states are valid only if the x-axis is chosen differently for electrophilic and nucleophilic reactants. This is necessary because any interaction stabilizes the $^1\Delta_g^+$ component of the $^1\Delta_g$ state and destabilizes the $^1\Sigma_g^+$ state relative to the $^1\Delta_g^-$ and $^3\Sigma_g^+$ states. Consequently, when an electrophilic species approaches $O_2(^1\Delta_g)$ it 'points' its filled π_x^* orbital at the species, giving rise to a stabilizing interaction, whereas $O_2(^1\Sigma_g^+)$ directs its empty orbital at the electrophile, producing a repulsive interaction.

It should be emphasized that the above description is a chemical interpretation of a general quantum mechanical result in perturbation theory calculations which is often described as a 'repulsion' between states which are mixed by a perturbation. It therefore does not take into account other possible interactions that are capable of shifting the positions of the ground state and the other component of the singlet delta state which have wave functions characteristic of free radicals. Nevertheless, there is little doubt that when singlet oxygen shows a high reactivity relative to the ground state it is the $^1\Delta_g^+$ component which is responsible. On the other hand, we would not expect $O_2(^1\Sigma_g^+)$ to react as a free radical or undergo processes similar to $O_2(^1\Delta_g)$. The fact that $O_2(^1\Sigma_g^+)$ is readily relaxed to $O_2(^1\Delta_g)$ could, however, lead to an observed reactivity when the $^1\Sigma_g^+$ state is formed as the first species in any system.

THE LIFETIME OF SINGLET OXYGEN

Isolated molecules of oxygen in the $^1\Delta_g$ state spontaneously undergo a transition to the ground state, principally through the transition:

$$O_2(^1\Delta_g) \rightarrow O_2(^3\Sigma_g^-) + h\nu(12\,700\text{ Å})$$

for which $k = 0{\cdot}935\ \text{h}^{-1}$ or the half-life is 45 min.[15] This is a magnetic dipole transition. Collisions with other molecules can shorten this lifetime in two ways: (a) they can induce an electric-dipole transition at the same wavelength; (b) they can induce a radiationless transition to the ground state. For example, in the presence of 1 atm of O_2 the radiative half-life becomes 10 min[15] and the non-radiative half-life becomes 14 ms. This makes the dark process almost 10^4 times faster than the radiative process. It is clear that radiative decay will not determine the actual lifetime of singlet oxygen below the stratosphere. However, it is worth noting that this shortening of the radiative lifetime in collisions means that a given 12 700 Å emission intensity from solution is indicative of a much smaller singlet oxygen concentration than the same intensity from the gas phase.

The ability of molecules to quench singlet oxygen in the gas phase and in solution is the subject of several papers in this volume, and exact rate constants are available for many quenchers. From these values we learn that in the gas phase at 1 atm the lifetime of singlet oxygen can vary between 1 and 10^{-5} s, depending on the nature of the gas. In solution the lifetime varies between 1 ms (in Freon 11) and 2 μs (in water) at room temperature. The mechanism of the relaxation in both phases is reasonably well understood and can be used to predict lifetimes in new media.[16,17].

COOPERATIVE TRANSITIONS AND ENERGY POOLING

Spectroscopic studies of singlet oxygen generated both chemically and electrically have revealed some novel radiative and energy disproportionating processes.[18–21]. It was found that pairs of singlet oxygen molecules could combine their electronic excitation to bring about processes which require the energy from up to four molecules of $O_2(^1\Delta_g)$. The most important of these are described by the following equations:

$$2O_2(^1\Delta_g)\begin{cases} \xrightarrow{k_1} 2O_2(^3\Sigma_g^-) + h\nu & (1) \\ \xrightarrow{k_2} O_2(^3\Sigma_g^-) + O_2(^1\Sigma_g^+) & (2) \\ + A \rightarrow 2O_2(^3\Sigma_g^-) + A^* & (3) \end{cases}$$

The first process gives rise to the so-called 'dimol' emission at 6340 Å and 7030 Å which is proportional to the square of the $O_2(^1\Delta_g)$ concentration, $k_1 = 0{\cdot}28\ \text{l mol}^{-1}\ \text{s}^{-1}$. Emission from this cooperative transition provided the first direct evidence for chemically generated singlet oxygen in the chlorine–peroxide reaction.[18] The same transition has also been used to excite singlet oxygen, providing direct evidence for some singlet oxygen reactions.[22] The second process [equation (2)] can be considered an energy disproportionation and is possible because the energy of two singlet delta molecules (45 kcal) is sufficient to excite the singlet sigma state (37·5 kcal), $k_2 = 1{\cdot}3 \times 10^3$ l mol^{-1} s^{-1}. One consequence of this process is the presence of small amounts of the higher singlet state when the lower (delta) state is formed.[19] The third process [equation (3)] was proposed to account for the luminescence from aromatic molecules excited by singlet oxygen.[20,23]. The detailed mechanism by which such co-operative excitation occurs is still controversial.

STATE CORRELATION DIAGRAMS

The two principal reactions characteristic of singlet oxygen are cycloaddition and the 'ene' reaction [reactions (4) and (5), respectively]:

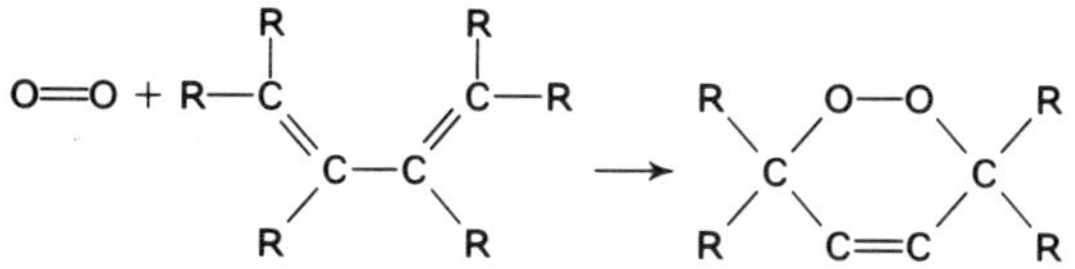

(4)

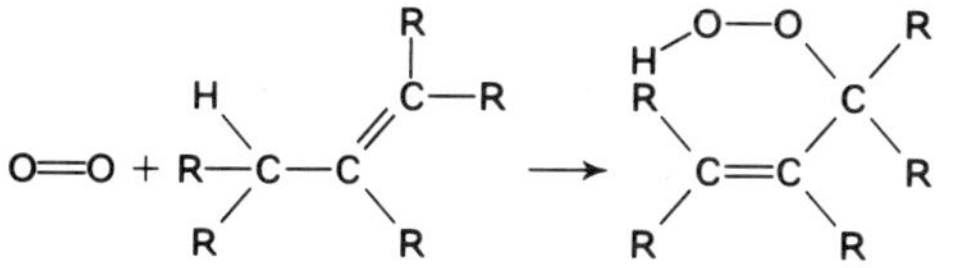

(5)

The formation of endoperoxides shown in reaction (4) requires a conjugated double bond. The formation of hydroperoxides as shown in reaction (5) requires an allylic hydrogen. When the double bonds are surrounded by electron-donating groups, the activation energy for both reactions can be reduced to nearly zero and the rate constant then approaches its maximum value of about 10^8 l mol^{-1} s^{-1}.[24] The formation of dioxetanes in singlet oxygen reactions appears to have a much higher activation energy and is observed chiefly when other paths are blocked.[25,26] This is consistent with the predictions of the Woodward–Hoffman selection rules for concerted cycloaddition reactions.[27] Such orbital symmetry considerations and more quantitative calculations[27] indicate that the $^1\Delta_g^+$ component is the only one which correlates directly with the ground-state products in reaction (4). These correlations are illustrated by the potential energy curves shown in Figure 2. It is likely that the curves also apply to the 'ene' reaction since there is a topological similarity between the two processes. They show similar rate constants, temperature dependences and variations with substituents around the double bond(s). It has also been suggested that peroxirane intermediate may be important in the 'ene' reaction.[27] Although such an intermediate has not been identified, some MINDO/3 calculations by Dewar and Thiel support this possibility.[28] Potential energy curves based on these calculations are presented in Figure 3 and remain

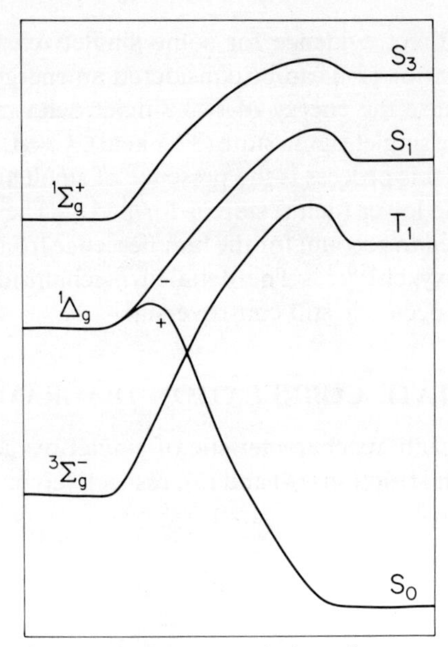

Figure 2 State correlation diagrams for the reactions of the three low-lying states of molecular oxygen with a diene to produce endoperoxides in triplet (T) and singlet (S) states

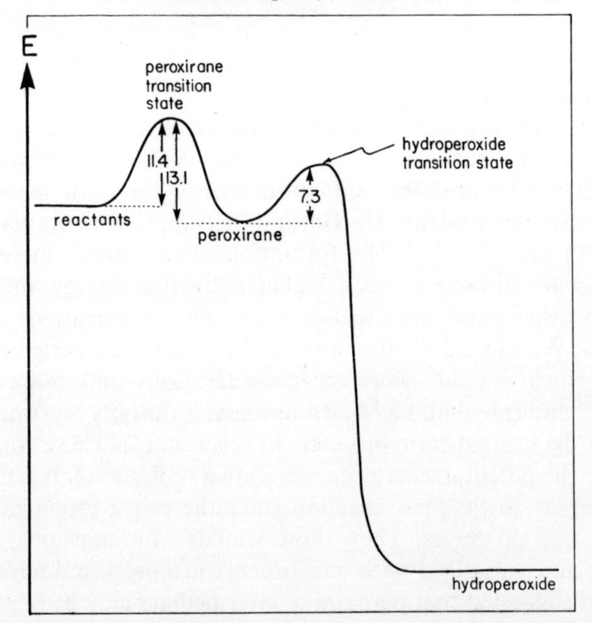

Figure 3 Calculated [28] potential energy curve for the reaction of singlet oxygen with propylene to form a hydroperoxide

an alternative path for the formation of hydroperoxides. The curves shown were calculated for the reaction with propylene and the procedure is not sensitive to changing substituents near the double bond.[28] More recent *Ab Initio* calculations question the importance of the peroxirane intermediate.[28]

REFERENCES

1. H. C. Urey, *The Planets: Their Origin and Development*, Yale University Press, New Haven, 1952.
2. P. H. Abelson, *Ann. N.Y. Acad. Sci.*, **69**, 276 (1957).
3. L. V. Berkner and L. C. Marshall, *Discuss. Faraday Soc.*, **37**, 122 (1964).
4. R. S. Mulliken, *Nature, Lond.*, **122**, 505 (1928).
5. W. H. J. Childe and R. Mecke, *Z. Physik*, **68**, 344 (1931).
6. G. Herzberg, *Nature, Lond.*, **133**, 759 (1934).
7. H. Kautsky and H. de Bruijn, *Naturwissenschaften*, **19**, 1043 (1931).
8. H. Kautsky, *Chem. Ber.*, **66**, 1588 (1933).
9. H. Kautsky, *Trans. Faraday Soc.*, **35**, 216 (1939).
10. A. Schönberg, *Justus Liebgis Ann. Chem.*, **518**, 299 (1935).
11. G. O. Schenk, *Naturwissenschaften*, **35**, 28 (1948).
12. C. S. Foote and S. Wexler, *J. Amer. Chem. Soc.*, **86**, 3879 (1964).
13. E. J. Corey and W. C. Taylor, *J. Amer. Chem. Soc.*, **86**, 388 (1964).
14. E. A. Ogryzlo, *Photophysiology*, **5**, 35 (1970).
15. R. M. Badger, A. C. Wright, and R. F. Whitlock, *J. Chem. Phys.*, **43**, 4345 (1965).
16. F. Wilkinson, Chapter 5.
17. E. A. Ogryzlo, Chapter 2.
18. R. J. Browne and E. A. Ogryzlo, *Proc. Chem. Soc.*, **1964**, 117.
19. S. J. Arnold and E. A. Ogryzlo, *Can. J. Phys.*, **45**, 2053 (1967).
20. E. A. Ogryzlo and A. E. Pearson, *J. Phys. Chem.*, **72**, 2913 (1968).
21. E. W. Gray and E. A. Ogryzlo, *Chem. Phys. Lett.*, **3**, 658 (1969).
22. D. F. Evans, *Chem. Commun.*, **1969**, 367.
23. A. U. Khan and M. Kasha, *J. Amer. Chem. Soc.*, **92**, 3293 (1970).
24. R. D. Ashford and E. A. Ogryzlo, *J. Amer. Chem. Soc.*, **97**, 3604 (1975).
25. P. D. Bartlett and A. P. Schaap, *J. Amer. Chem. Soc.*, **92**, 3223 (1970).
26. S. Mazur and C. S. Foote, *J. Amer. Chem. Soc.*, **92**, 3225 (1970).
27. D. R. Kearns, *J. Amer. Chem. Soc.*, **91**, 6554 (1969).
28. M. J. S. Dewar and W. Thiel, *J. Amer. Chem. Soc.*, **97**, 3978 (1975).
29. L. B. Harding and W. A. Goddard III, *J. Amer. Chem. Soc.* **99**, 4521 (1977).

3

Lifetime of the $^1\Sigma_g^+$ Oxygen State Produced in a Corona Discharge and other Measurements on $^1\Sigma_g^+$ and $^1\Delta_g$ States in an Electron-beam Controlled Discharge

F. BASTIEN, M. LECUILLER

Laboratoire de Physique des decharges ER 114 du CNRS, Ecole Superieur d'Electricité, Plateau du Moulton, 91190 Gif sur Yvette, France

G. FOURNIER, D. PIGACHE, D. PROUST †

Office National d'Etudes et de Recherches Aérospatiales, 92320 Chatillon sous Bagneux, France

INTRODUCTION

The singlet $^1\Delta_g$ and $^1\Sigma_g^+$ states of molecular oxygen are at present of great interest by reason of their possible applications in chemistry. However, difficulties in designing efficient and pure sources of singlet oxygen at a high production rate and at a high (atmospheric) pressure still remain to be overcome.

Spectroscopic measurements have been made on two types of electrical discharges in order to observe the light emission due to singlet molecular oxygen states created in those discharges.

CORONA DISCHARGE[1]

The corona discharge is produced in a pure (99.998%) oxygen flow between a positive high voltage point (radius 35 μm) and an earthed plane which for these experiments are both made of gold; the electrode gap is 1 cm. The discharge light is detected by an S20 cathode photomultiplier (Hamamatsu R372) at the slit of a Jobin and Yvon (HRS 2) grating monochromator. In order to reduce the thermal noise, the photomultiplier is cooled to −40 °C. The spectrum is recorded graphically at the output of a picoammeter.

Identification of the $^1\Sigma_g^+ \rightarrow {}^1\Sigma_g^-$ transition

Both the P and R branches of the (0, 0) band of the $^1\Sigma_g^+ \rightarrow {}^3\Sigma_g^-$ transition can easily be observed (Figure 1). However, other bands are not sufficiently intense to be visible. Under the best conditions, the (0,0) band has been displayed with high dispersion which is actually sufficient to resolve the P branch (Figure 2).

† Permanent address: Observatoire de Meudon, 92190 Meudon, France.

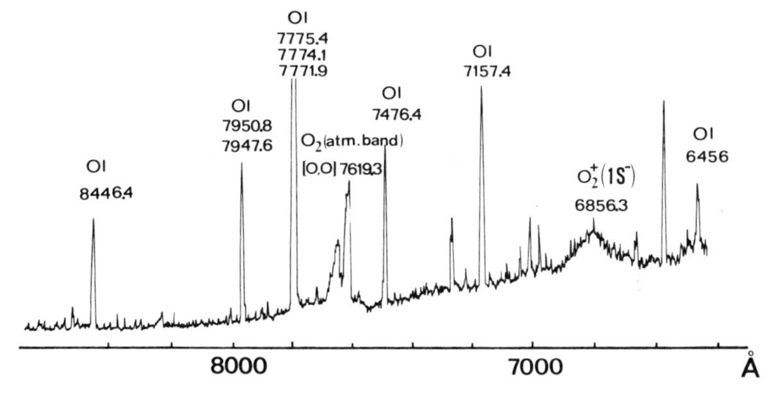

Figure 1 Emission spectrum of the discharge from 620 to 850 nm

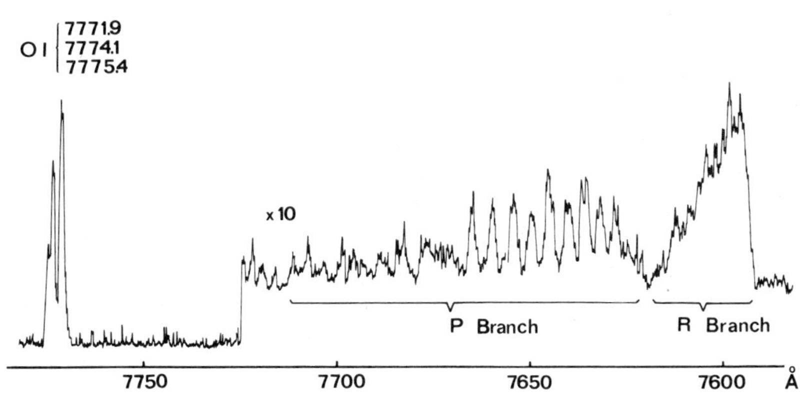

Figure 2 Atmospheric (0,0) band of the transition $^1\Sigma_g^+ \rightarrow {}^3\Sigma_g^-$ from 758 to 780 nm

Time dependence of the emission from the $^1\Sigma_g^+$ state

Since the discharge current is naturally periodic (characteristic of a corona discharge), each pulse can be used as an origin of time in the sampling method. The circuit for time resolved measurement is shown in Figure 3. By this method, it was found that the lifetime of the $^1\Sigma_g^+$ state at 760 mmHg is 2 μs. The known reaction rate with pure oxygen would lead to a lifetime of 170 μs at the same pressure. This great difference can be explained by the presence of species other than O_2. We have shown that the ozone concentration produced in the discharge is high enough to explain the observed difference.

$^1\Delta_g$ state

We attempted to observe the $^1\Delta_g \rightarrow {}^3\Sigma_g^-$ transition by using an InAs photovoltaic cell. The $^1\Delta_g$ state could not be observed in this way.

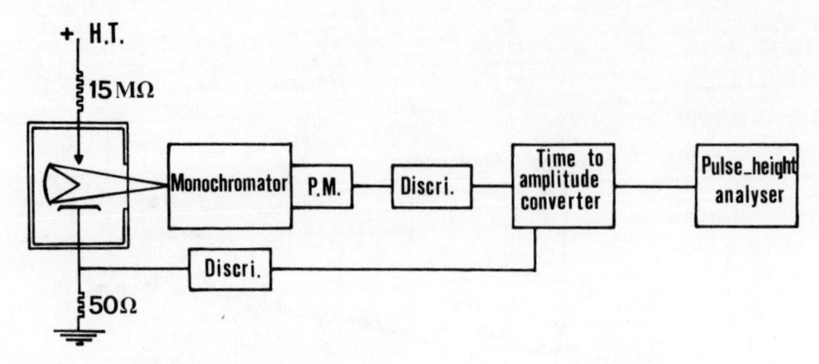

Figure 3 Circuit for time-resolved measurement

ELECTRON BEAM CONTROLLED DISCHARGE[2]

This discharge, which has been extensively investigated for gas lasers, has the following properties: it can be operated at high pressure (>200 mmHg) for long periods and with better homogeneity than the others and the electric field can be adjusted to any value below the breakdown threshold. Accordingly, it is anticipated that the generation of ozone (producing important quenching reactions) may be reduced.

The experimental arrangement is shown in Figure 4. Details of the electron gun and of the discharge chamber can be found in ref. 2. The chamber is evacuated to 10^{-2} mbar before being filled at 200 mbar with pure oxygen (99.998%) at room temperature. The electron beam (120 keV, 0.4 mA cm^{-2}) enters the chamber through a 20-μm aluminium foil with an area of 15×5 cm. It generates an electron density of a few multiples of 10^9 cm^{-3}. Typical values for the discharge parameters in these experiments are as follows: pulse duration, 50–200 μs; repetition rate, 3–25 Hz; discharge current, 450–600 mA; discharge voltage, 3000–7200 V; and gap, 26 mm. The values of the discharge current density and the ratio of the electric field to the gas density are given in the figure captions. The cathode potential is 1800 V.

It is well known that $^1\Delta_g$ singlet oxygen at high pressure emits visible radiation at 635 nm according to the simultaneous transition $2(^1\Delta_g) \rightarrow 2(^3\Sigma_g^-)$. This emission was observed simultaneously with that at 760 nm due to the $^1\Sigma_g^+ \rightarrow$

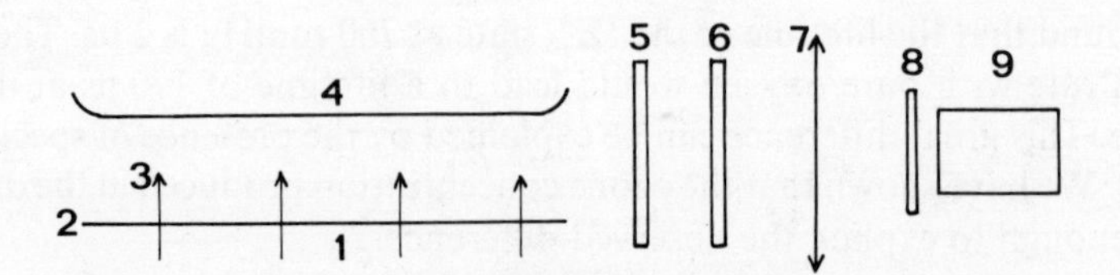

Figure 4 Experimental arrangement. 1 = Electron gun; 2 = electron window; 3 = discharges; 4 = movable anode; 5 = window of the discharge chamber; 6 = X-ray protection window; 7 = lens; 8 = filter; 9 = IR detector or monochromator with photomultiplier

$^3\Sigma_g^-$ transition. The same measurement technique as for the corona discharge is used here. The photomultiplier was screened from residual X-rays by a lead casing. In Figures 5 and 6 the radiation from the electron beam controlled discharge is compared with that from a corona discharge.[1] The transition band $2(^1\Delta_g) \rightarrow 2(^3\Sigma_g^-)$ is shown in Figure 5 and that of the transition $^1\Sigma_g^+ \rightarrow {}^3\Sigma_g^-$ in Figure 6. The reference to an atomic oxygen line shows that the radiation from

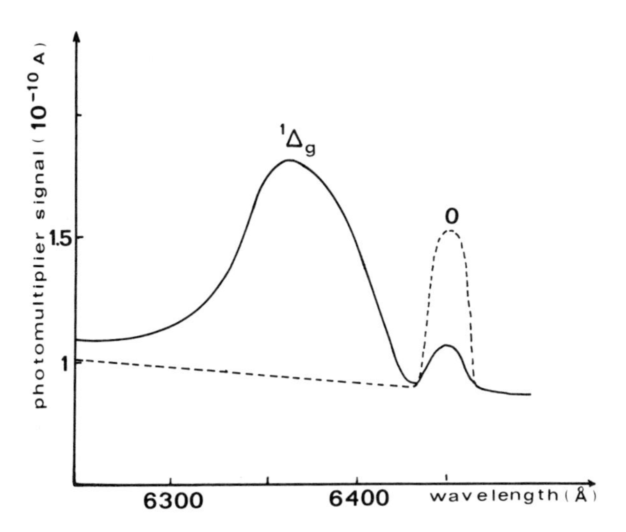

Figure 5 The transition band $2(^1\Delta_g) \rightarrow 2(^3\Sigma_g^-)$ and an atomic oxygen line under the following conditions: pulse duration, 50 μs; repetition rate, 25 Hz; discharge current density, 6 mA cm^{-2}; electric field/gas density, 1×10^{-16} V cm^{-2}. Dots are results from a corona discharge at atmospheric pressure

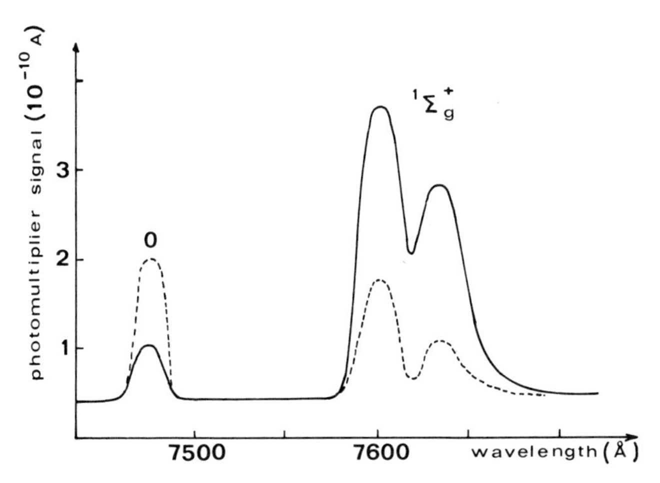

Figure 6 The transition band $^1\Sigma_g^+ \rightarrow {}^3\Sigma_g^-$ and an atomic oxygen line. Conditions as in Figure 5

metastable species is much more important with the electron beam discharge.

The work reported here is still preliminary and much remains to be done. The identification of the $^1\Delta_g \rightarrow {}^3\Sigma_g^-$ transition is being prepared as a better resolution of the visible spectra. A parametric study with quantitative interpretation is planned.

REFERENCES

1. F. Bastien and M. Lecuiller, *J. Phys. B, Atom. Molec. Phys.*, **8**, 1981 (1975).
2. J. Bonnet, *Note Technique*, No. 1975–4, ONERA (1975). Local report ed. Office National d'Etudes et de Recherches Aerospatiales, Chatillon sous Bagneux, France.

4

Physical Quenching of Singlet Oxygen

E. A. OGRYZLO

Department of Chemistry, University of British Columbia, Vancouver, Canada

INTRODUCTION

In this paper we shall restrict our attention to the physical quenching of singlet oxygen in the gas phase, where more detailed information about the mechanism can be obtained. In contrast to the condensed phase, where singlet oxygen has been identified exclusively with the $O_2(^1\Delta_g)$ state, in the gas phase both the $O_2(^1\Delta_g)$ and $O_2(^1\Sigma_g^+)$ states are important because of their interconvertibility through the following reactions:[1,2]

$$O_2(^1\Delta_g)+O_2(^1\Delta_g) \rightarrow O_2(^1\Sigma_g^+)+O_2(^3\Sigma_g^-) \quad (1)$$

$$O_2(^1\Sigma_g^+)+Q \rightarrow O_2(^1\Delta_g)+Q \quad (2)$$

where Q is any quenching species present in the system. The steady-state concentrations of these species in any system depend on the absolute concentration of singlet oxygen and the pressure of the quenching gas. However, in a typical discharge-generated stream of singlet oxygen where the pressure of $O_2(^3\Sigma_g^-)$ is about 2 torr, the pressure of $O_2(^1\Delta_g)$ is about 0.1 torr and the oxygen atoms have been almost eliminated with HgO, the $O_2(^1\Sigma_g^+)$ assumes a steady-state concentration which is only about 6×10^{-4} of the $O_2(^1\Delta_g)$ concentration, even in the absence of any wall deactivation of $O_2(^1\Sigma_g^+)$.[3] This relatively small $O_2(^1\Sigma_g^+)$ concentration may be further reduced by the addition of a good $O_2(^1\Sigma_g^+)$ quencher such as H_2O, but it is important to note that the high steady-state concentration will rapidly re-establish itself if the quencher is condensed out of the stream.

INTERACTIONS WITH OZONE

Although the quenching of both $O_2(^1\Delta_g)$ and $O_2(^1\Sigma_g^+)$ by ozone appears to involve more than simple physical interactions, they will be discussed briefly because of the question of whether ozone is present in discharge-generated singlet oxygen.

The two relevant reactions can be written as [2,4,5]

$$O_2(^1\Sigma_g^+)+O_3(^1A_1) \rightarrow 2O_2(^3\Sigma_g^-)+O(^3P) \qquad (3)$$
$$k_3=1{\cdot}4\times10^{10}\ \mathrm{l\ mol^{-1}\ s^{-1}};\ \Delta H°=-10{\cdot}5\ \mathrm{kcal}$$

and

$$O_2(^1\Delta_g)+O_3(^1A_1) \rightarrow 2O_2(^3\Sigma_g^-)+O(^3P) \qquad (4)$$
$$k_4=2\times10^{6}\ \mathrm{l\ mol^{-1}\ s^{-1}};\ \Delta H°=+2{\cdot}8\ \mathrm{kcal}$$

Both rate constants are the largest reported for each species in the gas phase. Taking into account the two other reactions which form and remove ozone:[6]

$$O+O_2+O_2 \rightarrow O_3+O_2 \qquad (5)$$
$$k_5=2{\cdot}2\times10^{8}\ \mathrm{l\ mol^{-1}\ s^{-1}}$$

and

$$O_3+O \rightarrow 2O_2 \qquad (6)$$
$$k_6=5\times10^{6}\ \mathrm{l\ mol^{-1}\ s^{-1}}$$

it can easily be shown[3] that when the oxygen-atom concentration is decreased below 10^{-3} torr (by the introduction of mercury, for example) a steady-state O_3 concentration is quickly established such that $[O_3]=0{\cdot}02[O]$, i.e. the ozone concentration is only 2% of the oxygen-atom concentration. It is clear that in a discharge-flow system where the oxygen atom concentration has been reduced to an insignificant concentration, ozone need not be considered a significant contaminant.

PHYSICAL QUENCHING OF $O_2(^1\Sigma_g^+)$

Although our primary interest lies in $O_2(^1\Delta_g)$, we initially concentrated on a study of the quenching of $O_2(^1\Sigma_g^+)$ because more accurate and complete data exist for $O_2(^1\Sigma_g^+)$, owing largely to the greater ease with which it can be relaxed. This relaxation can be written as

$$O_2(^1\Sigma_g^+)+Q \rightarrow O_2(^1\Delta_g)+Q \qquad (7)$$

in which 15·0 kcal $\mathrm{mol^{-1}}$ (5238·5 $\mathrm{cm^{-1}}$) of electronic energy must be transformed into nuclear motion in $O_2(^1\Delta_g)$ and the quencher. The data that have been obtained for this process are summarized in Figure 1. The point on each tear-drop surrounding each quencher indicates the best experimental value currently available.[7] The quenchers are grouped, somewhat arbitrarily, along the horizontal axis to facilitate comparison of structurally similar molecules.

Several general observations can be made. The ionization energy of the quencher is of no consequence; unsaturation and paramagnetism have no significant effect on quenching ability; deuteration decreases the quenching rate constant and the best quenchers are those containing hydrogen. In 1972, when much of these data appeared in the literature, it was noted by two laboratories[7,8] that the quenching abilities of the diatomic molecules correlate well

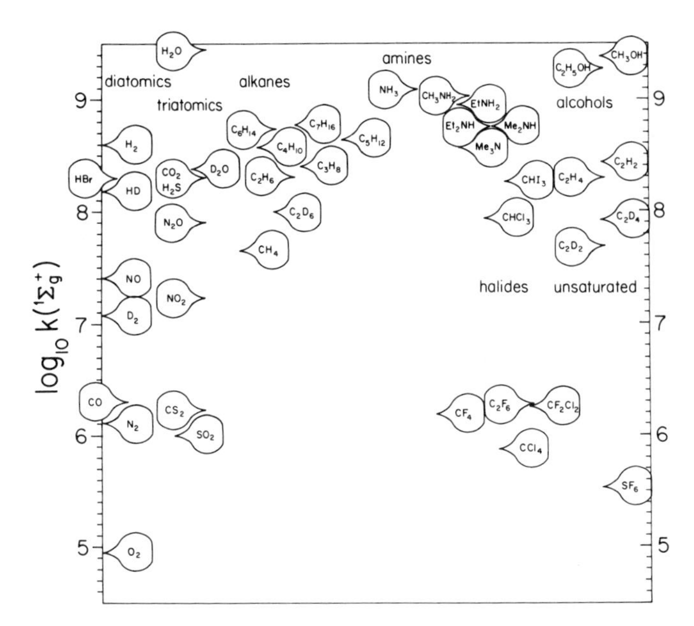

Figure 1 Logarithm of the rate constants for the quenching of $O_2(^1\Sigma_g^+)$ in units of $l\ mol^{-1}\ s^{-1}$

with the fundamental vibrational frequencies of the quenchers. This correlation led us to suggest that the effectiveness of a quencher is determined by its ability to become vibrationally excited by the 15 kcal liberated in the O_2 transition. Our initial attempts to use the 'Fermi golden rule' to calculate these rate constants, especially for homonuclear diatomics, were unsatisfactory and it was clear that a more sophisticated theoretical model was required.[7, 9]

There are two distinct ways in which the transitions on O_2 and the quencher can be coupled. The first involve short-range interactions (S.R.I.), where the perturbation which causes the relaxation is the short-range repulsive potential due to the actual collision between the two molecules. The alternative mechanism involves long-range interactions (L.R.I.), where the transition is brought about by the interaction between the transition quadrupole on oxygen with the transition dipole or quadrupole on the quenchers. The actual collision between the two molecules (if it occurs) serves only to limit the distance of closest approach.

The magnitude of the perturbation is different in each instance and this effect plays a critical role in determining how the two different relaxation rates can be calculated. In S.R.I. the perturbation energy is determined by the relative velocities of the two molecules and hence is about 200 cm^{-1} at room temperature. The only successful method of dealing with such a large perturbation has been scattering theory in the 'distorted-wave approximation'. In 1975, Kear and Abrahamson[10] applied the theory to the quenching of $O_2(^1\Sigma_g^+)$ by some diatomic molecules, and their results can be compared with the experimental values in Table 1. It can be seen that their values are about two orders of magnitude too low.

Now let us consider the contribution from long-range interactions (L.R.I.). Because the perturbation is due to multipolar interactions which are of the order of 10 cm^{-1}, a semi-classical calculation can be carried out. Vibrations, rotations and electronic motion are treated quantum mechanically while translational motion is treated classically. Standard time-dependent perturbation theory then yields a transition probability given [in the case of $O_2(^1\Sigma_g^+)$ quenching by HBr] by:

Table 1 Rate constants for the quenching of $O_2(^1\Sigma_g^+)$ at 300 K

Quencher	Rate constant ($l\ mol^{-1}\ s^{-1}$) Experimental[a]	Calculated (Kears and Abrahamson[10])
H_2	$8{\cdot}0 \times 10^8$ (33)	$4{\cdot}3 \times 10^7$
	$2{\cdot}4 \times 10^8$ (34, 35, 43)	
	$6{\cdot}8 \times 10^8$ (36)	
	$5{\cdot}5 \times 10^8$ (37)	
HD	$1{\cdot}9 \times 10^8$ (33)	3×10^6
	$1{\cdot}1 \times 10^8$ (35)	
D_2	$1{\cdot}2 \times 10^7$ (35,38)	$9{\cdot}2 \times 10^5$
	$1{\cdot}1 \times 10^7$ (33)	
CO	$1{\cdot}8 \times 10^6$ (39)	$3{\cdot}1 \times 10^4$
	$2{\cdot}6 \times 10^6$ (38)	
	$2{\cdot}0 \times 10^6$ (40)	
	$1{\cdot}5 \times 10^6$ (41)	
N_2	$1{\cdot}3 \times 10^6$ (34)	1×10^4
	$1{\cdot}2 \times 10^6$ (39)	
	$1{\cdot}8 \times 10^6$ (35)	
	$1{\cdot}1 \times 10^6$ (38)	
O_2	$9{\cdot}0 \times 10^4$ (34, 39)	$3{\cdot}4 \times 10^1$
	$2{\cdot}7 \times 10^5$ (38)	
	$6{\cdot}6 \times 10^4$ (44)	

[a] Information about the source of these experimental values can be found in reference 7 or 11.

$$P_{QQ} = \sum_{j_1 j_2} \frac{56\mu}{10h^2b^8kT} |\langle 0|O_2|0\rangle|^2 |\langle 2|HBr|0\rangle|^2 C^2(j_1 2 j_1'; 00) C^2(j_2 2 j_2'; 00) I_4 \quad (8)$$

where μ is the reduced mass, b is the impact parameter, $\langle 0|O_2|0\rangle$ is the electronic matrix element connecting the initial and final states on O_2, and $\langle 2|HBr|0\rangle$ is the vibrational matrix element connecting the initial and final states on HBr. The next two terms are Clebsch–Gordon coefficients for the rotational transitions in O_2 and HBr, and I_4 is the Fourier transform interaction potential evaluated at the net defect frequency, i.e. that corresponding to the energy transformed into translational motion. The probability was calculated with equation (8) by averaging over both impact parameter and a Boltzmann distribution of molecular velocities. The computer calculation then yields a series of perhaps several hundred individual vibrational–rotational channels, each with its own probability which has been adjusted to take into account the initial Boltzmann distribution in the rotational levels of the reactants. The overall rate is obtained by adding these individual probabilities.

The results of our calculations of k at 25 °C are presented in Table 2.[11] It can be seen that the agreement with experiment is excellent for H_2, HD and D_2. Such agreement in the absence of any adjustable parameters leads us to the

Table 2 Rate constants for the quenching of $O_2(^1\Sigma_g^+)$ at 300 K

Quencher	Rate constant ($l\ mol^{-1}\ s^{-1}$) Calculated (this work)	Experimental[a]	Major product (95%) O_2	Q	Energy discrepancy (cm^{-1})
H_2	$4{\cdot}6 \times 10^8$	$8{\cdot}0 \times 10^8$ (33)	$v=0$	$v=1$	1038
		$2{\cdot}4 \times 10^8$ (34, 35, 43)			
		$6{\cdot}8 \times 10^8$ (36)			
		$5{\cdot}5 \times 10^8$ (37)			
HD	$1{\cdot}0 \times 10^8$	$1{\cdot}9 \times 10^8$ (33)	$v=1$	$v=1$	−85
		$1{\cdot}1 \times 10^8$ (35)			
D_2	$1{\cdot}2 \times 10^7$	$1{\cdot}2 \times 10^7$ (35, 38)	$v=1$	$v=1$	722
		$1{\cdot}1 \times 10^7$ (33)			
NO	$9{\cdot}3 \times 10^6$	$3{\cdot}3 \times 10^7$ (34)	$v=1$	$v=2$	124
		$2{\cdot}4 \times 10^7$ (35)			
		$2{\cdot}5 \times 10^7$ (38)			
CO	$2{\cdot}5 \times 10^6$	$1{\cdot}8 \times 10^6$ (39)	$v=2$	$v=1$	118
		$2{\cdot}6 \times 10^6$ (38)			
		$2{\cdot}0 \times 10^6$ (40)			
		$1{\cdot}5 \times 10^6$ (41)			
HCl	$5{\cdot}3 \times 10^6$	$4{\cdot}0 \times 10^7$ (37)	$v=0$	$v=2$	−469
		$2{\cdot}5 \times 10^7$ (42)			
HBr	$3{\cdot}4 \times 10^8$	$2{\cdot}3 \times 10^8$ (42)	$v=0$	$v=2$	−98
N_2	$9{\cdot}12 \times 10^5$	$1{\cdot}3 \times 10^6$ (34)	$v=2$	$v=1$	69
		$1{\cdot}2 \times 10^6$ (39)			
		$1{\cdot}8 \times 10^6$ (35)			
		$1{\cdot}1 \times 10^6$ (38)			
O_2	~10	$9{\cdot}0 \times 10^4$ (34, 39)	$v=2$	$v=1$	705
		$2{\cdot}7 \times 10^5$ (38)	$v=1$	$v=2$	629
		$6{\cdot}6 \times 10^4$ (44)			

[a] Information about the source of these experimental values can be found in reference 7 or 11.

conclusion that it is long-range multipolar interactions that lead to the physical quenching of $O_2(^1\Sigma_g^+)$. With this conclusion, we can inspect the computer output to determine the most important quenching channel in each instance. The vibrational levels of O_2 and the quencher excited in most of the quenching acts are shown in Table 2. The most important rotational change in these quenching reactions depends on whether the energy discrepancy is positive or negative (see Table 2). If it is positive, i.e. excess energy is produced which can be dissipated in rotational motion, the most important rotational change which occurs is $\Delta J = +2$. This is true for H_2 and D_2 where a quadrupole–quadrupole interaction allows such a transition.

Although the results for NO are low (possibly because we could not consider the contribution from the unknown transition–quadrupole moment on NO), the results for CO, HBr and N_2 are within the combined experimental error. The calculated results for O_2 are very low because all transitions are far from resonance. However, since the quenching constants for Ar and He are not much different from the value for O_2, it appears possible that for these species S.R.I. provides the mechanism by which relaxation occurs. In view of this possibility there is a need for another method of distinguishing between the S.R.I. and L.R.I. mechanisms. One possibility is the temperature dependence of the quenching constant. Although quenching due to S.R.I. should always show a positive temperature dependence because the magnitude of the perturbation increases with increasing relative kinetic energy of the particles, transitions induced by L.R.I. could show no temperature dependence or even a negative dependence.

Braithwaite *et al.*[12] measured this temperature dependence for quenching with H_2 and obtained the results shown as circles in Figure 2. The results of S.R.I. and L.R.I. calculations are also shown.[12] The S.R.I. values are higher than those obtained by Kear and Abrahamson,[10] because the possibility of

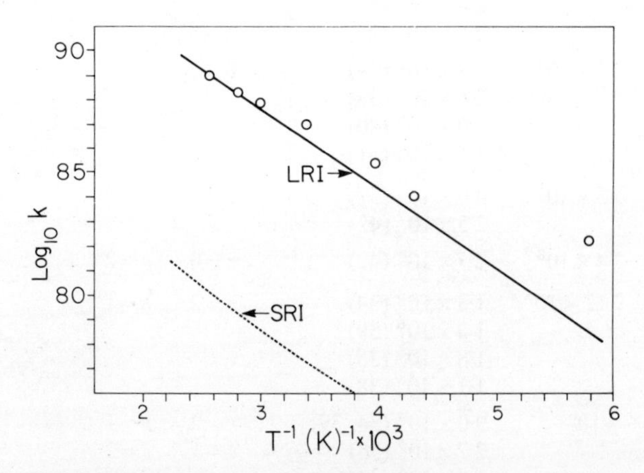

Figure 2 Rate constants for the quenching of $O_2(^1\Sigma_g^+)$ by H_2 as a function of temperature. Experimental points are represented by circles

rotational excitation was included in our calculation. However, a value of 0·33 was assumed for the electronic matrix element on O_2 since this value is more consistent with the quenching rates observed for the inert gas.[7] It can be seen that although L.R.I. calculations are again more successful at reproducing the absolute values of the rate constants, both mechanisms have the same temperature dependence, and this effect is due to the fact that when H_2 is excited to the first vibrational level 1080 cm^{-1} of energy remain to be dissipated. Consequently, the temperature dependence is determined principally by changing rotational populations which at high temperatures can better accommodate this excess energy.

It is clear that in order to differentiate between the two mechanisms we require a system which is closer to resonance. One possibility is the reaction

$$O_2(^1\Sigma_g^+) + HBr \rightarrow O_2(^1\Delta_g) + HBr\ (v=2) \tag{9}$$

which is only 120 cm^{-1} exothermic. The experimental and theoretical results for this system are shown in Figure 3.[13] Once again, LRI calculations are better at reproducing the experimental values. However, it would appear that S.R.I.

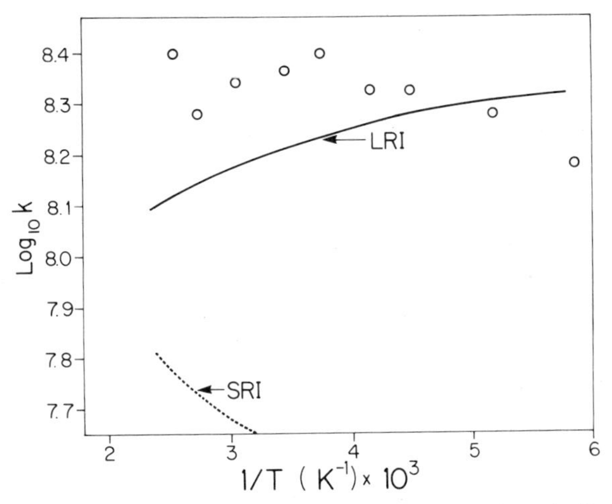

Figure 3 Rate constants for the quenching of $O_2(^1\Sigma_g^+)$ by HBr as a function of temperature. Experimental points are represented by circles.

could contribute to the quenching at high temperatures since a better fit is obtained when the rate constants from L.R.I. and S.R.I. are added.

Work is currently being carried out on triatomic molecules. Because of the greater flexibility in these molecules they can come closer to resonance and hence they will provide a more definitive test. Although calculations are not yet complete on these systems, we feel it is significant that the experimental quenching rates decrease slightly with increasing temperature.[14] Such a trend could result from an L.R.I. quenching mechanism, but would not result from an S.R.I. mechanism.

Before ending our discussion of $O_2(^1\Sigma_g^+)$ quenching, it is worth pointing out that the importance of rotational and vibrational excitation of the quencher in determining the quenching constant is made clear when one considers the halides in Figure 1. It can be seen that as soon as the last hydrogen atom is removed from the molecule, the quenching constant drops by about two orders of magnitude, which is consistent with the removal of the higher frequencies associated with the presence of the hydrogen atom.

PHYSICAL QUENCHING OF $O_2(^1\Delta_g)$

Rate constants for the relaxation of $O_2(^1\Delta_g)$ in the gas phase by 23 quenchers are presented in Figure 4. The short horizontal tails on the 'drops' surrounding each quencher give the 'best' current experimental values.[7] Although the data are not as complete as for $O_2(^1\Sigma_g^+)$, several similarities and differences are immediately apparent and can be attributed to the nature of the quenching reaction, which can be written as

$$O_2(^1\Delta_g) + Q \rightarrow O_2(^3\Sigma_g^-) + Q + 22{\cdot}54 \text{ kcal} \qquad (10)$$

The fact that the reaction is spin-forbidden with a 'singlet' quencher and furthermore must dissipate 6·5 kcal more than the $O_2(^1\Sigma_g^+)$ relaxation makes the quenching constants for species such as H_2, H_2O, CO_2, C_3H_8 and NH_3 about five orders of magnitude smaller than the corresponding $O_2(^1\Sigma_g^+)$ values. However, the relative values of these quenching constants indicate that, as with

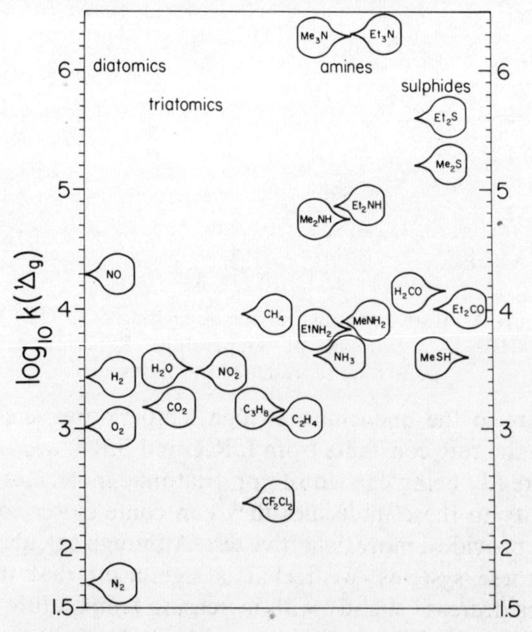

Figure 4 Logarithm of the rate constants for the quenching of $O_2(^1\Delta_g)$

$O_2(^1\Sigma_g^+)$, the $O_2(^1\Delta_g)$ quenching rates are governed by the ability of the quencher to become vibrationally and rotationally excited in the process. The restriction imposed by the 'spin–flip' is clear from the fact that molecules such as NO, NO_2 and O_2 can formally relax the spin restrictions and are consequently much more effective relative to the five singlet species mentioned above.

The amines also behave distinctly differently in the quenching of $O_2(^1\Sigma_g^+)$ and $O_2(^1\Delta_g)$. As $O_2(^1\Sigma_g^+)$ quenchers the amines show very little difference in their quenching abilities, with the high frequency of the N—H bond leading to slightly more effective quenching by the least substituted amines. In contrast, the $O_2(^1\Delta_g)$ quenching ability of the amines increases markedly with ethyl or methyl substitution, and the same appears to be true for the sulphides. This rapidly increasing quenching ability has been related to a decreasing ionization energy with increasing substitution.[15] Since $O_2(^1\Delta_g)$ has a high electron affinity, it has been proposed that a charge-transfer interaction occurs, which can not only bring about a stronger interaction between the two species but more importantly assist the 'spin–flip' through spin–orbit interactions in the complex.[15]

CONCLUSION

It is possible to consider the charge–transfer and spin–spin interactions of $O_2(^1\Delta_g)$ with some quenchers to be a weak 'chemical-quenching' process. In this view, the strictly physical quenching of both $O_2(^1\Sigma_g^+)$ and $O_2(^1\Delta_g)$ may be due to an identical mechanism. Our recent calculations lead us to believe that in most instances this is a simple 'long-range' interaction between the transition quadrupole on $O_2(^1\Sigma_g^+)$ or the transition magnetic dipole on $O_2(^1\Delta_g)$ with the transition quadrupole and/or dipole on the quencher. The relative efficiencies of such physical quenchers are determined largely by the magnitude of the transition multipole on the quencher and the proportion of the energy that must appear as relative translational motion. Higher transition probabilities are found when little of the electronic energy ends up as translational motion.

REFERENCES

1. S. J. Arnold and E. A. Ogryzlo, *Can. J. Phys.*, **45**, 2053 (1967).
2. R. G. O. Thomas and B. A. Thrush, *J. Chem. Soc., Faraday Trans II*, **8**, 664 (1975).
3. E. A. Ogryzlo, in *Singlet Oxygen*, Organic Chemistry Series, Vol. 00, (Ed. H. H. Wasserman), Academic Press, New York, 1977.
4. D. R. Snelling, *Can. J. Chem.*, **52**, 257 (1974).
5. M. Yaron, A. Von Engel and P. H. Vedaud, *Chem. Phys. Letts.*, **37**, 159 (1976).
6. R. F. Hampson, Jr., and D. Garvin, *N.B.S. Tech. Note*, No. 886, U.S. Dept. of Commerce, Washington, D.C., 1975.
7. J. A. Davidson and E. A. Ogryzlo, in *Chemiluminescence and Bioluminescence* (Ed. M. J. Cormier, D. M. Hercules, and J. Lee), Plenum Press, New York, 1973, p. 111.
8. K. Kear and E. W. Abrahamson, *Abstr. 2nd Noyes Photochem. Symp.*, Austin, Texas, 1972.
9. J. A. Davidson and E. A. Ogryzlo, *Can. J. Chem.*, **52**, 240 (1974).
10. K. Kear and E. W. Abrahamson, *J. Photochem.*, **3**, 409 (1974/75).

11. M. Braithwaite, J. A. Davidson, and E. A. Ogryzlo, *J. Chem. Phys.*, **65**, 771 (1976).
12. M. Braithwaite, E. A. Ogryzlo, J. A. Davidson, and H. I. Schiff, *J. Chem. Soc., Faraday Trans. II*, (in press).
13. M. Braithwaite, E. A. Ogryzlo, J. A. Davidson, and H. I. Schiff, *Chem. Phys. Lett.*, **42**, 158 (1976).
14. J. A. Davidson and H. I. Schiff, personal communication.
15. E. A. Ogryzlo and C. W. Tang, *J. Amer. Chem. Soc.*, **92**, 5034 (1970).

5

Physical Properties of Singlet Oxygen in Fluid Solvents

F. WILKINSON

School of Chemical Sciences, University of East Anglia, Norwich, NR4 7TJ, England

INTRODUCTION

Many of the unique properties of molecular oxygen arise because its highest occupied double degenerate pair of molecular orbitals, which are of π character, contain only two electrons. This electron configuration gives rise to only three electronic states which conform to Pauli's principle and these states can be assigned group theoretical symmetry labels and their spin multiplicities denoted by a superscript. Thus, the lowest or ground electronic state of molecular oxygen, which has zero angular momentum about the internuclear axis (Σ state) in agreement with Hund's rule, has a spin multiplicity of three and is therefore the $^3\Sigma_g^-$ state. The two electronically excited states which arise from this same electron configuration, i.e. they do not involve electron promotion to higher orbitals, are the $^1\Delta_g$ state and the $^1\Sigma_g^+$ state, which have zero-point energies 95 and 158 kJ mol^{-1}, respectively, above that of the $^3\Sigma_g^-$ ground state.

The electronic transitions $^1\Delta_g \leftarrow {}^3\Sigma_g^-$ and $^1\Sigma_g^+ \leftarrow {}^3\Sigma_g^-$, although highly forbidden, are readily observed in absorption and emission in the upper atmosphere and estimated radiative lifetimes of 64 min[1] and ca. 10 s[2,3], respectively, have been reported. Under most conditions, however, the lifetime of these excited singlet states of molecular oxygen are much shorter than these values both in the gas phase[4] and in condensed media,[5] so much so that special rapid reaction techniques are often required in order to study their properties. In fact, the $^1\Sigma_g^+$ state is so easily quenched, often thereby undergoing a non-radiative transition to give the $^1\Delta_g$ state, that very little is known about its properties under any conditions other than in the gas phase.

Because of limitations of space it will be necessary to be selective in this paper. Fortunately, there have been several excellent recent reviews[6–10] concerning various aspects of the properties of singlet oxygen. In view of the interests of the author and those of many of the participants at this Conference, it was decided to concentrate on a discussion of the kinetic methods which have been evolved for studying singlet oxygen decay and quenching of singlet oxygen by physical processes in fluid solution and to comment on the likely relevance of such work to problems encountered in more rigid polymer media. In this connection, it is important to recognize that there is considerable divergence of opinion as to the importance of singlet oxygen as an important intermediate in for example the photodegradation of polymers.[11,12]

MEASUREMENT OF THE LIFETIME OF SINGLET OXYGEN (O_2^*, $^1\Delta_g$) IN FLUID MEDIA

Indirect Methods

Many dye-sensitized photo-oxygenations of various acceptors, A, have been clearly shown to involve singlet oxygen ($^1\Delta_g$) since they give products and relative reactivities identical with those obtained following chemical production of singlet oxygen, for example from the reaction of metal hypochlorites with hydrogen peroxide. The following simple mechanism for photo-oxygenation by a type II mechanism[13] with a sensitizer, Sens, is well established:

$$\text{Sens} + h\nu \longrightarrow {}^1\text{Sens}^* \longrightarrow {}^3\text{Sens}^* \quad (\text{with yield } \phi_\text{T}) \tag{1}$$

$${}^3\text{Sens}^* + {}^3\text{O}_2({}^3\Sigma_g^-) \longrightarrow {}^1\text{O}_2^*({}^1\Delta_g) + {}^1\text{Sens} \tag{2}$$

$${}^1\text{O}_2^*({}^1\Delta_g) \xrightarrow{k_1} {}^3\text{O}_2({}^3\Sigma_g^-) \tag{3}$$

$${}^1\text{O}_2^*({}^1\Delta_g) + \text{A} \xrightarrow{k_\text{R}} \text{AO}_2 \text{ or other products} \tag{4}$$

From this mechanism, it follows that

$$\phi_{\text{AO}_2} = \frac{\phi_\text{T} k_\text{R}[\text{A}]}{k_1 + k_\text{R}[\text{A}]} \tag{5}$$

or

$$\frac{1}{\phi_{\text{AO}_2}} = \frac{1}{\phi_\text{T}}\left(1 + \frac{\beta}{[\text{A}]}\right) \tag{6}$$

where $\beta = k_1/k_\text{R}$. From the slope and intercept of the linear plots of $\phi_{\text{AO}_2}^{-1}$ *versus* $[\text{A}]^{-1}$, values of β can be obtained and thus relative reactivities under identical conditions where k_1 will be constant. Some compounds[14] have β-values as low as 10^{-4} mol dm^{-3} and, since k_R cannot exceed the diffusion-controlled rate constant ($k_\text{d} \approx 10^{10}$ dm^3 mol^{-1} s^{-1}), it follows that $k_1 < 10^6$ s^{-1}. Foote and Denny[15] were able to improve on this value since they found that very low concentrations of β-carotene, C, inhibited photo-oxidation by quenching singlet oxygen ($^1\Delta_g$). The mechanism of quenching was later shown to be due to energy transfer to produce triplet β-carotene,[16] viz.

$${}^1\text{O}_2^*({}^1\Delta_g) + \text{C} \longrightarrow {}^3\text{C}^* + {}^3\text{O}_2({}^3\Sigma_g^-) \tag{7}$$

Kinetic Spectroscopic Analysis of the Disappearance of an Oxidizable Acceptor following Pulse Excitation

The principle of this method involves producing singlet oxygen rapidly and allowing it to react with an acceptor, the absorption of which thereby decreases over a time period following the pulse excitation. Analysis of the kinetics of disappearance of the acceptor gives information concerning the unimolecular decay of singlet oxygen ($^1\Delta_g$) as well as the rate constant for its bimolecular reaction with the acceptor. This method was evolved by Adams and Wilkinson[17,18] using laser excitation of a sensitizer, e.g. methylene blue, to give its triplet state which, in fluid-aerated solution, is rapidly quenched by molecular oxygen to produce singlet oxygen, and the kinetics of its reaction with the acceptor, e.g. diphenylisobenzofuran (DPBF), was followed spectrophotometrically by analysing at the absorption maximum of this acceptor. Preliminary results were communicated to leading workers in the field interested in singlet oxygen, such as Foote *et al.*[19] and Kearns,[8] who quoted these results in their publications with due reference. Unfortunately, reference to the fact that Adams had fully described his method to Kearns in 1969 during a visit by Kearns to our laboratory was not made in later publications by Merkel and Kearns,[20–22] who used the same method with the same donor and the same acceptor. These workers, however, employed a different kinetic analysis treatment. As has been pointed out by Young *et al.*,[13] the kinetic analysis needs careful consideration, as explained below. Consider reactions (3) and (4) given above, in the absence of a quencher [as in step (7)], from which it follows that

$$-\frac{d[^1O_2^*]}{dt}=(k_1+k_R[A])[^1O_2^*] \tag{8}$$

and

$$-\frac{d[A]}{dt}=k_R[A][^1O_2^*] \tag{9}$$

If it is assumed that [A] does not alter substantially during the period of the analysis, e.g. by not more than 10%, then an average value, $[A]_{av}$, equal to $([A]_{t=0}+[A]_{t=\infty})/2$ can be substituted into equation (8), giving upon integration

$$[^1O_2^*]=[^1O_2^*]_{t=0}\exp\{-(k_1+k_R[A]_{av})t\} \tag{10}$$

Defining

$$k_D=k_1+k_R[A]_{av} \tag{11}$$

and substituting equations (10) and (11) into equation (9) gives

$$-\frac{d[A]}{dt}=k_R[A][^1O_2^*]_{t=0}\exp(-k_Dt) \tag{12}$$

Integration of equation (12) gives either

$$\ln\left(\frac{[A]}{[A_0]}\right) = \frac{k_R[^1O_2^*]_{t=0}}{k_D}\cdot\{\exp(-k_D t-1\} \tag{13}$$

or

$$\ln\left(\frac{[A]}{[A_\infty]}\right) = \frac{k_R[^1O_2^*]_{t=0}}{k_D}\cdot\exp(-k_D t) \tag{14}$$

Equation (13) was used by Adams and Wilkinson [18] and they did not observe a dependence of k_D on [A]. They therefore assumed that $k_D = k_1$ [see equation (11)]. However, Young *et al.*,[23] who analysed their data according to equation (14), have shown that there is a dependence of k_D on [A] as expected from equation (11), and this has been confirmed by Farmilo and Wilkinson.[16] The kinetic analysis employed by Merkel and Kearns [21] is different. Their treatment is equivalent to starting with equation (12), which already makes the approximation that [A] does not vary much, and integrating the left- and right-hand sides of equation (12) assuming [A] is constant on the right- but not on the left-hand side. This treatment gives

$$[A]_t - [A_\infty] = \frac{k_R[A][^1O_2^*]_{t=0}}{k_1 + k_R[A]}\cdot\exp\{-(k_1 + k_R[A])t\} \tag{15}$$

and the change in concentration is given by

$$[A]_t - [A]_\infty = \varepsilon l\, \Delta\text{Abs} \tag{16}$$

where ΔAbs is the change in absorbance. A plot of ln ΔAbs against time will, from equations (15) and (16), have a slope equal to $-(k_1 + k_R[A])$. By varying [A], values of k_1 and k_R can be determined. Henceforth we shall refer to this as the Kearns treatment.

We have carried out extensive comparisons of the methods of analysis, i.e. curve fitting to equation (13) or (14) as was done by Wilkinson and co-workers [16, 18] and Young *et al.* [23] and using the Kearns treatment. We have also, using an incremental multicalculation computer treatment, fitted the experimental curve without any assumption, i.e. accepting only that equations (8) and (9) apply, and derived the decay constants. The analysis of a single trace showed considerable divergence (up to 20%) in the value of k_D obtained from the different methods. However, the mean values of the fitted decay constants from twelve traces based on equation (13) or (14) were within ±5% of the mean values obtained from the much more time-consuming incremental method. The Kearns treatment was also shown to give agreement within 5% for the average value but it gave values for k_D which were consistently slightly higher than the incremental method. It follows, therefore, that careful curve fitting using equation (13) or (14) for several traces, or plots of ln ΔAbs against time, should give reliable values for k_D.

In the presence of a singlet oxygen quencher, Q, i.e. including the step

$$^{1}O_2^* + Q \xrightarrow{k_Q} \text{quenching} \tag{17}$$

equation (8) becomes

$$-\frac{d[^{1}O_2^*]}{dt} = (k_1 + k_R[A] + k_Q[Q])[^{1}O_2^*] \tag{18}$$

and we can define the decay constant in the presence of the acceptor as

$$k_D^1 = k_D + k_Q\,[Q] \tag{19}$$

Measurement of k_D^1 in the presence of variable concentrations of the quencher, Q, allows values of k_Q to be obtained.[16]

Kinetic Analysis following Energy Transfer to β-carotene

When β-carotene is subjected to direct laser photolysis, no triplet carotene is observed because of its low inter-system crossing yield. In deaerated solutions containing a triplet sensitizer and β-carotene, which has a triplet energy[16] of >95 kJ mol^{-1}, one observes the sensitized production of triplet β-carotene due to triplet–triplet energy transfer, i.e.

$$^{3}\text{Sens}^* + {}^{1}\text{C} \xrightarrow{k_{TC}} {}^{3}\text{C}^* + {}^{1}\text{Sens} \tag{20}$$

The triplet β-carotene which has a maximum absorption at 520 nm decays with a lifetime of 7·7 μs in benzene solution at room temperature. The decay can be written as

$$^{3}\text{C}^* \xrightarrow{k_{DC}} {}^{1}\text{C} \tag{21}$$

Truscott *et al.*[24] have shown that triplet β-carotene is quenched by oxygen with a rate constant of ca. 3×10^9 dm^3 mol^{-1} s^{-1} for the reaction

$$^{3}\text{C}^* + {}^{3}O_2 \xrightarrow{k_{CO}} \text{quenching} \tag{22}$$

and therefore in aerated benzene the decay time would be expected to be ca. 0·2 μs. However, Farmilo and Wilkinson[16] have shown, using triplet anthracene as the sensitizer in aerated benzene in the presence of relatively low concentrations of β-carotene, that the sensitized triplet β-carotene has a lifetime >2 μs. This arises because $k_T[O_2] \gg k_{TC}[C]$, i.e. almost all ($>90\%$) of the sensitizer triplet states are quenched by oxygen [step (2)] producing long-lived singlet oxygen which then transfers energy to β-carotene in a rate-determining step [reaction (7)]. Since the triplet β-carotene decay constant in aerated benzene is so large, the measured decay is equal to the decay of its precursor, namely singlet oxygen ($^{1}\Delta_g$), and thus triplet β-carotene is a probe for singlet oxygen decay. This can be demonstrated mathematically as shown below. According to the mechanism outlined above following the pulsed production of ^{3}Sens*, i.e. considering steps (2), (3), (7), (20), (21), and (22), gives

$$\frac{d[^3C^*]}{dt} = k_{\Delta C}[C][^1O_2^*] + k_{TC}[^3Sens^*][^1C] - k_{CO}[O_2][^3C^*] - k_{DC}[^3C^*] \quad (23)$$

$$-\frac{d[^1O_2^*]}{dt} = (k_1 + k_{\Delta C}[C])[^1O_2^*] - k_T[^3Sens^*][O_2] \quad (24)$$

$$-\frac{d[^3Sens^*]}{dt} = (k_T[O_2] + k_{TC}[C])[^3Sens^*] \quad (25)$$

The exact solution of these three equations is

$$[^3C^*] = P(e^{-k_A t} - e^{-k_C t}) + Q(e^{-k_B t} - e^{-k_C t}) \quad (26)$$

where

$$k_A = k_T[O_2] + k_{TC}[C] \quad (27)$$

$$k_B = k_1 + k_{\Delta C}[C] \quad (28)$$

and

$$k_C = k_{CO}[O_2] + k_{DC} \approx k_{CO}[O_2], \quad (29)$$

since k_{DC} is small compared with $k_{CO}[O_2]$.

$$P = \frac{[^3Sens^*]_{t=0}(k_{TC}[C](k_B - k_A) + k_{\Delta C}[C]k_T[O_2])}{(k_A - k_B)(k_A - k_C)} \quad (30)$$

and

$$Q = \frac{[^3Sens^*]_{t=0}k_{\Delta C}[C]k_T[O_2]}{(k_A - k_B)(k_C - k_B)} \quad (31)$$

At times greater than 1 μs, $e^{-k_A t}$ and $e^{-k_C t}$ become negligibly small and equation (26) becomes

$$[^3C^*] = Q\exp(-k_B t) \quad (32)$$

i.e. theory predicts a first-order decay of triplet β-carotene with a rate constant k_B which is equal to the singlet oxygen decay constant under these conditions. The linear dependence of k_B on [C] [equation (28)] was confirmed by Farmilo and Wilkinson [16] and the value of k_1 obtained from the analysis of the decay constants of singlet oxygen sensitized triplet β-carotene was, within experimental error, equal to that obtained by the method based on the kinetics of disappearance of an oxidizable acceptor following pulse excitation (see earlier).

In the presence of a quencher, equation (28) becomes

$$k'_B = k_1 + k_{\Delta C}[C] + k_Q[Q] \quad (33)$$

thus allowing the evaluation of k_Q. This method has also been extended recently to allow the yields of singlet oxygen formed during oxygen quenching of triplet sensitizers to be measured (see Garner and Wilkinson, Chapter 7).

LIFETIMES OF SINGLET OXYGEN IN VARIOUS SOLVENTS

The lifetime of singlet oxygen varies considerably (ca. 1000-fold) depending on the solvent, and a selection of literature values are given in Table 1. Merkel and Kearns[22] have shown that there is a correlation between the lifetime of singlet oxygen and the infrared spectral properties of the solvent. The greater the absorption near 8000 cm^{-1} the shorter is the lifetime of singlet oxygen, which suggests that direct conversion of the electronic energy from singlet oxygen ($^1\Delta_g$) into vibrational energy of the solvent is predominant in determining the decay of singlet oxygen. This is probably true in solvents such as water, alcohols and hydrocarbons which have overtones and combination bands that absorb relatively strongly in this spectral region, but in solvents such as carbon tetrachloride and Freon-11 it appears[25] that highly vibrational excited states of ground-state oxygen are produced during the decay of singlet oxygen. There is little indication of a heavy-atom effect on the lifetime of singlet oxygen (see Table 1).

Table 1 Lifetime of singlet oxygen in fluid solvents

Solvent	τ (μs)	Reference	Solvent	τ (μs)	Reference
H_2O	2	22	80% C_6H_6 / 20% C_2H_5OH	26±5	23
D_2O	20	26	C_6H_6	24	23
CH_3OH	7	22	C_6H_6	26±3	27
CH_3OH	10±1	23	C_6H_6	25±3	28
CH_3OH	9±1	27	C_6D_6	36±7	27
50% D_2O / 50% CH_3OH	11	22	C_6F_6	600±200	25
C_2H_5OH	12	22	C_5H_5N	33±15	23
C_2H_5OH	10±1	27	$C_4H_4O_2$	32±10	23
95% C_2H_5OH / 5% H_2O	5±2	28	$CHCl_3$	60±15	25
			$CDCl_3$	300±100	25
$CH_3(CH_2)_3OH$	19±3	23	CCl_4	700±200	25
$(CH_3)_3COH$	34±4	23	$CH_3(CH_2)_2Br$	10±2	27
CH_3COCH_3	26	22	$CH_3(CH_2)_2Br$	8±2	28
C_6H_{12}	17	22	CS_2	200±60	25
			Freon 11	1000±200	25

QUENCHING OF SINGLET OXYGEN ($^1\Delta_g$) BY NICKEL CHELATES WHICH ACT AS PHOTOSTABILIZERS OF POLYOLEFINS

Various compounds, including carotenoids and aliphatic and aromatic amines, quench singlet oxygen at least in part by non-chemical mechanisms. The literature is so extensive that space does not permit adequate discussion of this aspect. However, Carlsson and co-workers[29,30] have shown that certain metal(II) chelates, some of which act as photostabilizers of polyolefins, act as efficient singlet oxygen quenchers. We have measured singlet oxygen quenching constants for several nickel chelates and these, together with the values obtained by Carlsson *et al.*,[29] are given in Table 2. These nickel chelates have also been shown to quench the triplet state of benzophenone very efficiently in benzene solution.[31]

Table 2 Singlet oxygen quenching rate constants

Solvent	Quencher	k_Q ($dm^3\ mol^{-1}\ s^{-1}$)	Reference
Benzene[a]	β-Carotene	$(1.3 \pm 0{\cdot}2) \times 10^{10}$	16
Benzene[a]	β-Carotene	$(1.2 \pm 0{\cdot}1) \times 10^{10}$	16
Benzene	Negopex A(I)[b]	$(3{\cdot}1 \pm 0{\cdot}3) \times 10^{9}$	27
Benzene	Negopex B(II)[b]	$(2{\cdot}8 \pm 0{\cdot}3) \times 10^{9}$	27
Isooctane	Nickel(II) bis(2-hydroxy-5-methoxyphenyl-*N*-n-butylaldimine)	$3{\cdot}5 \times 10^{9}$	29
Benzene	Nickel(II) bis(*N*-tert-butylsalicylaldimine)	$(2{\cdot}6 \pm 0{\cdot}2) \times 10^{8}$	27
Isooctane	Nickel(II) n-butylamine-[2,2′-thiobis(4-tert-octyl)phenolate]	$2{\cdot}2 \times 10^{8}$	29
Benzene		$(1{\cdot}1 \pm 0{\cdot}1) \times 10^{8}$	27
Isooctane	D-α-Tocopherol	$(1{\cdot}2 \pm 0{\cdot}1) \times 10^{8}$	33
Isooctane	Nickel(II) bis[2,2′-thiobis-(4-tert-octyl)phenolate]	$1{\cdot}6 \times 10^{8}$	29
Benzene		$(1{\cdot}1 \pm 0{\cdot}1) \times 10^{8}$	27
Isooctane	1,4-Diazabicyclo[2.2.2]octane (DABCO)	$3{\cdot}5 \times 10^{7}$	27
Benzene	Nickel(II) bis(butyl-3,5-ditert-butyl-4-hydroxybenzylphosphonate)	$1{\cdot}4 \times 10^{7}$	27
Dichloromethane		$1{\cdot}2 \times 10^{7}$	29
Benzene	Ferrocene	$< 5 \times 10^{6}$	27
Benzene	Iodoethane	2×10^{4}	27

[a] Independent results obtained using the two different kinetic methods of analysis given in the text.
[b] Structure (I) with R = Me and (II) with $R = C_{11}H_{23}$:

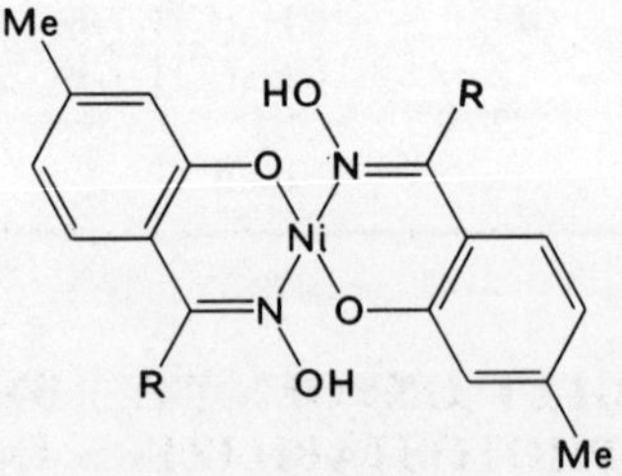

Gas-phase singlet oxygen has been shown to react with polyethylene to form hydroperoxides at the surface of the polymer[32] and several workers (e.g., ref. 12) have proposed that singlet oxygen is an important intermediate in the photo-oxidative degradation of various polymers. The fact that the polymer stabilizers shown in Table 2 quench singlet oxygen efficiently could account for their stabilizing properties. However, much more work on polymers is necessary before the mechanism of stabilization within photodegrading polymers can be established.

REFERENCES

1. R. M. Badger, A. C. Wright, and R. F. Whitlock, *J. Chem. Phys.*, **43** 4345 (1965).
2. W. H. J. Childs and R. Mecke, *Z. Physik*, **68**, 344 (1931).
3. L. Wallace and D. M. Hunten, *J. Geophys. Res.*, **73**, 4813 (1968).
4. R. P. Wayne, *Adv. Photochem.*, **7**, 311 (1969).
5. C. S. Foote, *Accounts Chem. Res.*, **1**, 104 (1968).
6. K. Gollnick, *Adv. Photochem.*, **6**, 1 (1969).
7. E. A. Ogryzlo, *Photophysiology*, **5**, 35 (1970).
8. D. R. Kearns, *Chem. Rev.*, **71**, 395 (1971).
9. J. F. Rabek, *Wiad. Chem.*, **25**, 293, 365, 435 (1971).
10. B. Stevens, *Accounts Chem. Res.*, **6**, 90 (1973).
11. R. P. R. Ranaweeva and G. Scott, *Chem. Ind.*, **1975**, 774.
12. B. Rånby and J. F. Rabek, *Photodegradation, Photo-oxidation and Photostabilization of Polymers*, Wiley, London, 1975.
13. R. H. Young, K. Wehnby, and R. L. Martin, *J. Amer. Chem. Soc.*, **93**, 5775 (1971).
14. K. Gollnick, *Adv. Photochem.*, **6**, 1 (1969).
15. C. S. Foote and R. W. Denny, *J. Amer. Chem. Soc.*, **90**, 6233 (1968).
16. A. Farmilo and F. Wilkinson, *Photochem. Photobiol.*, **18**, 447 (1973).
17. D. R. Adams, *PhD Thesis*, University of East Anglia, 1971.
18. D. R. Adams and F. Wilkinson, *J. Chem. Soc., Faraday Trans. II*, **68**, 586 (1972).
19. C. S. Foote, R. W. Denny, L. Weaver, Y. Chang, and J. Peters, *Ann. N.Y. Acad. Sci.*, **171**, 139 (1970).
20. P. B. Merkel and D. R. Kearns, *Chem. Phys. Lett.*, **12**, 120 (1971).
21. P. B. Merkel and D. R. Kearns, *J. Amer. Chem. Soc.*, **94**, 1029 (1972).
22. P. B. Merkel and D. R. Kearns, *J. Amer. Chem. Soc.*, **94**, 7244 (1972).
23. R. H. Young, D. Brewer, and R. A. Keller, *J. Amer. Chem. Soc.*, **95**, 375 (1973).
24. T. G. Truscott, E. J. Land, and A. Sykes, *Photochem. Photobiol.*, **17**, 43 (1973).
25. C. A. Long and D. R. Kearns, *J. Amer. Chem. Soc.*, **97**, 2018 (1975).
26. P. B. Merkel and D. R. Kearns, *J. Amer. Chem. Soc.*, **94**, 1030 (1972).
27. A. Farmilo and F. Wilkinson, unpublished results, using the method based on the disappearance of an oxidizable acceptor following pulse excitation.
28. A. Farmilo and F. Wilkinson, unpublished results, using triplet β-carotene as a probe for singlet oxygen.
29. D. J. Carlsson, G. D. Mendenhall, T. Suprunchuk, and D. M. Wiles, *J. Amer. Chem. Soc.*, **94**, 8960 (1972).
30. D. J. Carlsson, T. Suprunchuk, and D. M. Wiles, *Can. J. Chem.*, **52**, 3728 (1974).
31. A. Adamczyk and F. Wilkinson, *J. Amer. Polym. Sci.*, **18**, 1225 (1974).
32. M. L. Kaplan and P. G. Kelleher, *J. Polym. Sci. B*, **9**, 565 (1971).
33. S. R. Farenholtz, F. H. Doleiden, A. M. Trozzolo, and A. A. Lamola, *Photochem. Photobiol.*, **20**, 505 (1974).

6

The Mechanism of Quenching of Singlet Oxygen

R. H. YOUNG and D. R. BREWER

Union Carbide Corp., Bound Brook, N.J., USA
Sarah Lawrence College, Bronxville, N.Y., USA

INTRODUCTION

The quenching of singlet oxygen by definition involves the deactivation of the excited singlet states of the oxygen molecule. Although many papers have been written on the quenching of both $^1O_2(^1\Delta)$ and $^1O_2$ $(^1\Sigma)$ both in the gas phase and in solution, this paper will deal primarily with the quenching mechanisms of $^1O_2$ $(^1\Delta)$ in a solvent environment.

The deactivation of singlet oxygen $(^1\Delta)$ can involve either a route whereby the quencher undergoes no ultimate chemical change (physical quenching) or a chemical reaction resulting in a new product. In both cases similar intermediates may be involved in the reaction or quenching paths. Although chemical reaction (k_{rx}) with singlet oxygen is definitely one mechanism of quenching, for the purpose of clarity the term quenching in this paper will refer only to methods of physical quenching (k_q):

$$^1O_2\ (^1\Delta) \xrightarrow[k_q]{Q} {}^3O_2\ (^3\Sigma) + Q \quad \text{(physical quenching)} \tag{1}$$

$$^1O_2\ (^1\Delta) \xrightarrow[k_{rx}]{Q} Q\,O_2 \quad \text{(reaction)} \tag{2}$$

PRODUCTION OF SINGLET OXYGEN

There are numerous methods for the production of singlet oxygen, including direct laser excitation,[1] chemical methods[2–4] and decomposition of phosphite ozonides.[5–8] Perhaps the most common method is through dye-sensitized energy transfer to molecular oxygen:

$$^1\text{Dye}_0 \overset{h\nu}{\rightleftharpoons} {}^1\text{Dye} \xrightarrow{\text{isc}} {}^3\text{Dye}_1 \xrightarrow{^3O_2} {}^1\text{Dye}_0 + {}^1O_2 \tag{3}$$

For all methods of singlet oxygen production there are complications, and precautions must be taken. For dye-sensitized energy transfer the interactions of oxygen and the quencher with the excited singlet and triplet states of the dye sensitizer represent sources of complications.

INTERACTION OF OXYGEN WITH THE SINGLET STATE OF THE SENSITIZER

The quenching of singlet excited states by oxygen has been shown to result in enhanced inter-system crossing[9] at a rate which is diffusion controlled.[10] We were able to determine the rate constant for diffusion of oxygen in a number of solvents by a method involving the Stern–Volmer fluorescence quenching of the singlet state of a number (seven) of organic molecules.[11] The measured diffusion rate constants are given in Table 1.

Table 1 Rate constants[a] of diffusion of oxygen in a variety of solvents

Solvent	$k_{\text{Diffusion}} \times 10^{-10}$ ($\text{M}^{-1}\,\text{s}^{-1}$)
Acetone	3·8 ± 0·1
Benzene	2·6 ± 0·8
2-Butanol	1·7 ± 0·2
n-Butanol	1·9 ± 0·3
tert-Butanol	1·7 ± 0·3
Cyclohexanol	0·32 ± 0·13
Ethyl acetate	3·6 ± 0·5
Ethylene glycol	0·48 ± 0·01
n-Heptane	3·2 ± 0·7
Methanol	3·0
Pyridine	1·9 ± 0·6
2-Propanol	2·3 ± 0·3
Tetrahydrofuran	2·6 ± 0·4
Water	0·53 ± 0·11

[a] Average values calculated from data for four or more hydrocarbons and reported at the 90% confidence level.

INTERACTION OF OXYGEN WITH THE TRIPLET STATE OF THE SENSITIZER

Stevens[12] has suggested that the quenching of triplet states ($^3D_1$) by ground-state molecular oxygen yields singlet oxygen with a rate constant that is limited by the spin statistical factor of 1/9:

$$\text{Probability} = 1/9: \quad {}^3D_1 + {}^3O_2 \underset{k_{-\text{diff.}}}{\overset{k_{\text{diff.}}}{\rightleftharpoons}} {}^1({}^3D^3O_2) \xrightarrow{k_{\text{et}}} {}^1D_0 + {}^1O_2 \tag{4}$$

$$\text{Probability} = 3/9: \quad {}^3D_1 + {}^3O_2 \underset{k_{-\text{diff.}}}{\overset{k'_{\text{diff.}}}{\rightleftharpoons}} {}^3({}^3D^3O_2) \xrightarrow{k_{\text{isc}}} {}^1D_0 + {}^3O_2 \tag{5}$$

$$\text{Probability} = 5/9: \quad {}^3D_1 + {}^3O_2 \underset{k_{-\text{diff.}}}{\overset{k_{\text{diff.}}}{\rightleftharpoons}} {}^5({}^3D^3O_2). \tag{6}$$

Gizzeman *et al.*[13] have suggested that the overall rate constant for oxygen quenching of the triplet state of a sensitizer is affected by the viscosity and the polarity of the solvent. We now have shown that non-energy transfer quenching by oxygen can occur in solvents such as water and ethylene glycol where k_{isc} can compete with $k_{-diff.}$.[11]

Rate constants for the quenching of the triplet states of a number of dyes by oxygen were determined in a series of solvents using a flash photolytic technique with direct observation of the triplet moiety. These rate constants were then compared with the rate constants of diffusion from Table 1 and the results reported in Table 2. Examination of these results shows that the ratio is much closer to the 4/9, the predicted ratio when energy transfer *and* inter-system crossing *both* are modes of quenching of the triplet state in two of the solvent systems. This is the first time that quenching of the triplet states by oxygen has been observed by other than energy transfer.

Table 2 Ratio of rate constants of quenching of triplet states by oxygen to the rate constant of diffusion in different solvents

Solvent	Triplet state			
	Rose bengal	Methylene blue	Thionene	Erythrosin
Methanol	0·044	0·087	—	—
Cyclohexanol	0·058	0·120	0·15	0·11
Ethylene glycol	0·36	0·50	0·40	0·42
Water	0·41	0·47	0·35	—

1/9 = 0·111; 4/9 = 0·445

QUENCHING OF SINGLET OXYGEN

One of the most popular methods used to determine rate constants for either reaction or quenching of singlet oxygen measures the effect of a quencher on the rate of photo-oxidation of a reactive organic compound. This organic compound is often the very reactive diphenylisobenzofuran (DPBF) (**1**), which has been shown to undergo a very rapid reaction with singlet oxygen to yield a monomeric peroxide and/or *o*-dibenzoylbenzene, depending upon the reaction conditions:[14]

C_6H_5 C_6H_5 C_6H_5

$\xrightarrow{^1O_2}$ $\longrightarrow$ (7)

C_6H_5 C_6H_5 C_6H_5

(**1**)

The rate constant of this reaction has been involved in some controversy. However, it has been shown by ourselves and others[15] that the reaction occurs rapidly with no measurable physical quenching of singlet oxygen in most solvents.

The rate of interference of the reaction between the steady state produced singlet oxygen and DPBF by the quencher (Q) is a measure of the effectiveness of the quencher:

$$^1D_0 \underset{-h\nu}{\overset{h\nu}{\rightleftharpoons}} {}^1D_1 \longrightarrow {}^3D_1 \xrightarrow{^3O_2} {}^1O_2 \begin{cases} \xrightarrow{k_d} {}^3O_2 & (8) \\ \xrightarrow{DPBF} \text{DPBF-O}_2 & (9) \\ \xrightarrow[k_q]{k_{rx}} {}^3O_2 & (10) \end{cases}$$

$$-\frac{d[\text{DPBF}]}{dt} = k_{^1O_2}\left[\frac{k_{rx}[\text{DPBF}]}{k_d - k_{rx}[\text{DPBF}] + k_q[Q]}\right] \quad (11)$$

At high concentrations of DPBF this becomes a zero-order reaction and all of the singlet oxygen is captured by DPBF. At very low concentrations of DPBF through the use of a fluorescence technique, a first-order disappearance of DPBF is observed. A variation of a Stern–Volmer method can then be used directly to obtain the ratio (β value) of the rate of decay (k_d) to the rate constant (k_q) of quenching and/or reaction of a second component. If the rate of decay of singlet oxygen (k_q) is known, the absolute rate constant of quenching and/or reaction can be calculated.[16]

An alternative, more direct method, first reported by Kearns *et al.*[17] and developed independently by others,[18–19] can also be used. This method involves a laser flash excitation of the dye to produce a significant concentration of singlet oxygen. The disappearance of the singlet oxygen as a function of time can be monitored by the change in velocity of its reaction with DPBF:

$$-\frac{d[^1O_2]}{dt} = k_d[^1O_2] + k_{rx}[\text{DPBF}][^1O_2] + k_q[Q][^1O_2] \quad (12)$$

$$\ln \ln \left(\frac{[\text{DPBF}]_t}{[\text{DPBF}]_{t=\infty}}\right) = \ln\left(\frac{k_{rx}[^1O_2]_{t=0}}{\alpha}\right) - \alpha t \quad (13)$$

$$\alpha = k_{rx}[\text{DPBF}] + k_d(+k_q[Q]) \quad (14)$$

A more precise variation of this technique involving a least-squares fit of the data and fewer assumptions was used by ourselves to calculate α values. A plot of α against [DPBF] results in a straight line of slope equal to the rate constant of the reaction (k_{rx}) and an intercept equal to the rate constant of decay (k_d) of singlet oxygen for a given solvent.

This method was used to determine the rate constants of decay of singlet oxygen in a large number of solvents, rate constants of reaction (k_{rx}) and rate constants of quenching (k_q) for a large number of organic compounds.

THE MECHANISM OF QUENCHING OF SINGLET OXYGEN

A number of different mechanisms have been postulated for the quenching of singlet oxygen ($^1O_2$ $^1\Delta$), including energy transfer to vibrational energy (solvents), charge-transfer quenching, electronic energy transfer, quenching by excited triplet state molecules, and dimole interactions. Each of these quenching mechanisms will be discussed below.

Table 3 Lifetimes of singlet oxygen in solution

Solvent	Lifetime (μs)	Reference
Methanol	11 ± 1	11, 19
	7	17
	8·5	18
Ethanol	19 ± 3	11
	12	17
Carbon tetrachloride	460[a]	11
	700	17
Water	2·1[a]	11
	2·0	17
Glycol	6·1[a]	11
Benzene	27 ± 4	11
	24	17
Benzene–methanol (4:1)	26 ± 5	11, 19
	17[b]	16
Acetone	42 ± 8	11
	26	17
Acetonitrile	57 ± 9	11
	30	17
Chloroform	228 ± 90	11
	60	17
n-Butanol	19 ± 3	11, 19
	18·2[b]	16
tert-Butanol	34 ± 4	11, 19
	22[b]	16
Bromobenzene–methanol (4:1)	23 ± 4	11, 19
Bromobenzene	75 ± 30	11
2-Propanol	20 ± 1	11
Cyclohexanol	16 ± 2	11
2,2,2-Trifluoroethanol	44 ± 11	11
2,2-Dichloroethanol	47 ± 12	11
2,2,2-Trichloroethanol	51 ± 14	11
2-Fluoroethanol	18 ± 4	11
Epibromohydrin	46 ± 10	11
Pyridine	17 ± 4	11
Dioxan	32 ± 10	11, 19
Tetrahydrofuran	23 ± 2	11
Ethyl acetate	47 ± 15	11
Methylene chloride	140	11
Methyl benzoate	40 ± 16	11

[a] Extrapolated to zero concentration of co-solvent (CH_3OH).
[b] Re-calculated using corrected quenching rate constants for β-carotene.

Quenching of Singlet Oxygen by Solvents

Although the lifetime of $^1O_2$ ($^1\Delta$) has been obtained as 64 min in the gas phase,[20] in solution the lifetime has been shown[17–19] to vary from 10^{-3} to 10^{-6} s and is highly dependent upon the type of solvent. The lifetimes for $^1O_2$ in a number of solvents have been measured and are reported in Table 3. There is reasonable agreement among the lifetime results obtained by most workers. However, some caution should be exercised in using these results. In some cases the amount of reaction of the monitoring reagent, DPBF, was not considered, which will lead to low values for the lifetime of $^1O_2$. The effect of variation of the concentration of the acceptor should be evaluated carefully.[19] In addition, for some solvents which have a very low rate of decay (such as chloroform and carbon tetrachloride), the direct method discussed above results in a large error. However, excellent rate constants of reaction (or quenching) can still be obtained and in combination with β values (k_d/k_{rx}) measured by alternative methods can yield improved lifetimes.

A variation of the above method is used to measure the rate constants of decay for some solvents where either DPBF or the dye sensitizer is not soluble. This method involves the incorporation of a co-solvent (usually methanol). The rate constant of decay in the solvent can be calculated, allowing for the quenching effect of the methanol:

$$k_d\,(\text{total}) = k_d\,(\text{solvent}) + k_{d(CH_3OH)}[CH_3OH] + k_{rx}[DPBF] \tag{15}$$

This method normally assumes that the rate constant of decay of singlet oxygen by methanol can be calculated from the lifetime in neat methanol. The nearest neighbour effect at higher levels of quencher (methanol) may make this assumption invalid. To determine the validity of the assumption, the rate constants for quenching by a number of solvents were determined using these solvents as specific quenchers in carbon tetrachloride solution. The results were then compared with the rate constant of decay calculated from the lifetimes in the neat solvents (Table 4). In general, it can be seen that the rate constant for quenching calculated from pure solvents is only slightly higher than the real quenching rate constant for most systems. It was shown that the rate constant for reaction of DPBF in CCl_4 did not deviate over the range of quenchers (solvents) used for these experiments.

Kearns *et al.*[17] has developed an interesting theory to explain the three orders of magnitude difference in solvent quenching rate constants. They found that these rate constants were correlated with the optical densities of the solvents according to the empirical relationship

$$k_d \times 10^{-6} \propto 0{\cdot}5(\text{O.D.}_{7780\,\text{cm}^{-1}}) + 0{\cdot}05(\text{O.D.}_{6280\,\text{cm}^{-1}}) \tag{16}$$

where O.D. is the optical density of the solvent at the designated wavelength.

From this correlation, Kearns *et al.* developed a theory of decay in solution that is based on the transfer of the electronic energy of singlet oxygen into the vibrational modes of the solvent molecules. This theory was tested using the

Table 4 Rate constants of quenching of singlet oxygen by solvent molecules

Solvent	$k_q \times 10^3$ ($\text{M}^{-1}\,\text{s}^{-1}$)	
	Experimental[a]	Calculated[b]
Methanol	$3{\cdot}0 \pm 1{\cdot}0$	$3{\cdot}7 \pm 0{\cdot}2$
Ethanol	$1{\cdot}7 \pm 1{\cdot}1$	$3{\cdot}1 \pm 0{\cdot}2$
2-Propanol	$1{\cdot}7 \pm 0{\cdot}9$	$3{\cdot}8 \pm 0{\cdot}2$
n-Butanol	$2{\cdot}4 \pm 0{\cdot}6$	$4{\cdot}6 \pm 0{\cdot}7$
tert-Butanol	$1{\cdot}8 \pm 1{\cdot}5$	$2{\cdot}8 \pm 0{\cdot}4$
Pyridine	$1{\cdot}7 \pm 2{\cdot}2$	$4{\cdot}9 \pm 1{\cdot}3$
Dioxan	$0{\cdot}7 \pm 1{\cdot}5$	$2{\cdot}5 \pm 0{\cdot}9$
Tetrahydrofuran	$1{\cdot}9 \pm 1{\cdot}1$	$3{\cdot}6 \pm 0{\cdot}4$
Benzene	$2{\cdot}2 \pm 1{\cdot}5$	$2{\cdot}4 \pm 0{\cdot}5$
Bromobenzene	$1{\cdot}9 \pm 1{\cdot}3$	$1{\cdot}4 \pm 0{\cdot}6$
Methyl benzoate	$2{\cdot}9 \pm 1{\cdot}8$	$3{\cdot}2 \pm 1{\cdot}1$

[a] Determined in CCl_4 solution.
[b] Calculated from neat solvent, $k_q = k_d/[\text{Solvent}]$.

larger number of solvents reported in Table 3, with marginal success. In an attempt to improve the correlation, a best fit for the relative size of the Frank–Condon factors was obtained and was found to give only a slight improvement. Introduction of a term to allow for the difference in polarity of the solvents worked very well for a limited number of systems (alcohols).

$$\ln k_d = a + b\left[\frac{\text{Ionization potential}}{0{\cdot}5\ \text{O.D.}_{7880\,\text{cm}^{-1}} + 0{\cdot}05\ (\text{O.D.}_{6280\,\text{cm}^{-1}}) + \cdots}\right] \tag{17}$$

Charge-transfer Quenching of Singlet Oxygen

Different compounds have been suggested to quench singlet oxygen by a charge–transfer mechanism, including amines,[21–25], sulphides, [26–28] azomethine dyes,[29] and bilirubin.[30, 31] The suggested mechanism is

$$^1O_2 + A \underset{}{\overset{k_{\text{diff.}}}{\rightleftharpoons}} \left[\overset{\delta-}{O_2} \cdots \overset{\delta+}{A}\right] \longrightarrow {}^3O_2 + A \tag{18}$$

Evidence for the involvement of a charge–transfer or partial charge–transfer exciplex in the quenching of singlet oxygen is in several parts.

Solomon *et al.*[32] have suggested that the free energy (ΔF) of reaction between a donor and acceptor in an electron-transfer process is dependent upon the ionization potential (IP_d) of the donor, the singlet excitation energy of the acceptor (ΔE_a), the electron affinity of the acceptor (EA_a), and the coulombic energy gained from the interaction of the ionic species (C):

$$\Delta F \approx -\Delta E_a + IP_d - EA_a - C \tag{19}$$

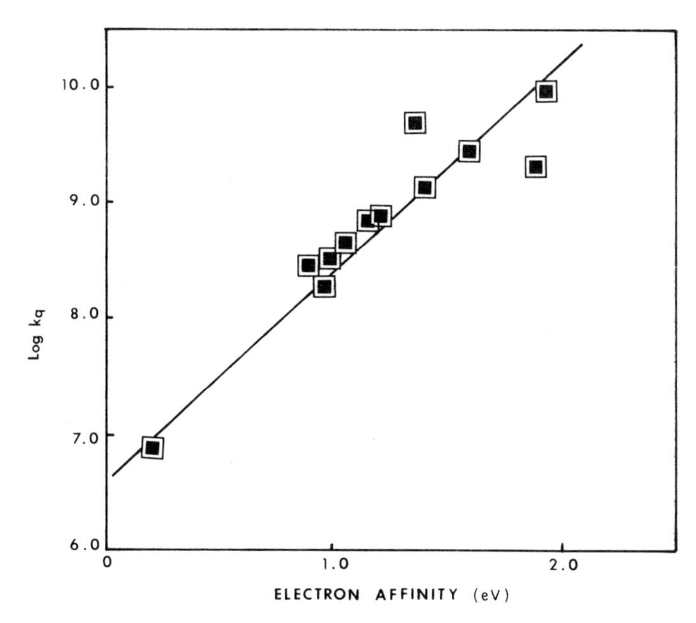

Figure 1 Quenching of the first excited state of aromatic compounds with DABCO (1,4-diazabicyclo[2.2.2]octane) in methanol

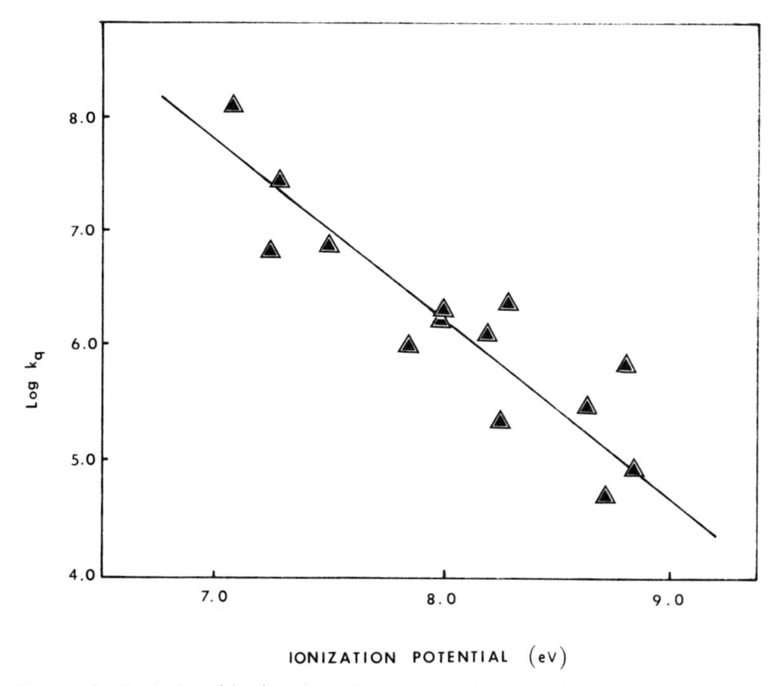

Figure 2 Relationship between log (rate of quenching) and ionization potential of amines for quenching of singlet oxygen in methanol

The reduction potential was used as a measure of the electron affinity along with measured values for the singlet energy for the first excited state of a number of molecules. A plot of the logarithm of the rate constant of quenching of the singlet state (fluorescence quenching) by amines against this measure of electron affinity of the excited singlet state (Weller function) was made for a large number of aromatic and polyaromatic compounds[23] and is shown in Figure 1. An excellent correlation was obtained, including the rate constant for the quenching of singlet oxygen by amine.

In addition, it has been shown[22,24,25] that there is a correlation between the rate constant of quenching of singlet oxygen and the ionization potential for a large number of amines (Figure 2). It has also been shown that rate constants with a number of substituted dimethylanilines gave a Hammett rho relationship with a rho value of $-1{\cdot}4$ to $-1{\cdot}7$ for the quenching of singlet oxygen. All of this evidence points towards the intermediacy of a partial charge-transfer exciplex in the quenching of the first excited singlet state of the oxygen molecule.

In a similar manner, it can be shown that a charge-transfer species is probably also involved in the quenching of singlet oxygen by sulphides. Figure 3 illustrates the correlation obtained between the logarithm of the rate constant of quenching

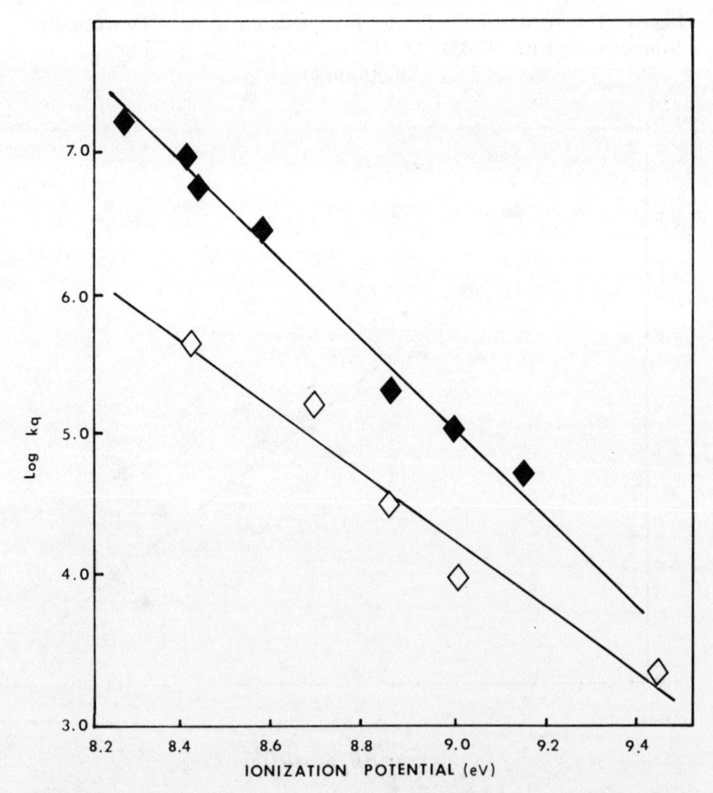

Figure 3 Plots of log (rate constant of quenching) *versus* ionization potential for gas phase values (◊) and values in methanol (♦)

and the ionization potential for a number of sulphides both in the gas phase[26] and in solution.[27] The difference in slope may reflect the relative effect of physical quenching (gas phase plus solution) plus the effect of reaction quenching (solution).

Reaction versus Quenching

There are a number of compounds which both physically quench and react with singlet oxygen, including amines,[11,33] sulphides,[34,35] bilirubin,[30,31] and tocopherols.[36–37] A summary of the rate constants for quenching and reaction of examples of these systems is given in Table 5. Foote and co-workers[34,35] have suggested that with sulphides the ratio of quenching to reaction is dependent upon the polarity of the solvent.

Table 5 Rate constants for physical quenching and reaction with singlet oxygen

Compound	k_q (s^{-1})	k_{rx}(s^{-1} M^{-1})	Solvent	Reference
Triethylamine	$2{\cdot}8 \times 10^6$	$5{\cdot}3 \times 10^5$	CH_3OH	11
Di-*n*-butylsulphide	$5{\cdot}0 \times 10^5$	$7{\cdot}8 \times 10^6$	CH_3OH	11
Bilirubin	$2{\cdot}1 \times 10^9$	$0{\cdot}4 \times 10^9$	$CHCl_3$	30
	$2{\cdot}1 \times 10^9$	$0{\cdot}2 \times 10^9$	CCl_4	31
α-Tocophenol	$6{\cdot}2 \times 10^8$	$4{\cdot}6 \times 10^7$	CH_3OH	38
	$2{\cdot}5 \times 10^8$	$2{\cdot}0 \times 10^6$	C_5H_5N	37
	$1{\cdot}7 \times 10^8$	—	C_6H_6	38

Quenching by Energy Transfer

A number of different compounds have been reported as effective quenchers of singlet oxygen in solution via an energy-transfer mechanism. One of the most effective of these is β-carotene, which has been reported by Foote and co-workers[38] to exhibit diffusion-controlled quenching. As can be seen from the calculated values for the rate constants of quenching in Table 6, the rate constants are actually about half of that for diffusion of oxygen.

Table 6 Rate constants of quenching of singlet oxygen by energy transfer

Compound	Solvent	$k_d/k_{rx} \times 10^6$	$k_d \times 10^4$	$k_q \times 10^{10}$	$k_{diff.} \times 10^{10}$	Reference
β-Carotene	CH_3OH	6·1	9·1	1·5	3·0	11, 12, 19
	C_6H_6–CH_3OH (4:1)	3·0	3·8	1·3	2·6	11, 16, 19
Polymethene pyrulium dye	—	—	—	3·0	—	43
Ni(II) bis-isooctane (di-*n*-butyl-dithiocarbonate)	iso-octane	—	—	1·7	—	41

Other compounds which have been suggested to involve energy-transfer quenching include the important class of transition metal chelates,[39,40] iodine atoms,[41] and a polymethane pyrulium dye.[42] The transition metal chelates have been shown to be effective as additives in protecting polymers from photo-oxidation degradations.[43–46]

Miscellaneous Mechanisms for Quenching of Singlet Oxygen

Other mechanisms have been proposed for specific quenching reactions of singlet oxygen. These include:

(*i*) quenching by triplet excited states:[47]

$$^1O_2\,(^1\Delta) + {}^3S_1 \longrightarrow {}^3O_2 + {}^1S_0 \tag{20}$$

(*ii*) energy pooling reactions:[48]

$$2\,^1O_2\,(^1\Delta) \longrightarrow 2\,O_2\,(^3\Sigma) + h\nu \tag{21}$$

$$2\,^1O_2\,(^1\Delta) \longrightarrow O_2\,(^1\Sigma) + O_2\,(^3\Sigma) \tag{22}$$

$$2\,^1O_2\,(^1\Delta) + {}^1\text{Rubene}_0 \longrightarrow O_2\,(^3\Sigma) + {}^1\text{Rubene}_1^* \tag{23}$$

REFERENCES

1. I. B. C. Matheson and J. Lee, *J. Amer. Chem. Soc.*, **94**, 3310 (1972).
2. C. S. Foote and S. Wexler, *J. Amer. Chem. Soc.*, **86**, 3879 (1964).
3. C. S. Foote and S. Wexler, *J. Amer. Chem. Soc.*, **86**, 3880 (1964).
4. E. McKeown and W. Waters, *J. Chem. Soc. B*, **1966**, 1040.
5. Q. E. Thompson, *J. Amer. Chem. Soc.*, **83**, 845 (1961).
6. E. Wasserman, R. W. Murry, M. L. Kaplan, and W. A. Yager, *J. Amer. Chem. Soc.*, **90**, 4160 (1968).
7. R. W. Murry and M. L. Kaplan, *J. Amer. Chem. Soc.*, **90**, 527 (1968).
8. R. W. Murry and M. L. Kaplan, *J. Amer. Chem. Soc.*, **90**, 4161 (1968).
9. B. Stevens and B. E. Algar, *Ann. N.Y. Acad. Sci.*, **171**, 50 (1970).
10. W. R. Ware, *J. Phys. Chem.*, **66**, 455 (1962).
11. D. R. Brewer, *PhD Thesis*, Georgetown University, 1974.
12. B. Stevens, *Org. Scintill. Liquid Scintill. Counting*, *Proc. Int. Conf.*, 117 (1971).
13. O. L. J. Gizzeman, F. Kaufman, and G. Porter, *J. Chem. Soc.*, *Faraday Trans. II*, **1973**, 708.
14. J. A. Howard and G. D. Mendenhall, *Can. J. Chem.*, **53**, 2199 (1975).
15. P. B. Merkel and D. R. Kearns, *J. Amer. Chem. Soc.*, **97**, 462 (1975).
16. R. H. Young, K. Wehrly, and R. L. Martin, *J. Amer. Chem. Soc.*, **93**, 5774 (1971).
17. D. R. Kearns, P. B. Merkel, and R. Nilsson, *J. Amer. Chem. Soc.*, **94**, 7244 (1972).
18. D. R. Adams and F. Wilkinson, *J. Chem. Soc.*, *Faraday Trans. II*, **1972**, 586.
19. R. H. Young, D. Brewer, and R. A. Keller, *J. Amer. Chem. Soc.*, **95**, 375 (1973).
20. R. M. Badger, A. C. Wright, and R. F. Whitlock, *J. Chem. Phys.*, **43**, 4345 (1965).
21. C. Quannes and T. Wilson, *J. Amer. Chem. Soc.*, **90**, 6527 (1968).
22. K. Furukawa and E. A. Ogryzlo, *J. Photochem.*, **1**, 163 (1972/73).
23. R. H. Young and R. L. Martin, *J. Amer. Chem. Soc.*, **94**, 5183 (1972).
24. R. H. Young, R. L. Martin, D. Feriozi, D. Brewer, and R. Kayser, *Photochem. Photobiol.*, **17**, 233 (1973).

25. R. H. Young, D. Brewer, R. Kayser, R. Martin, D. Feriozi, and R. A. Keller, *Can. J. Chem.*, **52**, 2889 (1974).
26. R. A. Ackerman, J. N. Pitts, and I. Rosenthal, *Symposium on Singlet Oxygen, Amer. Chem. Soc. Div. Pet. Chem. Prepr.*, **16**, No. 4, A25 (1971).
27. R. L. Martin, *PhD Thesis*, Georgetown University, 1972.
28. E. A. Ogryzlo and C. W. Tang, *J. Amer. Chem. Soc.*, **92**, 5030 (1970).
29. W. F. Smith, W. G. Kerkstroeter, and K. L. Eddy, *J. Amer. Chem. Soc.*, **97**, 2764 (1975).
30. C. S. Foote and T. Ching, *J. Amer. Chem. Soc.*, **97**, 6209 (1975).
31. B. Stevens and R. D. Small, Jr., *Photochem. Photobiol.*, **23**, 33 (1976).
32. B. S. Solomon, C. Steel, and Z. Weller, *Chem. Commun.*, **1969**, 927.
33. W. F. Smith, *J. Amer. Chem. Soc.*, **94**, 186 (1972).
34. C. S. Foote and J. W. Peters, *J. Amer. Chem. Soc.*, **93**, 3795 (1971).
35. C. S. Foote, R. W. Denny, L. Weaver, Y. Chang, and J. Peters, *Ann. N.Y. Acad. Sci.*, **171**, 139 (1970).
36. S. R. Fahrenholtz, F. H. Doleiden, A. M. Trozzolo, and A. A. Lamola, *Photochem. Photobiol.*, **20**, 505 (1974).
37. B. Stevens, R. D. Small, Jr., and S. R. Perez, *Photochem. Photobiol.*, **20**, 515 (1974).
38. C. S. Foote and R. W. Denny, *J. Amer. Chem. Soc.*, **90**, 6233 (1968).
39. D. J. Carlsson, T. Suprunchuk, and D. M. Wiles, *Can. J. Chem.*, **52**, 3728 (1974).
40. R. G. Derwent, and B. A. Thrush, *Faraday Discuss. Chem. Soc.*, **53**, 162 (1972).
41. P. B. Merkel and D. R. Kearns, *J. Amer. Chem. Soc.*, **94**, 7244 (1972).
42. A. Zweig and W. A. Henderson, Jr., *J. Polym. Sci.*, **13**, 717 (1975).
43. J. P. Guilory and C. F. Cooks, *J. Polym. Sci.*, **11**. 1927 (1973).
44. D. J. Carlson, T. Suprunchuk, and D. M. Wiles, *J. Polym. Sci., Polym. Lett. Ed.*, **11**, 61 (1973).
45. J. Flood, K. E. Russell, and J. K. S. Wan, *Macromolecules*, **65**, 669 (1973).
46. C. K. Duncan and D. R. Kearns, *Chem. Phys. Lett.*, **12**, 306 (1971).
47. U. Schurath, *J. Photochem.*, **4**, 215 (1975).
48. A. U. Khan and M. Kasha, *J. Chem. Phys.*, **39**, 2105 (1963).

7

The Use of β-Carotene as a Probe for Singlet Oxygen Yields and Lifetime

A. GARNER† and F. WILKINSON

School of Chemical Sciences, University of East Anglia, Norwich, NR4 7TJ, England

INTRODUCTION

The rate constants for oxygen quenching of an extensive range of aromatic triplet states ($^3M^*$) are known to approach a maximum value of 1/9th of the diffusion controlled rate in solution. These limiting rates are thought[1–3] to arise from spin statistical factors introduced during the collisional interaction of quencher and quenchee through the following mechanism:

$$^3M^* + {}^3O_2 \xleftarrow{k_d} \begin{cases} \xrightarrow{1/9\,k_d} {}^1[M\cdot\cdot O_2]^* \longrightarrow {}^1M + {}^1O_2^* & (1) \\ \xrightarrow{1/3\,k_d} {}^3[M\cdot\cdot O_2]^* \longrightarrow {}^1M + {}^3O_2 & (2) \\ \xrightarrow{5/9\,k_d} {}^5[M\cdot\cdot O_2]^* & (3) \end{cases}$$

Only process (1), which represents 1/9th of the total interactions, will lead to quenching of the triplet state, and consequently the yield of singlet oxygen ($^1O_2^*$) is unity.[4] As the energy gap between the initial and final states increases the observed quenching rate constant decreases as the Franck–Condon factors become less favourable for energy transfer;[1] however, this should not affect the singlet oxygen yield.

Our observations[5,6] that several triplet states are quenched by oxygen with rate constants greater than 1/9th diffusion controlled have led us to postulate the following modification to mechanisms (1) and (2) where charge-transfer interactions are thought to be energetically favourable:

$$3M^* + {}^3O_2 \underset{}{\overset{1/9\,k_d}{\rightleftharpoons}} {}^1[M\cdot\cdot O_2]^* \rightleftharpoons {}^1[M^{+\cdot}\cdot O_2^-] \longrightarrow {}^1[M\cdot\cdot{}^1O_2] \longrightarrow {}^1M + {}^1O_2^* \quad (4)$$

$$\downarrow\uparrow$$

$$3M^* + {}^3O_2 \underset{}{\overset{1/3\,k_d}{\rightleftharpoons}} {}^3[M\cdot\cdot O_2]^* \rightleftharpoons {}^3[M^{+\cdot}\cdot O_2^-] \longrightarrow {}^1M + {}^3O_2 \quad (5)$$

The observed rate constant for the oxygen quenching of the triplet state of

† Present address: Laboratory of Radiation and Biophysical Chemistry, School of Chemistry, The University, Newcastle upon Tyne, NE1 7RU, England.

N-methylindole in benzene solution agrees well with the theoretical limiting rate for this mechanism of 4/9th diffusion controlled.[6] In the absence of inter-system crossing between the almost degenerate singlet and triplet charge transfer states, the quantum yield of singlet oxygen would be 0·25, and yields between 0·25 and unity would be expected where inter-system crossing is efficient.

METHOD

In the presence of oxygen, benzene solutions of anthracene (A) containing β-carotene (C), when subjected to laser photolysis, have been shown to sensitize the formation of triplet carotene [7] as a result of the following mechanism:

$$^3A^* + {}^3O_2 \xrightarrow{k_{TO}} {}^1O_2^* + {}^1A \tag{6}$$

$$^3A^* + {}^1C \xrightarrow{k_{TC}} {}^3C^* + {}^1A \tag{7}$$

$$^1O_2^* + {}^1C \xrightarrow{k_{\Delta C}} {}^3C^* + {}^3O_2 \tag{8}$$

$$^1O_2^* \xrightarrow{k_1} {}^3O_2 \tag{9}$$

$$^3C^* + {}^3O_2 \xrightarrow{k_{CO}} {}^1C + {}^3O_2 \tag{10}$$

$$^3C^* \xrightarrow{k_{DC}} {}^1C \tag{11}$$

The experimental arrangement of the laser photolysis apparatus has been described.[8] Under the conditions of the experiment with $[O_2] \gg [C]$, process (6) dominates over (7), and the formation of triplet β-carotene is due mainly to process (8), which is thus confirmed as an energy-transfer process with a rate constant of about $1{\cdot}2 \times 10^{10}$ dm^3 mol^{-1} s^{-1}. Triplet β-carotene is produced as an observable transient with an absorption at 520 nm and an extinction in excess of 10^5 dm^3 mol^{-1} cm^{-1}.[9] Reaction (10) has also been demonstrated [10] with a rate constant of 3×10^9 dm^3 mol^{-1} s^{-1}.

The overall scheme for the sensitized triplet β-carotene formation may be represented as a sequential energy transfer path, $^3A^* \longrightarrow {}^1O_2^* \longrightarrow {}^3C^*$, and the kinetic analysis of this scheme has already been presented,[7] and gives

$$[^3C^*] = P(e^{-k_A t} - e^{-k_C t}) + Q(e^{-k_B t} - e^{-k_C t}) \tag{12}$$

where

$$k_A = k_{TO}[O_2] + k_{TC}[C] \tag{13}$$

$$k_B = k_1 + k_{\Delta C}[C] \tag{14}$$

and

$$k_C = k_{CO}[O_2] + k_{DC} \tag{15}$$

At times greater than 1 μs, $e^{-k_A t}$ and $e^{-k_C t}$ become negligible compared with $e^{-k_B t}$ and equation (12) reduces to

$$[^3C^*]=Qe^{-k_B t} \quad \text{with} \quad Q=\frac{[^3A^*]_0 k_{\Delta C}[C]k_{TO}[O_2]}{(k_A-k_B)(k_C-k_B)} \tag{16}$$

and it is observed that the decay of triplet carotene is first order and follows the decay of singlet oxygen according to the exponent $(-k_B t)$, i.e. step (8) is rate limiting. $[^3A^*]_0$ is the initial concentration of the anthracene triplet.

In laser photolysis experiments with *N*-methylindole and naphthalene (N) in benzene solution it was necessary [6] to sensitize the formation of their triplet states with higher energy triplet ketones, $^3S^*$. This introduces additional steps into the reaction sequence:

$$^3S^* + {}^1N \xrightarrow{k_{SN}} {}^3N^* + {}^1S \tag{17}$$

$$^3S^* + {}^3O_2 \xrightarrow{k_{SO}} {}^1O_2^* + {}^1S \tag{18}$$

$$^3S^* + {}^1C \xrightarrow{k_{SC}} {}^3C^* + {}^1S \tag{19}$$

$$^3S^* \xrightarrow{k_S} {}^1S \tag{20}$$

followed by

$$^3N^* + {}^3O_2 \xrightarrow{k_{NO}} {}^1O_2^* + {}^1N \tag{21}$$

$$^3N^* \xrightarrow{k_N} {}^1N \tag{22}$$

$$^3N^* + {}^1C \xrightarrow{k_{NC}} {}^3C^* + {}^1N \tag{23}$$

The overall energy transfer scheme can be represented as

$$^3S^* \longrightarrow {}^3N^* \longrightarrow {}^1O_2^* \longrightarrow {}^3C^* \tag{24}$$

and

$$^3S^* \longrightarrow {}^1O_2^* \longrightarrow {}^3C^* \tag{25}$$

and again it is possible to ignore the contributions of processes (19) and (23) to produce triplet β-carotene. Kinetic analysis of this scheme gives the following expression:

$$[^3C^*]=A(e^{-k_D t}-e^{-k_C t})+B(e^{-k_B t}-e^{-k_C t})+C(e^{-k'_A t}-e^{-k_C t}) \tag{26}$$

$$+D(e^{-k_B t}-e^{-k_C t})+P(e^{-k'_A t}-e^{-k_C t})+Q'(e^{-k_B t}-e^{-k_C t}) \tag{27}$$

where

$$k_D=k_{SN}[N]+k_{SO}[O_2]+k_{SC}[C]+k_S \tag{28}$$

and

$$k'_A = k_{NO}[O_2] + k_{NC}[C] + k_N \tag{29}$$

As in the case of anthracene sensitization of triplet β-carotene, after 1 μs only the exponent $e^{-k_B t}$ need to be considered, and consequently

$$[^3C^*] = (B + D + Q')\, e^{-k_B t} \tag{30}$$

with

$$B = \frac{[^3S^*]_0 k_{SN}[N] k_{NO}[O_2] k_{\Delta C}[C]}{(k_D - k'_A)(k_D - k_B)(k_B - k_C)} \tag{31}$$

$$D = \frac{[^3S^*]_0 k_{SN}[N] k_{NO}[O_2] k_{\Delta C}[C]}{(k_D - k'_A)(k'_A - k_B)(k_C - k_B)} \tag{32}$$

and

$$Q' = \frac{[^3S^*]_0 k_{SO}[O_2] k_{\Delta C}[C]}{(k_D - k_B)(k_C - k_B)} \tag{33}$$

Insertion of values of the rate constants into the above three terms gives the ratios $B:D:Q'$ as $-0{\cdot}18:1{\cdot}0:0{\cdot}09$, and the B and Q' terms are small and partly compensating. In general, for these schemes errors of less than 10% are produced by ignoring the B and Q' terms, and then

$$[^3C^*] = \frac{[^3S^*]_0 k_{SN}[N] k_{NO}[O_2] k_{\Delta C}[C]}{(k_D - k'_A)(k'_A - k_B)(k_C - k_B)} \cdot e^{-k_B t} \tag{34}$$

Differences in k_B arise when either S, N or A quench singlet oxygen, as is the case with *N*-methylindole ($k = 3 \times 10^6$ $dm^3\ mol^{-1}\ s^{-1}$); however, extrapolation of the first-order decay curves to zero time removes the dependence on the exponential term. Furthermore, the term D is insensitive to changes in k_B since both k'_A and k_C are very much greater than k_B.

RESULTS AND DISCUSSION

Rearrangement of equations (16) and (34) and insertion of the equality $[^3C^*]_{t=0} = O.D._{520}/\varepsilon_{520}$, where ε_{520} is the molar decadic extinction of triplet β-carotene at 520 nm and $O.D._{520}$ the optical density extrapolated to zero time, gives

$$k_{TO}[O_2] = \frac{(k_A - k_B)(k_C - k_B) O.D._{520}}{[^3A^*]_0 k_{\Delta C}[C] \varepsilon_{520}} \tag{35}$$

and

$$k_{NO}[O_2] = \frac{(k_D - k'_A)(k'_A - k_B)(k_C - k_B) O.D._{520}}{[^3S^*]_0 k_{SN}[N] k_{\Delta C}[C] \varepsilon_{520}} \tag{36}$$

The value of $k_{TO}[O_2]$ from the β-carotene sensitization is assumed to be equal to that of the oxygen quenching of $^3A^*$, i.e. the yield of singlet oxygen is assumed to be unity. Then $([^3A^*]_0 \varepsilon_{520})^{-1}$ can be evaluated from equation (35) and inserted into equation (36) where $[^3S^*]_0$ is arranged to be equal to $[^3A_0^*]$ in laser photolysis experiments; $k_{NO}[O_2]$ can be evaluated by inserting $O.D._{520}$ and the rate constants, measured independently, into equation (36).

The value of $k_{NO}[O_2]$ so obtained represents only that part of process (21) which gives rise to singlet oxygen. If this is represented as $k_{NO}^c[O_2]$, then the yield of singlet oxygen is equal to $\phi(^1O_2^*) = k_{NO}^c[O_2]/k_{NO}[O_2]$, where $k_{NO}[O_2]$ is the overall quenching rate by oxygen. This latter value is determined from the rate of quenching of $^3N^*$ by oxygen by observing the decay of the triplet–triplet absorption of N in laser photolysis experiments.[6]

Values of $\phi(^1O_2^*)$ so obtained are presented in Table 1, and low yields are observed for all cases except where naphthalene has been included. The results for triplet *N*-methylindole, where it is thought that quenching by processes (1) and (2) is occurring, show that the value of $\phi(^1O_2^*)$ lies between the limiting cases of 0·25 and unity discussed above. This may indicate that inter-system crossing between these states is competitive with internal conversion of the states.

Table 1 Yields of singlet oxygen from various triplet states

Sensitizer	Triplet quenched by oxygen	$O.D._{520}$ $(t=0)$	$k_{TO}[O_2]$ or $k_{NO}[O_2]$ (μs^{-1}): From O_2 quenching	From triplet β-carotene sensitization	$^1O_2^*$ yield[a]
	Anthracene	0·23	5·6	5·6[b]	1·0
Acetophenone	Naphthalene	0·143	3·7	3·9	1·05
Xanthone	Naphthalene	0·091	3·7	3·6	0·97
Acetophenone	*N*-Methylindole	0·079	27	15	0·56
Xanthone	*N*-Methylindole	0·042	27	11	0·40
	Acetophenone	0·082	6·8	4·4	0·65
	Benzophenone	0·067	4·8	2·7	0·56
	Xanthone	0·027(5)	9·5	5·0	0·53
	Stilbene	—	6·5[c]	—	0·13 ± 0·08[c]

[a] Errors of the order of ±15%.
[b] Assumed value (see text).
[c] Determined by pulse radiolysis.

Quenching of the triplet states of certain aromatic ketones by oxygen has been discussed before[5] in terms of a possible contribution of the spin forbidden process

$$^3[M\cdot\cdot O_2]^* \longrightarrow {}^1[M\cdot\cdot O_2]^* \qquad (38)$$

This should lead to a yield of singlet oxygen of unity even when quenching is greater than 1/9th diffusion controlled. The low yields reported here indicate that some decay path other than processes (1) and (38) is operating. Charge-transfer interactions may be implicated since Merkel and Kearns[11] have shown that quenching of the triplet states of several aromatic ketones is less than 1/9th diffusion controlled in CCl_4, which has a marginally lower dielectric constant than benzene. The lower yield of singlet oxygen from triplet *N*-methylindole when sensitized by xanthone, compared with sensitization by acetophenone, is not unexpected since we have observed[6] that quenching of triplet xanthone by *N*-methylindole does not give rise to 100% energy transfer.

We have also estimated the yield of singlet oxygen from triplet stilbene by a similar method using β-carotene in pulse radiolysis experiments, and the value is shown in Table 1. Stilbene is of special interest in connection with the geometry of its triplet excited state,[12] since the vertical energy gap from the triplet state is less than the energy between the torsionally relaxed ground and triplet states. In this case the low yield of singlet oxygen may indicate that process (2) is more efficient than (1) since Franck–Condon factors are favourable for the spin exchange process.

Finally, in view of the ability of certain singlet oxygen quenchers, including nickel chelates[7] and β-carotene, to reduce the photodegradation[13,14] of various polymers, it would be of interest to employ the method outlined above to determine the yields and lifetimes of singlet oxygen in solution in the presence of polymers.

REFERENCES

1. O. L. J. Gijzeman, F. Kaufman, and G. Porter, *J. Chem. Soc. Faraday Trans. II*, **69**, 708 (1973); O. L. J. Gijzeman and F. Kaufman, *J. Chem. Soc. Faraday Trans. II*, **69**, 721 (1973).
2. A. Adamczyk and F. Wilkinson, *Organic Scintillators and Liquid Scintillation Counting*, Academic Press, New York, 1971, p. 233.
3. K. Kawaoka, A. U. Khan, and D. R. Kearns, *J. Chem. Phys.*, **46**, 1842 (1967).
4. B. Stevens, *Accounts Chem. Res.*, **6**, 90 (1973).
5. A. Garner and F. Wilkinson, *J. Chem. Soc. Faraday Trans. II*, **72**, 1010 (1976).
6. A. Garner and F. Wilkinson, *Chem. Phys. Lett.*, **45**, 432 (1977); F. Wilkinson and A. Garner, *J. Chem. Soc. Faraday Trans., II*, **73**, 222 (1977).
7. A. Farmilo and F. Wilkinson, *Photochem. Photobiol.*, **18**, 447 (1973).
8. D. R. Adams and F. Wilkinson, *J. Chem. Soc. Faraday Trans. II*, **68**, 586 (1972).
9. E. J. Land, A. Sykes, and T. G. Truscott, *Photochem. Photobiol.*, **13**, 311 (1971).
10. T. G. Truscott, E. J. Land, and A. Sykes, *Photochem. Photobiol.*, **17**, 43 (1973).
11. P. B. Merkel and D. R. Kearns, *J. Chem. Phys.*, **58**, 398 (1973).
12. J. Saltiel and B. Thomas, *Chem. Phys. Lett.*, **37**, 147 (1976).
13. B. Rånby and J. F. Rabek, *Photodegradation, Photo-oxidation and Photostabilization of Polymers*, Wiley, New York, 1975.
14. J. F. Rabek and B. Rånby, *J. Polym. Sci. A1*, **12**, 273 (1974).

8

Kinetics of Singlet Oxygen Peroxidation

B. STEVENS
Department of Chemistry, University of South Florida, Tampa, FL. 33620, USA

Insofar as $O_2(^1\Delta_g)$ plays an intermediary role in the photosensitized peroxidation of unsaturated organic molecules,[1,2] elucidation of the reaction mechanism is concerned with the identity of those elementary processes in which $O_2(^1\Delta_g)$ is generated and consumed. Since $O_2(^1\Delta_g)$ is spectroscopically inaccessible in solution †, these processes must necessarily be deduced from the analysis of the dependence of the overall quantum yield on such experimental variables as acceptor inhibitor and dissolved oxygen concentration.[3]

In the simplest case where the excitation energy of $O_2(^1\Delta_g)$ (ca. 8000 cm^{-1}) exceeds the energy separation, ΔE_{ST}, of sensitizer singlet (S_1) and triplet (T_1) states, the energy transfer quenching processes (1) and (2) are prohibited by energetic and spin conservation requirements, respectively:

$$S_1 + O_2(^3\Sigma) \longrightarrow T_1 + O_2(^1\Delta) \quad (1)$$

$$S_1 + O_2(^3\Sigma) \longrightarrow S_0 + O_2(^1\Delta) \quad (2)$$

and the source of $O_2(^1\Delta)$ is limited to process (3)

$$T_1 + O_2(^3\Sigma) \rightarrow S_0 + O_2(^1\Delta) \quad (3)$$

which, except at very low oxygen concentrations,[4] competes effectively with triplet relaxation [process (4)]. The sensitizer triplet state in turn may be produced by

$$T_1 \rightarrow S_0 \quad (4)$$

inter-system crossing from, and oxygen quenching of, the sensitizer singlet state described by processes (5) and (6):

$$S_1 \rightarrow T_1 \quad (5)$$

$$S_1 + O_2(^3\Sigma) \rightarrow T_1 + O_2(^3\Sigma) \quad (6)$$

† $O_2(^1\Delta)$ has yet to be identified in absorption or emission in solutions undergoing photosensitized peroxidation.

If the fate of $O_2(^1\Delta)$ is limited to reaction with the acceptor or substrate molecule A [process (7)] and the solvent-induced electronic relaxation [process (8)], the overall photoperoxidation quantum yield

$$O_2(^1\Delta) + A \rightarrow AO_2 \tag{7}$$

$$O_2(^1\Delta) \rightarrow O_2(^3\Sigma) \tag{8}$$

is given by equation (9):

$$\gamma AO_2 = \left(\frac{\gamma_{IS} + K[O_2]}{1 + K[O_2]}\right)\left(\frac{[A]}{\beta_A + [A]}\right) \tag{9}$$

where $\gamma_{IS} = k_5\tau_s$, $K = k_6\tau_s$, and $\beta_A = k_8/k_7$, $k_4 \ll k_3[O_2]$, and τ_s is the unquenched lifetime of S_1. Equation (9) describes the general dependence of γ_{AO_2} on the concentration variables $[O_2]$ and $[A]$ and provides access to the parameters [5, 6]

$$\beta_A = (\gamma_{AO_2})_{[A]=\infty}\, d\gamma_{AO_2}^{-1}/d[A]^{-1} \tag{10}$$

and

$$\gamma_{IS} = (\gamma_{AO_2})_{[O_2]=0}(1 + \beta_A/[A]) \tag{11}$$

The former may be translated [7] into rate constants $k_7 = k_8/\beta_A$ using independent measurements [8, 9] of $k_8 = \tau_\Delta^{-1}$ for the appropriate solvent, whereas the latter is a characteristic property of the sensitizer and agrees well with independent measurements of γ_{IS} or with $1 - \gamma_F$ for sensitizers of known inter-system crossing (γ_{IS}) or fluorescence (γ_F) yields.

For these acceptors which exhibit photodimerization (e.g. anthracene and naphthacene) or which have solubilities much less than the optimum concentration $[A] \approx \beta_A$, the necessary use of competitive techniques involves selective excitation of a reference sensitizing acceptor, R, of known reactivity β_R in the presence of the sample acceptor, A. If the concentrations of both species are monitored spectroscopically at various exposure times t, relative reactivities may be obtained [10] from equation (12), whereas the reduction

$$\ln([R]_0/[R]_t) = (\beta_A/\beta_R)\ln([A]_0/[A]_t) \tag{12}$$

in (relative) quantum yield $\gamma_{RO_2}^0/\gamma_{RO_2}$ of reference acceptor peroxidation by the addition of the second acceptor A is described by equation (13):

$$(\gamma_{RO_2}^0/\gamma_{RO_2}{}^{-1})^{-1} = (\beta_A/[A])(1 + [R]/\beta_R) \tag{13}$$

which provides experimental access to β_A and β_R if the concentrations of both acceptors are varied independently (e.g. ref. 11).

Several questions arise in connection with this simple scheme, however. Firstly, the singlet–triplet splitting energy, ΔE_{ST}, of aromatic hydrocarbons exceeds the $O_2(^1\Delta)$ excitation energy and process (1) is not energetically prohibited in the self-sensitized peroxidation of these compounds. Secondly, a number of molecules (e.g. amines,[12] biliverdin,[13] tocopherols,[11,14] and β-carotene[15]) are capable of quenching $O_2(^1\Delta)$ in the physical sense [process (14)] by (reversible) electron or energy transfer:

$$O_2(^1\Delta)+A \rightarrow O_2(^3\Sigma)+A \text{ (or } ^3A) \tag{14}$$

In certain cases [bilirubin[13,16] and 1,3-diphenylisobenzofuran[17] (DPBF)], physical quenching and chemical reaction have been reported to proceed simultaneously and the effect of physical quenching by aromatic acceptors on the significance of reported reactivities β should be examined, particularly since DPBF and rubrene have been widely used as reference acceptors. Finally, the nature of the photosensitized oxidation of leuco-dyes in which singlet oxygen was first postulated as an intermediate[18] deserves further examination since the overall process involves electron transfer (cf. the photoreduction of dye to leuco-dye[19]) rather than characteristic $O_2(^1\Delta)$ addition.

It is with these aspects of the general mechanism that this contribution is concerned.

THE EFFECT OF PHYSICAL QUENCHING

The inclusion of process (14) modifies the $O_2(^1\Delta)$/acceptor addition efficiency to

$$\phi_A = k_7[A]/\{k_8+(k_7+k_{14})[A]\} \tag{15}$$

$$= \frac{\beta_A^0}{\beta_A}\left(\frac{[A]}{\beta_A^0+[A]}\right) \tag{16}$$

where the total (physical and chemical) reactivity index, $\beta_A^0 = k_8/(k_7+k_{14})$, cannot exceed the chemical reactivity index, $\beta_A = k_8/k_7$. Equation (10) now becomes

$$\beta_A^0 = (\gamma_{AO_2})_{[A]\rightarrow\infty}\, d\gamma_{AO_2}^{-1}/d[A]^{-1} \tag{17}$$

while equation (13) has the significance

$$(\gamma_{RO_2}^0/\gamma_{RO_2}^{-1})^{-1} = (\beta_A^0/[A])(1+[R]/\beta_R^0) \tag{18}$$

and equation (12) is unchanged. Methods are therefore available for the measurement of total reactivity indices β^0 [equations (17) and (18)] and of *relative* chemical reactivities β_A/β_R [equation (12)] and it may be argued[16] that physical quenching by both acceptors is unlikely if $\beta_A/\beta_R = \beta_A^0/\beta_R^0$ unless k_{14}/k_7 is independent of the acceptor. However, an unambiguous assessment of the extent of physical quenching requires the estimation of β_A^0/β_A for the same acceptor.

THE QUANTUM YIELD, γ_Δ, of $O_2(^1\Delta)$ FORMATION

Introduction of the energy fission process (1) leads to expression (19) for the quantum yield of $O_2(^1\Delta)$ formation at concentrations of dissolved oxygen $[O_2] \gg k_4/k_3$:

$$\gamma_\Delta = (\gamma_{IS} + \alpha K[O_2])/(1 + K[O_2]) \tag{19}$$

Here the Stern–Volmer constant for oxygen quenching of the sensitizer singlet state $K = (k_1 + k_6)\tau_s$ and the parameter $\alpha = 1 + k_1/(k_1 + k_6)$ has limiting values of 1 ($k_1 \ll k_6$) and 2 ($k_1 \gg k_6$). From equations (16) and (19), the overall quantum yield is expressed as [cf. equation (9)]

$$\gamma_{AO_2} = \left(\frac{\gamma_{IS} + \alpha K[O_2]}{1 + K[O_2]}\right)\left(\frac{[A]}{\beta_A^0 + [A]}\right)\frac{\beta_A^0}{\beta_A} \tag{20}$$

which, at constant acceptor concentration, provides experimental access to the quantities

$$\alpha\beta_A^0/\beta_A = (1 + \beta_A^0/[A])\,d\{\gamma_{AO_2}(1 + K[O_2])\}/dK[O_2] \tag{21}$$

and

$$\gamma_{IS}\beta_A^0/\beta_A = (1 + \beta_A^0/[A])(\gamma_{AO_2})_{[O_2] \to 0} \tag{22}$$

Values of these parameters for the sensitizer/acceptors listed are given in Table 1 (see refs. 4 and 11 for experimental details).

Table 1 Photoperoxidation parameters in benzene at 25 °C

Parameter	Rubrene	DMA[a]	DMBA[b]	Tetracene
$10^3\beta^0$ (M)	$1{\cdot}0 \pm 0{\cdot}1$[c]	$1{\cdot}1 \pm 0{\cdot}1$[d]	$1{\cdot}6 \pm 0{\cdot}2$[d]	$2{\cdot}4 \pm 0{\cdot}3$[d]
$\alpha\beta^0/\beta$	$1{\cdot}91 \pm 0{\cdot}24$	$1{\cdot}84 \pm 0{\cdot}22$	$1{\cdot}0 \pm 0{\cdot}2$	$1{\cdot}16 \pm 0{\cdot}2$
$\gamma_{IS}\beta^0/\beta$	$0{\cdot}08 \pm 0{\cdot}04$	$0{\cdot}12 \pm 0{\cdot}06$	$0{\cdot}66 \pm 0{\cdot}06$	$0{\cdot}61 \pm 0{\cdot}07$
γ_F^e	$0{\cdot}98 \pm 0{\cdot}02$	$0{\cdot}90 \pm 0{\cdot}04$	$0{\cdot}36 \pm 0{\cdot}03$	$0{\cdot}19 \pm 0{\cdot}02$
γ_{IS}			$\leqslant 0{\cdot}67$	$0{\cdot}68 \pm 0{\cdot}02$[f]
α	$\geqslant 1{\cdot}67$	$\geqslant 1{\cdot}62$	$\leqslant 1{\cdot}34$	$1{\cdot}04 \pm 0{\cdot}22$
k_1/k_6	$\geqslant 2{\cdot}0$	$\geqslant 1{\cdot}6$	$\leqslant 0{\cdot}39$	$\leqslant 0{\cdot}35$
β/β^0	$\leqslant 1{\cdot}29$	$\leqslant 1{\cdot}27$	$\leqslant 1{\cdot}12$	$1{\cdot}11 \pm 0{\cdot}14$
k_{14}/k_7	$\leqslant 0{\cdot}29$	$\leqslant 0{\cdot}27$	$\leqslant 0{\cdot}12$	$\leqslant 0{\cdot}33$

[a] 9,10-Dimethylanthracene.
[b] 9,10-Dimethyl-1,2-benzanthracene.
[c] Equation (17).
[d] Equation (18) with rubrene as reference acceptor.
[e] Ref. 21.
[f] Ref. 20.

THE LIMITS OF ENERGY FISSION AND OF PHYSICAL QUENCHING

For tetracene, for which a value of $\gamma_{IS} = 0{\cdot}68 \pm 0{\cdot}02$ has been reported for the solvent used here,[20] the quotient $\beta/\beta^0 = 1{\cdot}11 \pm 0{\cdot}14$ can be estimated directly from the quantity $\gamma_{IS}\beta^0/\beta$ obtained from equation (22), to provide $\alpha = 1{\cdot}04 \pm 0{\cdot}22$ from $\alpha\beta^0/\beta$ [equation (21)].

For the other acceptors, it is possible to assign limits to the quantities α and β/β_0 if the data are analysed subject to the conditions

$$1 \leqslant \alpha \leqslant 2; \qquad \beta/\beta_0 \geqslant 1$$

For DMBA

$$\frac{\beta}{\beta^0} = \gamma_{IS}/(\gamma_{IS}\beta^0/\beta) \leqslant (1-\gamma_F)/(\gamma_{IS}\beta^0/\beta) = 1{\cdot}12 \tag{23}$$

whence

$$\alpha = (\alpha\beta^0/\beta)(\beta/\beta^0) \leqslant 1{\cdot}28$$

Since estimates of γ_{IS} for rubrene and DMA from the fluorescence quantum yields of these compounds are of the order of the uncertainty in measurement of γ_F, a lower limit to α is obtained from $\alpha\beta^0/\beta$ and the condition $\beta/\beta^0 > 1$. This limit of α then provides an upper limit to β/β^0 from the same experimental parameter; these are listed in Table 1, together with the corresponding limits of the quotients k_1/k_6 and k_{14}/k_7.

It is concluded that within the limits of error these acceptors do not physically quench $O_2(^1\Delta)$ whereas the energy fission process (1) is largely responsible for oxygen quenching of the rubrene and DMA singlet states of high fluorescence quantum yield while fluorescence quenching by oxygen of tetracene and DMBA proceeds essentially by the energy transfer process (6). This difference in behaviour may be attributed to the relative energies of the complex states

$$^3\Gamma_g(T_n^3\Sigma) > {}^3\Gamma_i(S_1^3\Sigma) > {}^3\Gamma_f(T_1^1\Delta) \quad k_1 > k_6$$
$$^3\Gamma_i(S_1^3\Sigma) > {}^3\Gamma_g(T_n^3\Sigma) > {}^3\Gamma_f(T_1^1\Delta) \quad k_1 < k_6$$

which reflect the ordering of molecular states:

$$T_n > S_1 \gg T_1 \quad \gamma_{IS} \approx 0$$
$$S_n > T_n > T_1 \quad \gamma_{IS} \gg 0$$

Absolute values of the total reactivity indices β^0 and relative values of chemical reactivity indices β_R/β_{DPFB} for rubrene and 1,3-diphenylisobenzofuran in several solvents are listed in Table 2. Since $\beta_R/\beta_R^0 \leqslant 1{\cdot}29$ in benzene (Table 1) an upper limit to

$$\frac{\beta_{DPBF}}{\beta_{DPBF}^0} \leqslant \frac{\beta_R^0/\beta_{DPBF}^0}{\beta_R/\beta_{DPBF}} \times 1{\cdot}29 = 1{\cdot}25 \tag{24}$$

is estimated for this solvent. The independence of $\beta_R^0/\beta_{DPBF}^0 \approx \beta_R/\beta_{DPBF}$ of the solvent requires that k_{14}/k_7 is independent of both solvent and acceptor, or that physical quenching of $O_2(^1\Delta)$ by DPBF in Freon-113 cannot exceed 25% of the total reactivity exhibited by this acceptor; this is contrary to the findings of Matheson *et al.*[17] that $k_{14}/k_7 \approx 10$ in this acceptor–solvent system. It is of interest to note that the estimated lifetime, τ_Δ, of $O_2(^1\Delta)$ in Freon-113 (2·2 ms) is close to the lower limit computed from infrared absorption data.[17]

Table 2 Solvent dependence of reactivity indices for rubrene (R) and 1,3-diphenylisobenzofuran (DPBF)

Solvent	Freon-113	CCl_4	CS_2	C_6H_6
$10^5\beta_R^0$ (M)[a]	1·1	2·5	9·0	100
$10^5\beta_{DPBF}^0$ (M)[a]	0·054	0·16	0·53	6·00
β_R^0/β_{DPBF}^0	20	16	17	17
β_R/β_{DPBF}[b]	20	18	18	20
τ_Δ (μs)	2200[c]	700[d]	200[d]	24[d]

[a] From equation (17).
[b] From equation (12).
[c] Estimated from $(\tau_\Delta)_1/(\tau_\Delta)_2 = \beta_2^0/\beta_1^0$ for solvents 1 and 2 assuming common value for rate constants k_7 and k_{14}.
[d] Ref. 8.

THE QUENCHING OF $O_2(^1\Delta)$ BY LEUCOMALACHITE GREEN

The total reactivity of leucomalachite green (DH) towards $O_2(^1\Delta)$ was examined using competitive methods with 9,10-dimethylanthracene (DMA) [equation (18)] and 9-methylanthracene (MA) [equation (12)] as reference acceptors in toluene. For this solvent the experimental data

$$\beta_{DH}^0 = 3{\cdot}3 \pm 1{\cdot}0 \times 10^{-4}\ \mathrm{M}; \qquad \beta_{DMA}^0 = 1{\cdot}5 \pm 0{\cdot}2 \times 10^{-3}\ \mathrm{M} \tag{25}$$

and $\beta_{DH}/\beta_{MA} = 24 \pm 8$ are consistent with the parameters

$$\tau_\Delta = 18\ \mu\mathrm{s} \tag{26}$$

$$k_{14}(\mathrm{DH}) = 1{\cdot}7 \pm 0{\cdot}7 \times 10^8\ \mathrm{M}^{-1}\ \mathrm{s}^{-1} \tag{27}$$

$$k_7(\mathrm{DH}) = 1{\cdot}3 \pm 0{\cdot}6 \times 10^5\ \mathrm{M}^{-1}\ \mathrm{s}^{-1} \tag{28}$$

indicative of strong physical quenching of $O_2(^1\Delta)$ by the reduced dye, presumably via the electron transfer sequence

$$\mathrm{DH} + O_2(^1\Delta) \rightarrow \mathrm{DH}^+O_2^- \rightarrow \mathrm{DH} + O_2(^3\Sigma) \tag{29}$$

Production of the dye D^+, which was not observed in this solvent, would be expected to require ionic separation of the intermediate complex followed by deprotonation and disproportionation of the semiquinone,[19] viz.

$$\mathrm{DH}^+ \rightleftharpoons \mathrm{D} + \mathrm{H}^+ \tag{30}$$

$$\mathrm{DH}^+ + \mathrm{D} \longrightarrow \mathrm{D}^+ + \mathrm{DH} \tag{31}$$

if a counter ion is available to stabilize the dye cation. It is not immediately obvious how this sequence is promoted in the solid state[18] but it is of interest to note that the recombination of O_2^- and H^+ would produce the overall effect of hydrogen atom abstraction by $O_2(^1\Delta)$ which has been proposed in connection with the photosensitized oxidation of bilirubin to biliverdin.[13]

REFERENCES

1. C. S. Foote, *Accounts Chem. Res.*, **1**, 104 (1968).
2. D. R. Kearns, *Chem. Rev.*, **71**, 395 (1971).
3. B. Stevens and B. E. Algar, *Ann. N.Y. Acad. Sci.*, **171**, 50 (1970).
4. B. Stevens and B. E. Algar, *J. Phys. Chem.*, **72** 3468 (1968).
5. E. J. Bowen and D. W. Tanner, *Trans. Faraday Soc.*, **51**, 475 (1955).
6. B. Stevens and B. E. Algar, *Chem. Phys. Lett.*, **1**, 58, 219 (1967).
7. B. Stevens, S. R. Perez, and J. A. Ors, *J. Amer. Chem. Soc.*, **96**, 6846 (1974).
8. P. B. Merkel and D. R. Kearns, *J. Amer. Chem. Soc.*, **94**, 7244 (1972).
9. R. H. Young, D. Brewer, and R. A. Keller, *J. Amer. Chem. Soc.*, **95**, 375 (1973).
10. T. Wilson, *J. Amer. Chem. Soc.*, **88**, 2898 (1966).
11. B. Stevens, S. R. Perez, and R. D. Small, *Photochem. Photobiol.*, **19**, 315 (1974); **20**, 515 (1974).
12. R. H. Young, R. L. Martin, D. Feriozi, D. Brewer, and R. Karper, *Photochem. Photobiol.*, **17**, 233 (1973).
13. B. Stevens and R. D. Small, *Photochem. Photobiol.*, **23**, 33 (1976).
14. S. R. Fahrenholtz, F. H. Doleiden, A. M. Trozzolo, and A. A. Lamola, *Photochem. Photobiol.*, **20**, 505 (1974); C. S. Foote, T.-Y. Ching, and G. G. Geller, *Photochem. Photobiol.*, **20**, 511 (1974).
15. C. S. Foote, R. W. Denny, L. Weaver, T. Chang, and J. Peters, *Ann. N.Y. Acad. Sci.*, **171**, 139 (1970).
16. C. S. Foote and T.-Y. Ching, *J. Amer. Chem. Soc.*, **97**, 6209 (1975).
17. I. B. C. Matheson, J. Lee, B. S. Yamanashi, and M. L. Wolbarsht, *J. Amer. Chem. Soc.*, **96**, 3343 (1974).
18. H. Kautsky, H. de Bruijn, R. Neuwirt, and W. Baumeister, *Chem. Ber.*, **66**, 1588 (1933).
19. B. Stevens, R. R. Sharpe, and W. S. W. Bingham, *Photochem. Photobiol.*, **6**, 83 (1967).
20. A. Kearvell and F. Wilkinson, *Chem. Phys. Lett.*, **11**, 472 (1971).
21. B. Stevens and B. E. Algar, *J. Phys. Chem.*, **72**, 2582 (1968).

9

Quenchers of Singlet Oxygen—A Critical Review

D. BELLUŠ

Central Research Laboratories- CIBA-GEIGY AG,
4002 Basle, Switzerland

INTRODUCTION

The possibility that electronically excited, metastable singlet oxygen molecules might be involved as 'an active oxygenation species' in dye-sensitized reactions was first suggested by Kautsky and co-workers in the early 1930s.[1] Singlet molecular oxygen as a reactive species for organic syntheses was re-discovered by Foote and Wexler [2] and Corey and Taylor [3] in 1964. They demonstrated the chemical identity of 'active oxygenating species' generated by energy transfer from a photosensitizer with the singlet oxygen prepared either chemically (hypochlorite and H_2O_2 system)[4] or by microwave discharge.[5] As a result, very intensive research on the chemistry, physics and biology of this intriguing intermediate species was stimulated. Extensive reviews unequivocally suggested the involvement of singlet oxygen in organic oxygenation reactions,[6–20] in biological systems (photodynamic action),[21–24] in the photodegradation of polymers,[25–29] in the chalking of pigments,[30] in the phototendering of vat dyes,[31] and in atmospheric processes,[32] especially in the chemistry of polluted urban atmospheres.[33–35] Reviews of methods for the generation of singlet oxygen in the gas [13, 32] and in liquid phases [13, 26] as well as of its electronic states and spectroscopy [13, 32, 36] were published. However, the phenomenon of the quenching of singlet oxygen in solution was usually treated only marginally in those reviews (with two very recent exceptions [37, 225]). There is also some confusion in the definitions of the basic terms regarding quenching and quenchers of singlet oxygen. On account of the steadily increasing number of papers on singlet oxygen quenching in solution and in polymers, it became increasingly difficult to overlook this field.

It is the purpose of this review to fill this gap and to furnish an interdisciplinary review of molecules, known as quenchers of singlet molecular oxygen in liquid and in solid state. We shall focus our attention on the structural features of such quenchers in the hope that some structure–activity correlations might be drawn that were not evident previously. Detailed kinetic and mechanistic aspects of the quenching of singlet oxygen are beyond the scope of this review and will be discussed only where necessary. Available references up to September 1976 are included.

SINGLET STATES OF MOLECULAR OXYGEN

Spectroscopic studies have identified singlet oxygen in both of its excited singlet states, which are described symbolically in Table 1.[38–40]

Table 1 The electronic states of molecular oxygen and their properties

Oxygen molecule	Configurations of electrons in highest occupied M.O.s	Relative energy (kcal mol^{-1})	Lifetimes (s)	
			Gas phase	Liquid phase
Second excited state, $O_2(^1\Sigma_g^+)$	↑ ↓	37·5	7–12[a]	10^{-11}–10^{-9}
First excited state, $O_2(^1\Delta_g)$	↑↓ —	22·5	2700[a]	$2\times10^{-6}-10^{-3}$[b]
Ground state, $(^3\Sigma_g^-)$	↑ ↑	0	∞	∞

[a] Intrinsic radiative lifetimes. For $O_2(^1\Delta_g)$ in air at atmospheric pressure this is reduced to ≤ 100 ms and is due largely to the quenching process $O_2(^1\Delta_g)+O_2(^3\Sigma_g^-) \rightarrow 2\,O_2(^3\Sigma_g^-)$, for which a rate constant of $1{\cdot}2\times10^3$ M^{-1} s^{-1} has been reported.[41]
[b] Strongly solvent-dependent values, see Table 2.

Gaseous $O_2(^1\Sigma_g^+)$ is collisionally deactivated more rapidly than $O_2(^1\Delta_g)$ by all gases. Many tables of rate constants for gas-phase deactivation of singlet oxygen in both states have been published.[32,33,42,229]

Through collisional deactivation, the lifetimes are dramatically reduced in solution, but much more so for $O_2(^1\Sigma_g^+)$. Ackerman *et al.*[46] consider that the fundamental difference between the relative inefficiency of deactivation of the $^1\Delta_g$ state and the high efficiency of deactivation of the $^1\Sigma_g^+$ state may be attributed to the fact that $O_2(^1\Sigma_g^+)$ is transformed initially into $O_2(^1\Delta_g)$ by a spin-allowed process, where $O_2(^1\Delta_g)$ must undergo a spin-forbidden transition to reach the ground state. Thus, it is not surprising that for most singlet oxygen reactions in solution the $^1\Delta_g$ state is involved as the 'active oxygenating species' (reactions suggested to involve $O_2(^1\Sigma_g^+)$ were critically discussed by Kearns).[13] Consequently, all of the methods of detection and the quenching reactions discussed below refer to singlet molecular oxygen in its $^1\Delta_g$ states, denoted simply by $^1O_2$.

DETECTION OF QUENCHERS OF SINGLET MOLECULAR OXYGEN ($^1O_2$)

For an understanding of basic kinetic terms which we frequently employ in this review, it is essential to present a general kinetic scheme of the processes involved in oxygenations with $^1O_2$ in the presence of a quencher, Q. Scheme 1, which

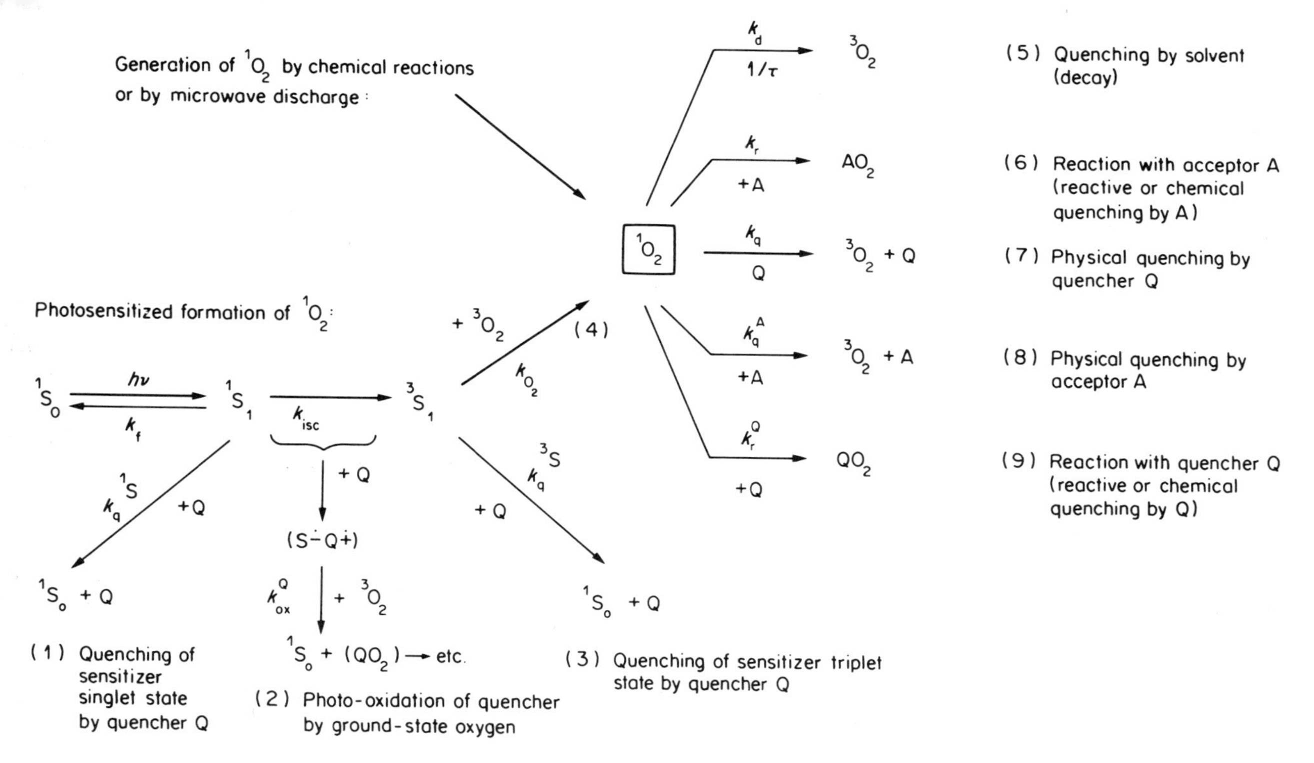

Scheme 1 General kinetic formulae of main reactions occurring in oxygenation reactions with singlet molecular oxygen $^1O_2$

Table 2 Lifetimes (τ) and pseudounimolecular rate constants (k_d) of decay of $^1O_2$ in various solvents

Solvent	Mixed solvents	$\tau \times 10^6$ (s)	$k_d \times 10^{-4}$ (s^{-1})	Method of measurement [a]	Reference
H_2O		2	50	A	49
	H_2O–CH_3OH (1:1)	3·5	28·6	A	49
	H_2O–CH_3OH (1:1)	3·6	27·7	B	197
D_2O		20	5	A	49
	D_2O–CH_3OH (1:1)	11	9·1	A	49
	D_2O–CD_3OD (1:1)	35	2·9	A	49
$HO(CH_2)_2OH$		2·1	47·6	B	197
2-Butoxyethanol		2·6	38	C	57, 100
CH_3OH		5	20	A	198
		5·5	18·2	N	51, 199
		7	14·3	A	49
		10; 11·4	8·8; 9·7	A	197
	CH_3OH–$HO(CH_2)_2OH$ (1:1)	8·3	12	B	197
$HCON(CH_3)_2$		7·1	14	A	198
n-Butanol		9·1	11	N	51, 199
		19	5·2	A	197
C_2H_5OH (95%)		5·6	18	A	198
C_2H_5OH		12	8·3	A	49
Hexadecane		11	9	C	57, 100
Benzene		12·5	8	A	198
		24	4·2	A	49
		24·4	4·1	A [b]	94
	Benzene–CH_3OH (4:1)	8·3	12	N	51, 199
	Benzene–CH_3OH (4:1)	26	3·8	A	197
tert-Butanol		13·5	7·4	N	51, 199
		34	3	A	197
Cyclohexane		17	5·9	A	49
Pyridine		17	5·9	G [c]	167
		33	3·1	A	197
CH_3SOCH_3		19·2	5·2	O	187
		30	3·3	G	201
Toluene		20	5	L	202
Isooctane		20; 21·3	5; 4·7	E	57, 100

Solvent	Mixture			Method	Ref.
Isooctane		25	4	D	57
CH_3COCH_3		26	3·8	A	49
CD_3COCD_3		26	3·8	A	49
CH_3CN		30	3·3	A	49
Dioxan		32	2·9	A	197
Bromobenzene		34	2·94	A	200
	Bromobenzene–CH_3OH (4:1)	23	4·3	A	197
1,2,4-Trichlorobenzene		50	2	L	202
$CHCl_3$		60	1·7	A	49
	$CHCl_3$–CH_3OH (9:1)	26	3·9	M	62
Chlorobenzene		91	1·1	F	100
CH_2Cl_2		105	0·95	K	100
		137	0·73	E	57, 100
	CH_2Cl_2–CH_3OH (11:5)	50	2	G	142
	CH_2Cl_2–CH_3OH (15:1)	91	1·1	G	142
CS_2		200	0·5	A	49 [d]
		250	0·4	H	179
$CDCl_3$		300	0·33	A	203
C_6F_6 (Hexafluorobenzene)		600	0·17	A	203
CCl_4		700	0·14	A	49
CF_3Cl (Freon-11)		1000	0·1	A	203

[a] A: Direct method employing laser pulse technique and 1,3-diphenylisobenzofuran (DPBF) as acceptor. DPBF reacts with $^1O_2$ essentially without physical quenching (i.e. $k_q^A \leqslant 0{\cdot}1k_r$).[49, 62, 204, 205] B: From photo-oxygenation of DPBF, using its known β-value. C: Calculated from rubrene oxygenation with $^1O_2$ from microwave discharge using k_q of 1,4-diazabicyclo[2.2.2]hexane (DABCO) as standard ($k_q = 3{\cdot}4 \times 10^7$ M^{-1} s^{-1}); $k_r/k_q^A \leqslant 20$ for rubrene has been reported.[151] D: From photosensitized photo-oxygenation of rubrene, using k_q of DABCO as in Method C. E: From rubrene oxygenation with $^1O_2$ formed by thermal decomposition of triphenyl phosphite ozonide. F: From photosensitized oxygenation of 2-methyl-2-pentene (assuming $k_r = 10^6$ M^{-1} s^{-1}). G: Method is not given. H: Direct method employing flash photolysis technique and DPBH as $^1O_2$ acceptor. K: As method D, but assuming k_r (rubrene) = $7{\cdot}3 \times 10^7$ M^{-1} s^{-1}. Actually, this k_r value is approximately an average of recently published[58, 200, 202] k_r values for rubrene. L: Calculated from equation (10). M: Calculated from the molar fractions of each solvent in the mixture using previously published k_d values. N: From photo-oxygenation experiment employing k_d/k_q value of β-carotene and diffusion constant of $^1O_2$ in calculation. O: By comparison of β-values of 2-methyl-2-pentene in CH_3OH and in dimethyl sulphoxide.

[b] Instead of DPBF oxygenation, decay of triplet excited anthracene in presence of β-carotene was recorded.

[c] After ref. 167 as personal communcation from D. Brewer.

[d] For effect of long lifetime of $^1O_2$ in CS_2 on photo-oxygenations of anthracene and 2-methyl-2-pentene in CS_2, see ref. 90 and references cited therein.

includes the most important kinetic terms, has intentionally been made complicated in order to illustrate the complexity of the situation which must be taken into account when speculating about, for example, the participation of $^1O_2$ or quenchers of $^1O_2$ in oxygenation reactions.

The rate constant of physical quenching, k_q is an absolute measure of the quenching efficiency of compound Q capable of deactivating $^1O_2$ to ground-state oxygen, $^3O_2$. Considering Scheme 1, it seems to be rather laborious to obtain a reliable value for k_q, and this is often the case. Fortunately, good methods are available for performing kinetic measurements (especially photosensitized experiments) in which many of the undesired reactions are eliminated or can be neglected. Nevertheless, many further complications are possible in special cases, e.g. quenching of $^1O_2$ with triplet-excited sensitizer,[47–49] quenching of $^1O_2$ with oxidized quencher, and competitive UV absorption by $^1S_0$ and/or quencher.[50] These alternatives could make Q appear to be a better quencher of $^1O_2$ than it really is.

The value of primary importance in the determination of other basic kinetic parameters of $^1O_2$ oxygenation reactions is the pseudo-unimolecular rate constant k_d of radiationless decay of $^1O_2$ in various solvents ('quenching by solvent'). The presently available values of K_d and the lifetimes of $^1O_2(\tau = 1/k_d)$ are listed in Table 2 in order of increasing lifetimes. Probably the most accurate values are those measured by direct methods A and H. Relative lifetimes of $^1O_2$ in a series of aromatic solvents have also been published.[51] In addition to Table 2, k_d values for a number of other solvents become known very recently and are published elsewhere in this book.[224,225] The most remarkable new result is the 2200-μs lifetime of $^1O_2$ in Freon–113, which represents the highest known lifetime of singlet oxygen in solution.[224]

Examination of the data in Table 2 clearly indicates that $^1O_2$ lifetimes cannot be correlated with most of the usual solvent properties. Polarity, viscosity, polarizability, ionization potential, and oxygen solubility, and also the presence of a heavy atom in the solvent molecule, appear to be unimportant factors in the radiationless decay of $^1O_2$ in solution.[49]

A simple theoretical interpretation of the solvent dependence of $^1O_2$ decay in terms of electronic to solvent vibrational energy transfer has been presented by Merkel and co-workers,[49,52] who related the solvent quenching efficiency (expressed as $1/\tau$) to the optical densities (O.D.) of solvents in 1-cm cells at 7880 and 6280 cm^{-1} corresponding to the $0 \rightarrow 0$ and $0 \rightarrow 1$ components of $O_2(^1\Delta_g) \rightarrow O_2(^3\Sigma_g^+)$ transition. Their expression

$$1/\tau \approx 0{\cdot}5(\text{O.D.}_{7880}) + 0{\cdot}05(\text{O.D.}_{6280})\,\mu\text{s}^{-1} \tag{10}$$

yields values that differ by not more than an order of magnitude from the experimental values of τ listed in Table 2 and explains the 10-fold increase in τ observed when D_2O (with lower O.D. at 7880 and 6280 cm^{-1}) replaces H_2O as the solvent. Since deuteration involves a very minor perturbation of the chemical properties of the solvent, the deuterium effect on the lifetime of $^1O_2$

becomes a powerful diagnostic tool for investigating the involvement of $^1O_2$ in various chemical and physical processes, especially in biological systems.

It is of interest that the effect of halogen-containing solvents on the lifetime of $^1O_2$ is opposite to that of those solvents on the fluorescence-quenching rate constants of singlet excited aliphatic ketones. For example, k_f ($\times 10^{-6}$, $\text{M}^{-1}\ \text{s}^{-1}$) for singlet acetone quenching increases in the order CH_2Cl_2 (1·4)$<$$CHCl_3$ (5·2)$\ll$$CCl_4$ (42·0).[53]

From the general kinetic formulae (Scheme 1), it follows that many complications in the kinetic treatment of oxygenations with $^1O_2$ can be encountered in photosensitized oxygenations. However, it is easier to satisfy the kinetically important requirement of a steady state for the $^1O_2$ concentration in photosensitized oxygenations (constant light intensity, constant partial pressure of oxygen, constant pulse lifetime, and homogeneity of the laser radiation, etc.) than in the experiments involving chemically generated $^1O_2$ or in experiments in the discharge-flow system, where the accuracy is limited by the small contact area between the two phases.[54] Furthermore, much information has been gathered in recent years, especially by Gollnick, about quantum yields of $^1O_2$ formation by energy transfer from excited sensitizers (especially dyes), singlet and triplet energies of sensitizers, undesired side-reactions, etc.[9,55] As a result, some experimental arrangements and kinetic schemes are known which permit an evaluation of the potential quenching efficiency of an unknown compound Q by photosensitized oxygenation techniques with a high degree of accuracy.

If only a 'yes–no' answer is sufficient, a combination of (*i*) oxygen consumption measurements and (*ii*) analysis of concentration of acceptor A decrease or AO_2 formation [events (*i*) and (*ii*) can be measured simultaneously] in the absence and in the presence of a compound Q for a given time of constant irradiation, *t*, completed with (*iii*) oxygenation(s) of A in the presence and in the absence of Q with chemically or by microwave-discharge generated $^1O_2$, should be carried out. In most cases these three simple experiments can either discover side-reaction paths of $^1O_2$, AO_2 or Q, respectively, or deliver a reliable 'yes–no' answer.[56] Provided that no complications occur, reasonable relative quenching efficiencies related to some known quencher Q (Tables 2–10) can very easily be obtained in this way.

If, however, exact values of the quenching constant, k_q, are desired, more sophisticated kinetic treatments are necessary, as follows.

Physical quenching only [*process* (*7*)]. For the system in which only processes $^1S_0 + h\nu \rightarrow {}^3S_1$ and processes (4), (5), (6), and (7) are operative, while other processes may reliably be neglected, the self-sensitized photo-oxygenation of rubrene (9,10,11,12-tetraphraphenylnapththacene) (S=A=Ru), followed by bleaching of rubrene absorption at 520 nm, was introduced by Carlsson *et al.*[57] as a simple method for the determination of k_q. From the steady-state treatment $[d(^1O_2)/dt = 0]$ of the above processes it follows that

$$k_q = [k_r(\text{Ru})_0 + k_d]C, \quad \text{where} \quad C = \frac{1}{(\text{Q})}\left\{\frac{[d(\text{Ru})/dt]^0}{[d(\text{Ru})/dt]^Q}\right\} - 1 \qquad (11)$$

$[d(Ru)/dt]^0$ and $[d(Ru)/dt]^Q$ are initial rates of loss of rubrene in the absence and in the presence of quencher, respectively, (Q) is the quencher concentration, k_r for rubrene is given by a–K in the first footnote in Table 2, k_d values for various solvents are summarized in Table 2, $(Ru)_0$ is the initial concentration of rubrene, and superscripts Q and 0 indicate the presence or the absence of quencher, respectively.

Under the usual spectroscopic conditions $[(Ru)_0 \approx 10^{-6}$–10^{-5}M], the contribution of the term $k_r(Ru)_0C$ to the total value of k_q may be rather small, e.g. 25% or less. This is another reason for the determination of exact values of k_d (Table 2).

The integrated forms of the above equation can be used for the calculation of k_q in quenching experiments with $^1O_2$ generated from $(PhO)_3PO_3$ or from a microwave-discharge system.[57]

Physical quenching or excited sensitizer quenching? [processes (1+3) or (7)?]. Very often the question arises of whether only the above processes or also the quenching of excited sensitizer S ($^1S_1$ or $^3S_1$) by quencher Q is involved. Zweig and Henderson[58] gave a useful kinetic treatment of this situation. Essentially, the amount of product AO_2 is given by the equation

$$\frac{1}{(AO_2)} = \frac{1}{(^1O_2)} + \left[\frac{k_q(Q)+k_d}{k_r}\right]\frac{1}{(^1O_2)}\cdot\frac{1}{(A)} \tag{12}$$

where $(^1O_2)$ is the total amount of $^1O_2$ generated, (A) is the acceptor (e.g. rubrene in self-sensitized photo-oxygenation) concentration and other terms are as specified above.

If the irradiation time and (Q) are kept constant, and if the amount of A converted into AO_2 is kept to less than ca. 10%, then a plot of $1/(AO_2)$ or $1/\Delta A$ against $1/(A)$ gives a straight line. The interceptor of this line is $1/(^1O_2)$ and its slope is $1/(^1O_2)[k_d + k_q(Q)/k_r]$. Provided that $(^1O_2)$, k_d and k_r are known, k_q can be calculated. On the other hand, a family of such curves for several different concentration of quenchers allows a distinction to be made between true physical $^1O_2$ quenching and two alternative processes: light absorption by the quencher and its quenching of excited sensitizer [processes (1+3)]. If these alternative processes are involved, the only parameter in equation (12) that will be affected is $(^1O_2)$, which decreases with increasing concentration of Q. The value of $1/(^1O_2)$ in equation (12) increases as the amount of quencher increases, and both the slope and intercept increase proportionally (Figure 1). If only physical $^1O_2$ quenching is involved, the amount of $^1O_2$ formed does not vary with increasing Q; the intercept of the plot does not change, and only the slope increases with increasing Q (Figure 2).

Quencher Q takes part simultaneously in processes (1), (2), (3), (4), (7), and (9). This kinetically very complicated situation occurs especially with amines as quenchers. Kinetic treatments are given in various original papers.[59–62] A great variety of other special kinetic situations and competitive experiments may be found in the references cited in Tables 2–10.

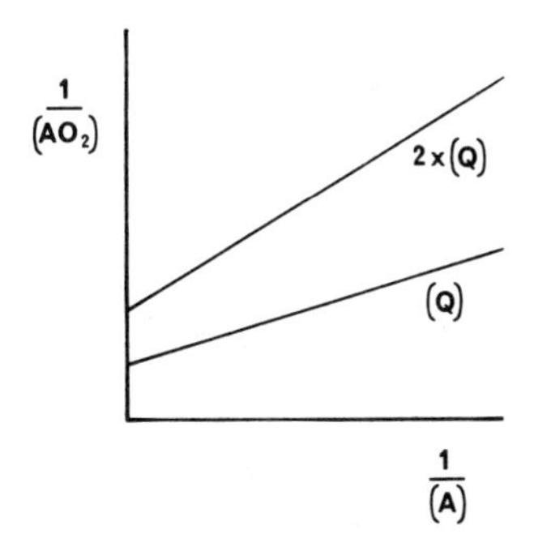

Figure 1 Exclusive quenching of excited sensitizer S ($^1S_1$ or $^3S_1$)

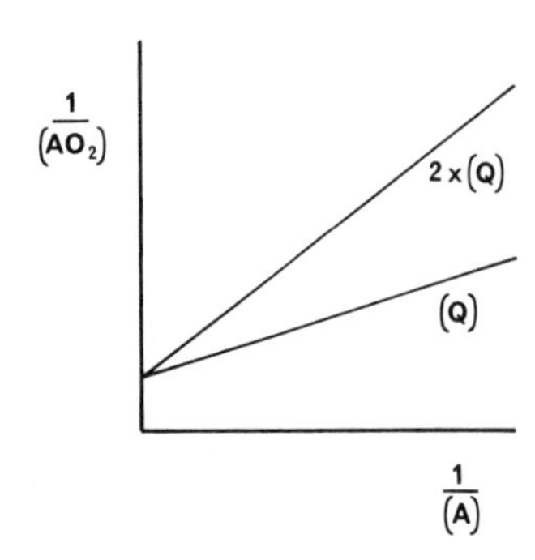

Figure 2 Exclusive physical quenching of $^1O_2$

REVIEW OF PHYSICAL QUENCHERS OF $^1O_2$

Which compound acts as a quencher of $^1O_2$? A common definition fails and there is some confusion in the literature on this point. In a broad sense, an $^1O_2$ quencher is a compound that accelerates the deactivation of $^1O_2$ in a given system without reacting with $^1O_2$†. Another condition is that the rate of $^1O_2$ deactivation with quencher Q [equation (13)] should be higher than the solvent-dependent decay of $^1O_2$ [equation (14)] in a given solvent:

$$[-d(^1O_2)/dt]_{\text{quencher}} = k_q(Q)(^1O_2) \tag{13}$$

$$[-d(^1O_2)/dt]_{\text{solvent}} = k(\text{solvent})(^1O_2) = k_d(^1O_2) \tag{14}$$

Since (13) > (14) is required, it follows that

$$k_q > k_d/(Q) \tag{15}$$

Therefore, a short, solvent-dependent (k_d, see Table 2) *definition of quencher Q* is as follows. In a given system, a quencher of $^1O_2$ is a compound satisfying the requirement $k_q > k_d/(Q)$ and acting mainly as a physical quencher ($k_q \gg k_r^Q$).

In other words, the probability of a bimolecular physical quenching process ($^1O_2 + Q \rightarrow {}^3O_2 + Q$) depends on the lifetime of $^1O_2$ in a given solvent. It has not been evident so far whether the absolute values of k_q are independent of the solvent or whether they are a function of some solvent property. The upper limit of k_q for pure physical quenching is about $3 \times 10^{10}\,\text{M}^{-1}\,\text{s}^{-1}$ (diffusion-controlled rate).

For use in practical and biological systems, i.e. in the presence of moisture or in aqueous solutions, Q should deactivate $^1O_2$ even more effectively than water ($k_d = 5 \times 10^5\,\text{s}^{-1}$). To achieve this result with concentrations of quenchers

† The requirement of 'inertness' of Q is justified but by definition an extremely elastic term and therefore it stays a sore point in each definition of Q.

as low as 10^{-2}–10^{-3} M, a value of $k_q \approx 5 \times 10^7$–$10^8$ M^{-1} s^{-1} seems to be the lowest limit for a compound with practical importance as a quencher of $^1O_2$. As will be seen in the following tables, many important naturally occurring and man-made quenchers have considerably higher, in fact nearly diffusion-controlled, values of k_q.

Carotenoids

The destruction of biopolymers in the living organism by photosensitized oxygenations was recognized long before the discovery of $^1O_2$ [63] and was called the 'photodynamic action'.[64] It is beyond the scope of this review to present all of the evidence for the involvement of carotenoid pigments in systems irradiated with visible light and containing a suitable photosensitizer and oxygen.

Carotenoids are by far the most extensive class of compounds which were recognized to be quenchers of singlet oxygen. The total production of carotenoids in nature has been conservatively estimated at about 10^8 tons per year.[65] Most of this huge output is in the form of four major carotenoids (fucoxanthin, lutein, violaxanthin and neoxanthin) containing nine or more conjugated double bonds, an inevitable prerequisite of high quenching activity (see later). The probable mechanisms by which carotenoid pigments protect components of the living organism against photo-sensitized oxygenation are depicted in Figure 3.

It should be obvious by now that the quenching of $^1O_2$ by carotenoids must play a prominent role in explaining the protective action of carotenoid pigments. In 1968, the first report of the interaction between carotenoids and $^1O_2$ was published by Foote and Denny,[67] who argued that the protective action of carotenoids against photo-oxygenations could not be adequately explained by the ability of carotenoids to quench triplet sensitizers. Although β-carotene quenches them effectively (e.g. chlorophyl, $E_T = 29$ kcal mol^{-1};[68] napthacene, $E_T = 29$ kcal mol^{-1};[69] zinc phthalocyanine, $E_T = 26{\cdot}1$ kcal mol^{-1};[70] pentacene, $E_T = 23$ kcal mol^{-1}; and many others),[71] this energy transfer reaction is diffusion controlled and should compete with the similarly controlled quenching of triplet sensitizer (e.g. chlorophyl) by $^3O_2$. (Essentially, carotenoids can also quench dye singlet states, e.g. of methylene blue [72]). Since both reactions occur at the same rate, the carotenoids could only protect if their local concentrations in biological systems greatly exceeded that of $^3O_2$. Therefore, the alternative possibility, that the carotenoids interact with $^1O_2$, was first proposed and experimentally verified by Foote and co-workers in 1968. They found [67] that extremely low concentrations of β-carotene ($< 10^{-4}$ M) effectively quenched sensitized photo-oxygenations as well as oxygenations involving $^1O_2$ generated by reaction of NaOC1 and H_2O_2, and calculated from their data that one β-carotene molecule may quench as many as 1000 $^1O_2$ molecules.[73] To account for this extraordinarily high quenching efficiency, they proposed the following mechanism in which electronic excitation energy is transfered from $^1O_2$ to β-carotene, generating triplet-state β-carotene and ground-state oxygen:

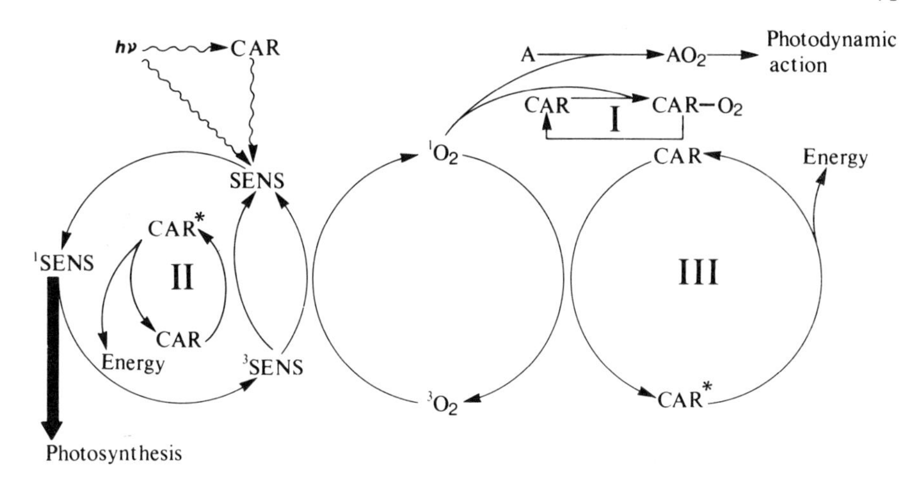

Figure 3 The three mechanisms whereby carotenoid pigments (CAR) protect cells against photosensitized oxidation. Light (wavy line) excites the photosensitizer (SENS), which in the case of photosynthetic organisms is deactivated primarily through photosynthesis. Those singlet excited sensitizer (^{1}SENS) molecules which are not deactivated by photosynthesis can be converted into the long-lived triplet excited sensitizer (^{3}SENS). These molecules can react with ground-state oxygen ($^3O_2$) to form singlet excited oxygen ($^1O_2$), which can react with a suitable receptor molecule (A) to form an oxidized product (AO_2) capable of leading to a photodynamic action. In mechanism I, the CAR act as preferred substrates and can be oxidized to CAR–O_2, which may be regenerated by a dark, enzymic process. In mechanism II, CAR can quench ^{3}SENS, presumably going to an excited state, CAR*, and then being deactivated to CAR and energy. In mechanism III, CAR can quench $^1O_2$ forming an excited state, CAR*, which can again be deactivated to CAR and energy. In addition, the accessory pigment function of carotenoid is depicted by the direct excitation of CAR by light and the transfer of this radiant energy to the SENS. [Reprinted, with permission, from N. I. Krinsky, in *Carotenoids*, Ed. O. Isler, Birkhäuser Verlag, Basle, 1971, p. 684]

$$^1O_2 + {}^1\beta\text{-carotene} \xrightarrow{k_q} {}^3O_2 + {}^3\beta\text{-carotene} \qquad (16)$$

This mechanism is feasible if the lowest triplet state of β-carotene is either nearly degenerate with (indication: triplet β-carotene can be quenched by ground-state oxygen [74]) or energetically below the $^1O_2$ level (22·5 kcal mol^{-1}). Actually, the E_T value for β-carotene was later measured and found to be between 17 and 25 kcal mol^{-1} (see Table 3). Additional support for this quenching mechanism is provided by the observation that other long-chain polyenes (with eight or more conjugated double bonds) are also very efficient quenchers, whereas the shorter polyenes (seven or fewer double bonds), which possess higher triplet energies, are much poorer quenchers of $^1O_2$ reactions. In agreement with this, Foote *et al.*[75] showed that the protective action of four carotenoids against chlorophyll-a photobleaching (also shown in Figure 4) drops most sharply between seven and nine conjugated double bonds. The quenching efficiencies and triplet energies of various carotenoids are compiled in Table 3.

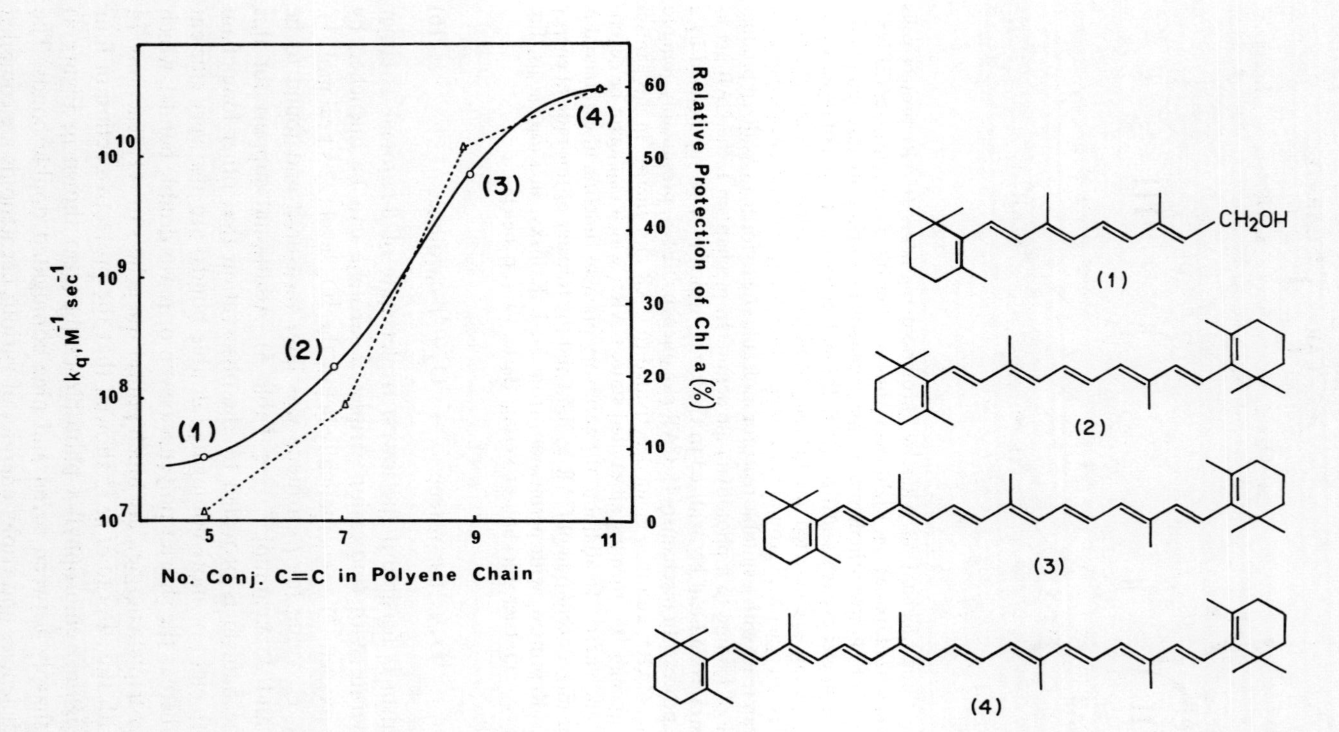

Figure 4 $^1O_2$ quenching rates (k_q) (○) and protective action against photobleaching of chlorophyll-a (ref. 76) (△) as a function of the number of conjugated double bonds as well as structures of quenchers (**1**)–(**4**). [Reprinted, with permission, from C. S. Foote, Y. C. Chang, and R. W. Denny, *J. Amer. Chem. Soc.*, **92**, 5216 (1970)]

The data in Table 3 reveal that emphasis was put on the determination of k_q for β-carotene. Since β-carotene is frequently used as a diagnostic tool for involvement [77–85,227] or non-involvement [86,87] of $^1O_2$ in organic or biological [88,89] oxygenation reactions, and further as a standard for the determination of k_d,[51,90] of relative reaction rate constants,[91] and of relative quenching constants,[85,92,93] a knowledge of the exact value of its k_q value is of great importance. Direct measurements of the rate constant k_q in benzene (2×10^{10} $M^{-1}\ s^{-1}$ by Merkel and Kearns [49] and $1{\cdot}3 \times 10$ $M^{-1}\ s^{-1}$ by Farmilo and Wilkinson [94]), in CS_2 ($2{\cdot}6 \times 10^{10}$ $M^{-1}\ s^{-1}$ by Foote *et al.*[90]) and in 3:2 benzene–methanol ($2{\cdot}3 \times 10^{10}$ $M^{-1}\ s^{-1}$ by Mathews-Roth *et al.*[95]) have yielded values which are remarkably close to that (3×10^{10} $M^{-1}\ s^{-1}$) assessed by Foote *et al.*[75] In contrast to the above values, Matheson and Lee [96] measured k_q for β-carotene in Freon–113 and reported a value of $1{\cdot}4 \times 10^9$ $M^{-1}\ s^{-1}$, which was only one-twentieth of that of the diffusion-controlled reaction in this solvent. This value was critized,[62,71] because it seems unlikely that the change of solvent would shift the energy levels (of either β-carotene or $^1O_2$) sufficiently to observe these differences. However, the recently published values of k_q in CH_2Cl_2 [97] and CCl_4 [98] and the revised value in Freon–113[99] (see Table 3) surprisingly suggest that there is actually a loose correlation between k_q for β-carotene and the $^1O_2$ lifetime in the given solvents. As follows from panel discussions during the 1st EUCHEM Conference on Singlet Oxygen Reactions with Polymers in Lidingö (Stockholm), September 2–4, 1976, and from two papers presented there,[225,231] the most reasonable value of k_q for β-carotene seems to be $1{\cdot}5 \pm 0{\cdot}5 \times 10^{10}$ $M^{-1}\ s^{-1}$. Thus, owing to the great importance (see above) of the k_q value of β-carotene and in order to facilitate a more reliable utilization of β-carotene in quenching experiments, these apparent differences should be resolved by precise experimentation.

In spite of the high carotenoid protective effect in both natural biopolymers and man-made polymer matrices (demonstrated, e.g. by the inhibition of self-sensitized oxygenation of tetracene in polystyrene [93] or by the measured rate constant $k_q = 4{\cdot}3 \times 10^8$ $M^{-1}\ s^{-1}$ in polystyrene [226]), β-carotene proved to be ineffective as a UV stabilizer in polypropylene films under conditions of xenon-arc irradiation. The possible explanation is that the polyene chain of β-carotene is quickly destroyed via radical or peroxide routes [89] during the processing of the polymer or, under prolonged irradiation, it may even react with $^1O_2$.[78,101–103] Thus, at present the only practical use of carotenoids is in the treatment of human patients suffering from light-sensitive diseases.[104]

Amines

Amines are less efficient quenchers of singlet oxygen than carotenoids. Typically, the known rate constants, k_q, for quenching of $^1O_2$ by aliphatic and cycloaliphatic amines range from 10^5 to 4×10^7 $M^{-1}\ s^{-1}$ and, for quenching by aromatic amines, from 5×10^6 to $1{\cdot}5 \times 10^9$ $M^{-1}\ s^{-1}$. Quenching data for $^1O_2$ by amines in solution together with experimental conditions for their determination

Table 3 Carotenoids: $^1O_2$ quenching rate constants (k_q) and triplet energy levels (E_T)

Carotenoid[a]	Number of conjugated double bonds	E_T (kcal mol^{-1})	$k_q \times 10^{-9}$ (M^{-1} s^{-1})	Experimental conditions for measurement of k_q			Reference
				Solvent	$^1O_2$ acceptor	Origin of $^1O_2$[b]	
Sarcina phytoene	3	(47)[k]	≤0·019	C_6H_6–CH_3OH (3:2)	Rubrene	A-1	95
All-*trans*-retinol (**1**)	5		0·027[c]	C_6H_6–CH_3OH (4:1)	2-methyl-2-pentene (2-MP)	A-1	75
Sarcina phytofluene	5		≤0·1	C_6H_6–CH_3OH (3:2)	Rubrene	A-1	95
Synthetic C_{30} carotene analogue (**2**)	7	25·3[n]	0·17[c]	C_6H_6–CH_3OH (4:1)	2-MP	A-1	75
P-422[h]	8		12	C_6H_6–CH_3OH (3:2)	Rubrene	A-1	95
Synthetic C_{35} carotene analogue (**3**)	9	20·4[n]	5·7[c]	C_6H_6–CH_3OH (4:1)	2-MP	A-1	75
P-438[h]	9		31	C_6H_6–CH_3OH (3:2)	Rubrene	A-1	95
β-apo-8′-Carotenol	9		25[d]	C_6H_6–CH_3OH (4:1)	2-MP	A-1	73
Lutein	10		21	C_6H_6–CH_3OH (3:2)	Rubrene	A-1	95
β-Carotene (**4**)	11	17·1[n]	30[e]	—	—	—	75
		21–25[o]	30	Toluene	Rubrene	A-2	58
		<22·8[p]	26[j]	CS_2	Anthracene	A-3	90
			23	C_6H_6–CH_3OH (3:2)	Rubrene	A-1	95
			20	C_6H_6	DPBF[f]	B-1	49
			13	CH_2Cl_2	Rubrene	A-2	100
			13	C_6H_6	DPBF	B-2	94
			13	C_6H_6–C_2H_5OH (8:1)	THMP[q]	A-4	83
			10	—	—	—[m]	57
			9·1	Bromobenzene	Rubrene	D	92
			8·5	CH_2Cl_2	Rubrene	A-2	97
			6·5	Pyridine	Rubrene	A-2	167
			6·2	CCl_4	Rubrene	A-2	98
			3	Freon-113	DPBF	C	99
			1·4	Freon-113	DPBF	C	96
Lycopene	11		(~30)[d]	C_6H_6–CH_3OH (4:1)	2-MP	A-1	73, 75
Isozeaxanthin	11		29	C_6H_6–CH_3OH (3:2)	Rubrene	A-1	95
Synthetic C_{50} carotenoid	15	13[n]	~30	—	—	—[g]	75[i]
Synthetic C_{60} carotenoid	19	11[n]	~30	—	—	—[g]	75[i]

[a] Structural formulae can be found in ref. 206.

[b] A: Competitive photo-oxygenation with sensitizer: 1, methylene blue; 2, rubrene; 3, zinc tetraphenylporphine; 4, Rose Bengal. B: Direct method employing laser pulse excitation of sensitizer: 1, methylene blue; 2, anthracene. C: Direct $^3O_2$ excitation with the 1·06-μm continuous wave output of Nd laser. D: See first footnote, F, in Table 9.

[c] Calculated from k_q of β-carotene (3×10^{10} M^{-1} s^{-1}).

[d] k_q could not be determined accurately because substantial destruction of carotenoid occurred under the experimental conditions.

[e] This value was assumed, because diffusion-controlled reactions involving molecular oxygen of high diffusibility occur in benzene at room temperature with a rate constant of 3×10^{10} M^{-1} s^{-1}.[207]

[f] 1,3-Diphenylisobenzofuran.

[g] Experimental conditions are not given.

[h] Mutant strain from *Sarcina lutea*.[95]

[i] Footnote 10a in ref. 75.

[j] Calculated, assuming $k_d = 4 \times 10^3$ s^{-1} for CS_2.

[k] E_T of *trans*-1,3,5-hexatriene.[208]

[m] Given as calculated value from the modified Debye equation.

[n] After Fig. 3 in ref. 209.

[o] Ref. 71.

[p] Ref. 94.

[q] 2,2,6,6-Tetramethyl-4-hydroxypiperidine [see Formula (**10**)].

Table 4 Amines: solution $^{1}O_2$ quenching rate constants (k_q) and ionization potentials (I_p)

Amine	Ionization potential (eV) [a]	$k_q \times 10^{-6}$ ($M^{-1} s^{-1}$)	Experimental conditions for measurement of k_q: Solvent	$^{1}O_2$ acceptor	Origin of $^{1}O_2$ [b]	Reference
Aliphatic and cyclophatic amines:						
Benzylamine	8·64					
Hexamethylenetetramine	8·26 [c]	$\lesssim 1$ (Benzylamine to Cyclohexylamine, piperidine)	CH_3OH	2,5-Diphenylfuran	A-1	60
Isopropylamine, *tert*-butylamine						
Cyclohexylamine, piperidine						
1,4-Piperazine	8·21	1·23	CH_3OH	2,5-Diphenylfuran	A-1	60
Quinuclidine	8·02; 7·70 [p]	1·76	CH_3OH	2,5-Diphenylfuran	A-1	60
Diethylamine	8·01	1·88	CH_3OH	2,5-Diphenylfuran	A-1	60
Pyrrolidine	8·30	2·14	CH_3OH	2,5-Diphenylfuran	A-1	60
Nicotine (**5**)		4·3 [j]	CH_3OH	α-Terpinene	A-1	107
1,4-Diazabicyclo[2.2.2]octane	7·52	4·5	*n*-Butanol	2,5-Diphenylfuran	A-1	109
(DABCO) (**6**)	7·23 [k]	7·3	CH_3OH	2,5-Diphenylfuran	A-1	60, 109
		14 [m]	Pyridine	Rubrene	A-1	59
		15 [h]	CH_3OH	Anthracene	A-4	90
		16	C_6H_6–CH_3OH (4:1)	2-Methyl-2-pentene	A-2	73
		29 [i]	CS_2	Anthracene	A-4	90
		30	Isooctane	Rubrene	A-3	100
		33	CH_2Cl_2	Rubrene	B	57, 100
		35; 39	Isooctane	Rubrene	B	57, 100
		44 [g]	C_6H_6	Anthracene	A-4	90
		310	Pyridine	Rubrene	A-3	167
		670	Toluene	Rubrene	A-3	58
Triethylamine	7·85 [f]	2·1	$C_2Cl_3F_3$ (Freon-113)	Tetracyclone [e]	C	210
	7·50	7·8 [m]	Pyridine	2-Methyl-2-pentene	A-1	59
		9·30	CH_3OH	2,5-Diphenylfuran	A-1	60
		12	— [d]	2,5-Diphenylfuran	A-1	109
Aromatic amines:						
Diphenylamine	7·25	6·1	CH_3OH	2,5-Diphenylfuran	A-1	60
N-Methylaniline	7·84 [f]	26·6	CH_3OH	2,5-Diphenylfuran	A-1	60
	7·10	0·038	$C_2Cl_3F_3$ (Freon-113)	Tetracyclone [e]	C	210
2-Aminonaphthalene		180 [n]	C_2H_5OH	2,5-Dimethylfuran	A-1	173

1-Isopropylamino-4-phenylaminobenzene		210; 330	Isooctane	Rubrene	A-3	97, 100
		400	Hexadecane	Rubrene	D	100
		1500 [n]	C_2H_5OH	2,5-Dimethylfuran	A-1	173
4,4′-Diaminobiphenyl		550 [n]	C_2H_5OH	2,5-Dimethylfuran	A-1	173
o-Phenylenediamine		600 [n]	C_2H_5OH	2,5-Dimethylfuran	A-1	173
1-Cyclohexylamino-4-phenylaminobenzene		1100 [n]	C_2H_5OH	2,5-Dimethylfuran	A-1	173
Substituted NN-dimethylanilines: *						
NN-Dimethylaniline, *p*–CN		2	CH_3OH	2,5-Diphenylfuran	A-1	60
p–CHO		3	CH_3OH	2,5-Diphenylfuran	A-1	60
m–Cl		14	CH_3OH	2,5-Diphenylfuran	A-1	60
p–Br		32	CH_3OH	2,5-Diphenylfuran	A-1	60
m–OCH_3		60	CH_3OH	2,5-Dimethylfuran	A-1	60
H	7·14 [f]	0·2	$C_2Cl_3F_3$ (Freon-113)	Tetracyclone [e]	C	210
	7·30	73	—	1,3-Diphenylisobenzofuran	A-1	197
		130	CH_3OH	2,5-Diphenylfuran	A-1	60
p-CH_3		181	CH_3OH	2,5-Diphenylfuran	A-1	60
p-OCH_3		209	CH_3OH	2,5-Diphenylfuran	A-1	60
p–$N(CH_3)_2$	6·54 [p]	675	CH_3OH	2,5-Diphenylfuran	A-1	60

[a] After J. L. Franklin *et al.* (cited in ref. 60), unless otherwise stated.
[b] A: competitive photo-oxygenation with sensitizer: 1, Rose Bengal; 2, methylene blue; 3, rubrene; 4, zinc tetraphenylporphine. B: thermal decomposition of $(PhO)_3PO_3$. C: direct $^3O_2$ excitation with 1·06-μm continuous wave output of Nd laser. D: microwave discharge.
[c] Ref. 211.
[d] Solvent is not given (probably methanol).
[e] Tetraphenylcyclopentadienone.
[f] Ref. 212.
[g] Calculated, assuming $k_d = 4{\cdot}2 \times 10^4$ M^{-1} s^{-1} for benzene.
[h] Calculated, assuming $k_d = 10^5$ M^{-1} s^{-1} for methanol.
[i] Calculated, assuming $k_d = 5 \times 10^3$ M^{-1} s^{-1} for CS_2.
[j] After ref. 59, assuming k_q (nicotine) $\approx 0{\cdot}15 k_q$ (DABCO) and k_q (DABCO) $= 3 \times 10^7$ M^{-1} s^{-1}.
[k] Ref. 213.
[m] Calculated, assuming $k_d = 3{\cdot}1 \times 10^4$ M^{-1} s^{-1} for pyridine.
[n] Calculated from $k_r = 10^8$ M^{-1} s^{-1} for 2,5-dimethylfuran,[214] $k_d = 10^5$ M^{-1} s^{-1} for ethanol (Table 2) and from the data in ref. 173, as given in ref. 26.
[p] Ref. 215.

and some ionization potentials are compiled in Table 4. Values of k_q for amines in gas phase were determined by Ogryzlo and co-workers.[105,106]

Several observations are apparent when examining the data presented in Table 4, as follows.

(i) Nicotine (**5**) is the first compound for which it has been clearly demonstrated (Schenck and Gollnick [107]) that $^1O_2$ (or, as originally proposed,[107] the kinetically equivalent complex $Sens^{rad}.O_2$) can be quenched with an amine. Careful kinetic analysis showed that $^1O_2$ is quenched by nicotine at 5·4 times the rate it reacts with this amine. Later calculations [59] showed that (**5**) is about one-seventh as reactive a quencher as (**6**) (DABCO).

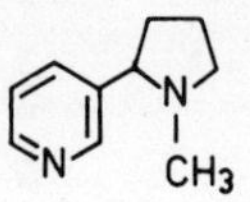

(5) (nicotine) 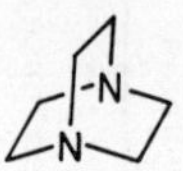 (6) (DABCO) 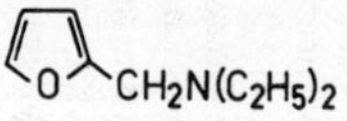 (7)

(ii) The most repeatedly determined value of k_q is that of 1,4-diazabicyclo [2.2.2] octane (**6**, DABCO). In 1968, Ouannès and Wilson [108] discovered that (**6**) is a powerful inhibitor of $^1O_2$ reactions in solution and then demonstrated that it also quenches $^1O_2$ in the gas phase. In contrast to **5**, **6** does not react chemically with $^1O_2$, either in the gas phase or in solution ($k_q > 100k_r$ [59]). Structural reasons for this behaviour will be explained later. The published values of k_q for DABCO are scattered and show no evident solvent effect. Formerly, very small solvent effects were reported, although with a limited amount of data.[51,90,109]

DABCO is only moderately effective as a $^1O_2$ quencher. The average of all but the last two k_q values from Table 4 is $2{\cdot}4 \times 10^7\ \text{M}^{-1}\ \text{s}^{-1}$, which is about 1000 times less than k_q for β-carotene. However, DABCO is an oxidatively stable, water-soluble compound and does not absorb light of wavelengths longer than 300 nm. Therefore, it is widely used as a diagnostic test for the participation,[31,78,79,81,82,84,88,89,91,110–118,228] or non-participation [86,87,119] of $^1O_2$ in oxygenation reactions as well as a standard for the determination of relative quenching efficiencies.[56,92] Anomalies in the tests with DABCO were also reported: rate acceleration due to the reaction with acceptor [84,222] and enhancement of $^1O_2$ dimole 634-nm emission from aqueous solution in the presence of DABCO.[120]

(iii) Amines with low ionization potentials are better quenchers (see also Figure 5). This results in the sequence tertiary > secondary > primary amines. Ogryzlo and Tang [105] explained this sequence as being the result of the formation of a charge-transfer (C-T) intermediate. A small spin–orbit coupling between the singlet and triplet states in the C-T intermediate would allow a spin flip to occur and hence a facile inter-system crossing from singlet to triplet oxygen:

$$^1O_2 + NR_3 \rightleftarrows [^1O_2^- \ ^+NR_3] \xrightarrow[\text{coupling}]{\text{Spin–orbit}} [^3O_2^- \ ^+NR_3] \longrightarrow {}^3O_2 + NR_3 \quad (17)$$

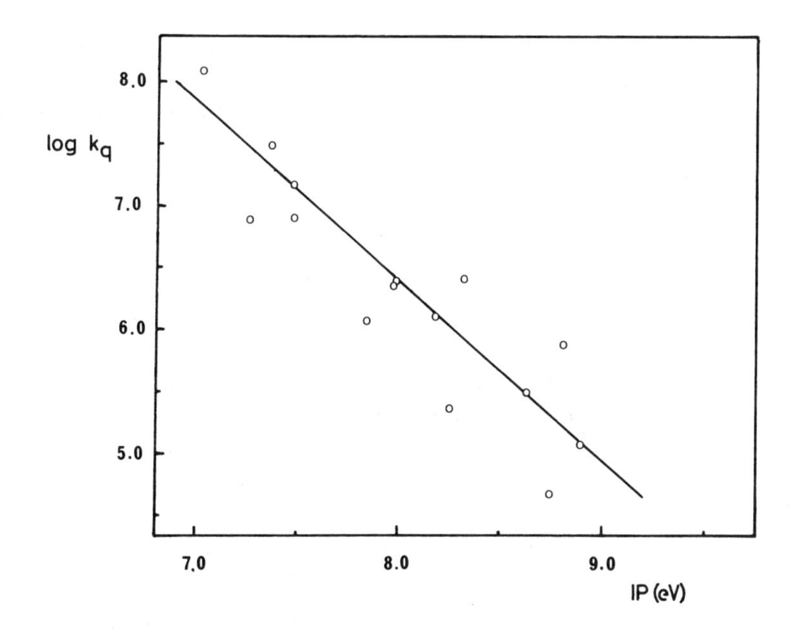

Figure 5 Relationship between log (rate of quenching) and ionization potentials of amines for quenching of singlet oxygen (methanol solution). [Reprinted, with permission, from R. H. Young, R. L. Martin, D. Feriozi, D. Brewer, and R. Kayser, *Photochem. Photobiol.*, **17**, 233 (1973)]

The results of Young *et al.*[60] seem to indicate that only a partial C-T complex is responsible for the quenching of $^1O_2$. Equation (17) should therefore be rewritten as

$$^1O_2 + NR_3 \rightleftarrows [^1O_2^- \ldots {}^+NR_3] \longrightarrow {}^3O_2 + NR_3 \qquad (18)$$

Alkyl substitution of the α-carbon of the amine reduces the quenching rate below that predicted by the ionization potential, thus indicating that $^1O_2$ quenching is sensitive to steric effects.[218]

Young *et al.*[60] also obtained the quenching rate constants of the dye-sensitized photo-oxygenation of 1,3-diphenylisobenzofuran for a series of substituted *NN*-dimethylanilines (see compounds with asterisks in Table 4). A Hammett plot was made and a moderate rho value of $-1{\cdot}39$ was obtained. This confirms the previous suggestion that a complex which is C-T in nature is responsible for the quenching of $^1O_2$. In other work, Young and Martin[109] obtained a good correlation between the logarithm of the rate constant of quenching of the first excited singlet state of aromatic compounds with DABCO and a measure of the electron affinity of the excited singlet state.

Gollnick and Lindner[121] observed that there is a correlation between the reaction rate of photo-oxidation of amines with $^1O_2$ and their quenching rates; i.e. the reaction rate is also a function of ionization potential (similar observations were made during the investigations of photoreductive processes by amines[122]). With this and all other previous findings in mind, they proposed that both the quenching and the reaction of $^1O_2$ with amines pass through the same intermediate (**8**) [irrespective of the extent of charge separation and bonding in (**8**)]:

(1) R = alkyl:

$$R_3N + {}^1O_2 \longrightarrow \underset{(\mathbf{8})}{[R_3N^+{-}OO^-]} \longrightarrow R_3N + {}^3O_2 \quad \text{(Physical quenching)}$$

Exclusive physical quenching occurs if (a) substituent R bears no α-hydrogen (e.g. tert-butylamine); (b) there are α-hydrogen atoms present, but the nitrogen atom is in a bridge-head position of a cage molecule (e.g. DABCO) and thus the formation of any C=N bond is restricted for structural reasons ('Bredt's rule'[123]).

(2) R_1, R_2, R_3 = alkyl and/or hydrogen:

(a)

$$\underset{\text{Primary or secondary amine}}{H{-}\overset{|}{\underset{|}{C}}{-}N{-}H} + {}^1O_2 \longrightarrow \left[\underset{H\quad OO^-}{{-}\overset{|}{\underset{|}{C}}{-}\overset{|}{\underset{|}{N^{\pm}}}{-}H} \rightleftarrows \underset{H\quad OOH}{{-}\overset{|}{\underset{|}{C}}{-}\overset{|}{\underset{|}{N}}}\right] \longrightarrow {}^3O_2 + \text{amine} \quad \text{(Physical quenching)}$$

$$H_2O_2 + \;\rangle C{=}N{-} \xrightarrow{+H_2O} \;\rangle C{=}O + H_2N{-} \quad \text{('Reactive' quenching)} \qquad (19)$$

(b)

$$\underset{\text{Tertiary Amine}}{{-}\overset{|}{\underset{\underset{H}{|}}{C}}{-}\overset{|}{N}{-}} + {}^1O_2 \longrightarrow {}^3O_2 + \text{amine} \quad \text{(Physical quenching)}$$

$$\left[\underset{H\qquad {}^-O{-}O}{\rangle C{-}N^{\pm}} \longrightarrow \;\rangle C{=}\overset{+}{N}\langle \cdot {}^-OOH \rightleftarrows \underset{OOH}{{-}\overset{|}{\underset{|}{C}}{-}\overset{-}{N}{-}} \longrightarrow \underset{OO^-\quad H}{\overset{|}{\underset{|}{C}}{-}\overset{|}{\underset{|}{N^{\pm}}}}\right] \qquad (20)$$

$$\xrightarrow{+H_2O} \;\rangle C{=}O + HN\langle \; + H_2O_2$$

or the amide is formed when the carbon atom bears another hydrogen.

According to this mechanism, it is clear that quaternary salts can neither quench nor react with $^1O_2$. It must be pointed out at this stage that in each sensitized photo-oxidation of amines (=Q), there is most probably an inherent competition between two primary processes: one involving singlet oxygen [process (9) in Scheme 1] and another one, involving an interaction of the excited sensitizer molecule with amine, followed by the reaction of the α-amino radical (or radical ion?) with ground-state oxygen [process (2) in Scheme 1].[61] For example, Bartholomew and Davidson[124] have shown that some of the photo-oxidations were more efficient at lower oxygen pressure, under which conditions the triplet dye was less efficiently quenched by oxygen [process (4) in Scheme 1] and therefore more available for abstracting hydrogen from the amine. Furthermore, the quenching of the excited singlet[72] and triplet states[116,122,217] of the dye by amine is important when high concentrations of amines are used ($\geq 5 \times 10^{-2}$ M). The probability of predomination of processes [(1), (2), and (3) *versus* (4), (5), and (9); Scheme 1] in the sensitized photo-oxidation of amines depends on the rate constants of each process and the concentrations of reactants.

Competitive experiments applied to the determination of k_q values reflect the probability of trapping and deactivation of $^1O_2$ either by an acceptor or by a quencher. Since k_q is a bimolecular rate constant, such a probability depends on the relative concentrations of both acceptor and quencher (e.g. amine). If, however, the local concentration of quencher in the proximity of acceptor may be increased, e.g. by an intramolecular attachment to the $^1O_2$-active site of the acceptor molecule, thus allowing the quencher to encounter *nearly all* $^1O_2$ molecules approaching the acceptor, the efficiency of the $^1O_2$ quenching process is expected to increase dramatically. This effect was precisely demonstrated by Atkinson *et al.*[125]: in a series of cleverly constructed molecules the ability of the intramolecularly bonded tertiary amino groups to quench the three main types of photo-oxygenation reactions was much higher than in the external quenchers such as triethylamine or even DABCO. For example, 100% of 2-methylfuran (5×10^{-2} M in CH_3OH) was consumed in methylene blue sensitized photo-oxygenation during 1 h, 90% in the presence of triethylamine (5×10^{-2} M) and 85% in the presence of DABCO (5×10^{-2} M). However, only 10% of 'intra-molecularly quenched' *NN*-diethylfurfurylamine (7) was oxygenated with $^1O_2$ during the same time. Other interesting examples of extremely efficient intra-molecular quenching are given in the original paper.[125]

$$(9) \xrightarrow[k_r^Q]{+^1O_2} (10) \xrightarrow[k_r^Q]{+^1O_2} (11) \xrightarrow[k_q]{+^1O_2} (11) + {}^3O_2 \qquad (21)$$

(9) $\xrightarrow{+^1O_2,\ k_q}$ (9) + $^3O_2$; (10) $\xrightarrow{+^1O_2,\ k_q}$ (10) + $^3O_2$

(9) R, N–CH_3 (10) R, N–H   (11) R, N–O·

Piperidines (**9**) and (**10**), and nitroxide (**11**) are recognized as being very efficient light stabilizers for polyolefins.[126] Whatever the reasons may be for the activity of these UV-transparent, sterically hindered piperidines,[126] their ability to quench singlet oxygen will nevertheless now be considered.

Despite various methods of measurement, the calculated values of apparent quenching constants 'k_q' of 1,2,2,6,6-pentamethylpiperidines (**9**) are in good agreement and reveal that compounds (**9**) are even better quenchers of $^1O_2$ than

Table 5 Apparent quenching constants of 1,2,2,6,6-pentamethylpiperidines (**9**)

(**9**)	$(k_q+k_r^Q)\times 10^{-7}$ ($M^{-1}\,s^{-1}$)	Experimental conditions for measurement of $(k_q+k_r^Q)$: Solvent	Acceptor	Generation[a] of $^1O_2$	Ref.
$R=O_2CC_8H_{17}$	4·3	C_6H_6–C_2H_5OH (1:1)	rubrene	A	127
R=OH	5	CH_2Cl_2	naphtalene	B	83
R=OH	3	C_6H_6–CH_3OH (1:1)	9,10-$(CH_3OCH_2)_2$-anthracene	C	56
Codeine + $^1O_2$ → norcodeine	≃1	*tert*-C_4H_9OH–H_2O (5:1)	—	D	128

[a] A: Methylene blue as sensitizer. Irradiation with 633-nm light of He/Ne gas laser. $(k_q+k_r^Q)$ value was calculated from Figures 4 and 5 in ref. 127, using k_q(DABCO)$=2{\cdot}4\times 10^7$ $M^{-1}\,s^{-1}$. B: Competitive experiment using Rose Bengal as sensitizer. C: Rose Bengal as sensitizer. $(k_q+k_r^Q)$ calculated as under A. D: Rose Bengal as sensitizer. $(k_q+k_r^Q)\approx 0{\cdot}02k_r$(2,5-dimethylfuran). Average value of k_r(2,5-dimethylfuran)$=5\times 10^8$ $M^{-1}\,s^{-1}$ was taken.

DABCO (Table 5). The ionization potential of (**9**) (R=H), $I_p=7{\cdot}65$ eV,[126] is also in the range of DABCO. However, in refs. 83 and 127, the exclusively inhibitive effect of (**9**) on the photo-oxygenation of $^1O_2$ acceptor was measured. Simultaneous measurements of oxygen uptake revealed,[56] however, that (**9**) is self-oxidized at a rate similar to that of quenching. In additional photo-oxygenation experiments, with (**9**) R=OH) as the sole acceptor, we found a very clean conversion (**9**) → (**10**) (>98% yield by gas–liquid chromatography). Similarly, Lindner *et al.*[128] reported smooth *N*-demethylation of codeine to nor-codeine under conditions of photosensitized as well as microwave-discharge $^1O_2$ oxygenations (see also refs. 111, 129, and 130). Thus, in our opinion, the observed quenching effect of (**9**) must be attributed to a large extent to the 'reactive' quenching [process (9) in Scheme 1] and the determined rate constants 'k_q' actually represents the sum of k_q and k_r^Q (see Table 5).

2,2,6,6-Tetramethylpiperidines (**10**) were found to be very poor quenchers of $^1O_2$.[56, 127] The measured value of k_q [(**10**), R=OH] of $\leqslant 5\times 10^5$ $M^{-1}\,s^{-1}$ [83, 218] is in fact lower than that of unsubstituted piperidine ($k_q=9\times 10^5$ $M^{-1}\,s^{-1}$ [60]). The ionization potential of (**10**) (R=OH) is 7·85 eV.[126]

It was recently reported that (**10**) undergoes dye-sensitized photo-oxidation producing stable nitroxide radicals (**11**).[83, 131] The rate of this photo-oxidation is low, $k_r=2{\cdot}9\times 10^3$ $M^{-1}\,s^{-1}$, and it can be reduced by known $^1O_2$ quenchers such as β-carotene and nickel dibutyldithiocarbamate.[83] Similar formation of nitroxides was observed by a sensitive e.s.r. method in photo-oxidized solutions of homotropine, 1,1′- and 1,2′-diadamatylamine, and 4,4′-dimethoxy-, 4,4′-dimethyl-, and 4,4′-di-*tert*-octyldiphenylamine.[83, 131].

We found[56] that paramagnetic nitroxides, which are stable end products of the unique $^1O_2$ oxidation sequence (**9**) → (**10**) → (**11**), are nearly as effective as DABCO in the quenching of sensitized photo-oxygenations of 2-methyl-2-pentene and 9,10-dimethoxymethylanthracene. The value of k_q [(**11**), R=OH] calculated from relative quenching rates is 2×10^7 M^{-1} s^{-1}. The $^1O_2$ quenching effect of nitroxide (**11**) (R=OH) was further substantiated by experiments with $^1O_2$ generated from NaOCl–H_2O_2 reagent, in which the yield of 3-hydroxy-2, 3-dimethyl-l-butene from 2,3-dimethyl-2-butene decreased by 1·6 in the presence of 10^{-2} mol of (**11**). Compound (**11**) is stable against further photo-oxidation and against H_2O_2, and, in turn, does not destroy H_2O_2.

The $^1O_2$ quenching by (**11**) is interesting, because several quenching modes can be taken into account. The adiabatic first ionization potential of (**11**) is 6·73 eV,[132] a value similar to that of the best amine $^1O_2$ quenchers operating by a C-T mechanism. Some radicals and radical ions quench singlets by the Förster transfer (long-range interaction) mechanism,[133] however, for (**11**) such a process appears to be extremely unlikely because of its very small long-wave extinction coefficient. In addition, the exchange-induced inter-system crossing caused by paramagnetism in the quenching radical (**11**) may also be operative.[134]

It would be an attractive idea to speculate that the $^1O_2$ quenching efficiency of the nitroxide formed, as well as its ability to quench excited triplet states[135] (well documented also with other structurally similar, sterically hindered nitroxides[136–138]) are responsible for the UV-protective effect of 2,2,6,6-tetramethylpiperidines in polymers. However, a number of arguments testify that these processes are not significant in polymers:

(a) small $^1O_2$ quenching rates of (**11**) [as well as of (**9**) and (**10**)], which may be expected to become even smaller in the polymer matrix;[93]

(b) low reaction rate of process (**10**) → (**11**);

(c) extensive formation of $^1O_2$ in polymers—although possible by several ways[26]—has not yet been unambiguously established;

(d) low additive [e.g. (**9**) or (**10**)] levels usually applied in commercial polymers.

This would strongly suggest that (**9**), (**10**), or (**11**) must act in a polymer according to a complex mechanism(s) other than the exclusive $^1O_2$ or excited state quenching. Besides extensive trapping of free radicals, essential in the degradation of polymers, by (**11**) (see also ref. 139), the $^1O_2$ 'reactive' quenching by (**9**) or (**10**) as well as physical $^1O_2$ quenching by (**9**), (**10**), or (**11**) probably make some rather small contribution to the overall mechanism of UV protection.

Another example may demonstrate the importance of additional protection modes: while 3-hydroxyquinuclidine is an efficient $^1O_2$ quencher in solution (about 75% of DABCO activity), its C_{18} ester shows no effect as a light stabilizer in polypropylene.[135]

Nitrogen-containing Compounds other than Amines

This section includes various types of compounds, the only common features being good to excellent quenching efficiency and the presence of one or more nitrogen atom(s) in the molecule.

Nitroxides

These compounds were dealt with in the previous section.

Nitrones, Azo Dioxides ('Dinitrones') and Nitroso Compounds

In 1975, Foote and Ching[140] sought possible 1,3-dipolar adducts of $^1O_2$ with stable 1,3-dipolar reagents (**12**) and (**14**). Neither (**12**) nor (**14**) reacted in the expected manner. While (**12**) underwent a quantitative $^1O_2$ ene reaction, (**12**) → (**13**), (**14**) was found to quench $^1O_2$ efficiently even though it failed to react at all. The rates of chemical reaction (k_r) and physical quenching (k_q) of $^1O_2$ with (**12**) and (**14**) were determined by competitive inhibition of photo-oxygenation of 1,3-diphenylisobenzofuran.

It is suggested[140] that the mechanism of this quenching process is pobably a C-T interaction analogous to that proposed for electron-rich compounds such as amines, although the formation of an unstable 1,3-dipolar adduct followed by its breakdown to initial compounds is not precluded.

Although 3,3,4,4-tetramethyl-1,2-diazetine-1,2-dioxide (**15**) is known as an useful triplet quencher,[141] neither (**15**) nor a series of similar 1,2-dioxides quench $^1O_2$.[142] However, the α-chloro-substituted (**16**) was found to inhibit strongly the reaction of rubrene with $^1O_2$, generated from $(C_6H_5O)_3PO_3$.[142]

(12) $+ {}^1O_2 \longrightarrow$ (13) $(k_r + k_q) = 2.1 \times 10^7\ M^{-1}\ s^{-1}$

(14) $+ {}^1O_2 \longrightarrow$ (14) $+ {}^3O_2$ $k_q = 5 \times 10^7\ M^{-1}\ s^{-1}$

	(15)	(16)	(17)	(18)	(19)
$k_q \times 10^{-9}\ M^{-1}\ s^{-1}$	0.080	9.3 or 12.0		5.3	9.3

It was estimated that about 0·7% of (**16**) is in equilibrium with (**17**) and at least part of the $^1O_2$ quenching is due to (**17**). Using this assumption, k_q for (**17**) was calculated as $9{\cdot}2 \times 10^9\,\text{M}^{-1}\,\text{s}^{-1}$ or $1{\cdot}2 \times 10^{10}\,\text{M}^{-1}\,\text{s}^{-1}$, depending on the experimental conditions.

Quenching studies with both (**18**) and (**19**) demonstrated that nitroso compounds are indeed very efficient $^1O_2$ quenchers.[142] Comparison of the quenching constants with that of β-carotene suggests that quenching of $^1O_2$ occurs by an energy transfer mechanism. The exceptionally low effective triplet energy of these nitroso compounds ($\leqslant$23 kcal mol^{-1}) may result from the coupling of energy transfer with the relaxation of the bent nitroso triplets to a low-energy linear configuration, isoelectronic with $^3O_2$.[142]

It is not yet known if compounds (**14**)–(**19**) have any practical applications, although, besides their $^1O_2$ quenching ability, nitrones such as (**14**) can effectively trap R· radicals[143] and cyclic azo dioxides similar to (**16**) can photochemically expel NO,[144] thus producing cyclic nitroxides in both instances.

Azomethine Dyes

Normally dyes are not associated with the quenching of $^1O_2$ but rather with the sensitized formation of this species. Azomethine dyes (**20**) are ineffective as photosensitizers for $^1O_2$ but instead constitute a new class of efficient solution quenchers of $^1O_2$.[145] Compounds (**20**) may be considered as *para*-substituted anilines, and as such are expected to quench $^1O_2$. However, (**20**) are much better quenchers than would be predicted on the basis of Hammett constants for this type of *para*-substituted anilines.

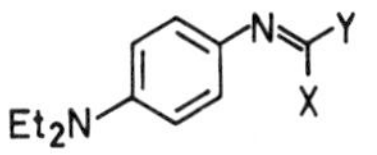

(20) 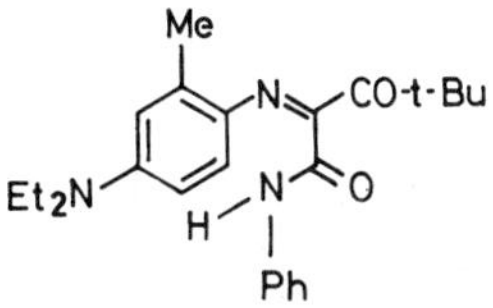 (21) (22)

$k_q = 2.6 \times 10^7\,\text{M}^{-1}\,\text{s}^{-1}$

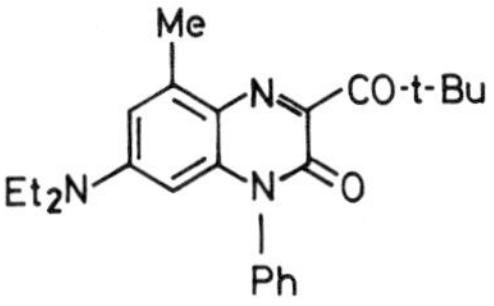

Using Rose Bengal or azine (**22**) as sensitizer, the $^1O_2$ quenching efficiencies of several azomethine dyes were determined by measuring the rate of inhibition of the reaction of $^1O_2$ with 2-methyl-2-pentene. Quenching rate constants, k_q, span approximately two orders of magnitude and, surprisingly, they are solvent dependent. For example, they are about 10-fold higher in acetonitrile solution ($k_q \approx 4 \times 10^8$–$3{\cdot}9 \times 10^{10}\,\text{M}^{-1}\,\text{s}^{-1}$) than in either benzene or pyridine ($k_q \approx 10^7$–$6 \times 10^9\,\text{M}^{-1}\,\text{s}^{-1}$). The quenching efficiency of (**20**) depends on the *ortho*-substituent: steric crowding *ortho* to the azomethine linkage generally increases k_q, whereas steric crowding *ortho* to the dialkylamino group may result in a decreased k_q. The influence of these and of a number of other structural factors is discussed in detail in the original paper.[145] No evidence was found for chemical reaction of any of the azomethine dyes with $^1O_2$.

In this connection, it is interesting to compare the $^1O_2$ sensitizing and quenching efficiencies of azomethine dye (**21**) with azine dye (**22**). The latter differs from the former solely by the presence of a sigma bond between the amide nitrogen atom and the aromatic ring. This bond prevents rotation about the C=N bond. As a consequence, azine (**22**) is a sensitizer, whereas dye (**21**) is a quencher of $^1O_2$. Compound (**22**) has a triplet lifetime of 80 μs, whereas that of (**21**) is less than 10 μs. Dyes (**20**) are also capable of quenching very low-lying triplets, e.g. that of violanthrene ($E_T \approx 25$ kcal mol^{-1}), at almost the diffusion-controlled rate.[145]

For the $^1O_2$ quenching process the authors propose a mechanism involving the formation of a dye–$^1O_2$ excited complex, the binding energy of which is increased by energy transfer and C-T contribution.[145]

Amino Acids and Proteins

Dye-sensitized photo-oxygenation of amino acids in proteins or of bases in nucleic acids is believed to be the basis for most of the phenomena classified under 'photodynamic action' (see Figure 3). $^1O_2$ is probably the intermediate in some but not all of these photodynamic reactions.

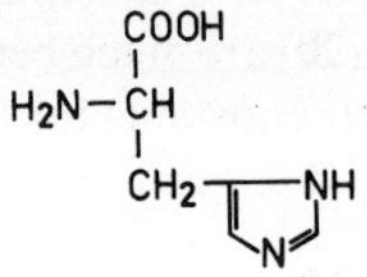

(23) (histidine)

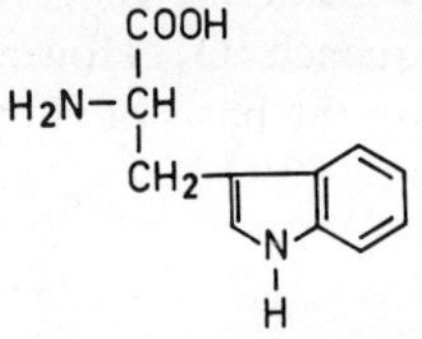

(24) (tryptophan)

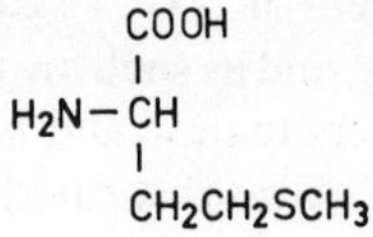

(25) (methionine)

In 1966, Glad and Spikes[146] showed that 4×10^{-3} M of free histidine (**23**) significantly protects trypsin against photoinactivation, in the presence of either methylene blue or eosin Y as sensitizer. Some years later, Nilsson *et al.*,[116] Matheson *et al.*[147] and Schmidt and Rosenkranz[148] showed that the protective action of histidine (**23**) as well as some other amino acids and proteins (Table 6) consists in the physical (k_q) and 'reactive' (k_r^Q) quenching of $^1O_2$ and of the quenching of the sensitizer triplet state (k_q^{3S}). Whichever reaction prevails depends on the structure of the amino acid as a separate molecule or as a constituent unit of protein. By careful kinetic and quenching (e.g. with N_3^-) experiments in normal and perdeuterated solvents the $^1O_2$ oxygenation mechanism was established and the pertinent rate constants measured (Table 6).

The k_q values (calculated) of proteins were estimated by summing the contributions from (**23**)–(**25**), assuming that all other amino acids have no effect; this seems reasonable since their amino groups would be tied up in formation of the peptide bond. The results suggest that (**23**)–(**25**) in these proteins are (within a rough factor of 2–3) as accessible for $^1O_2$ as when they are free in solution.

Table 6 Amino acids and proteins: rate constants of quenching processes

Compound	$k_q^{3S} \times 10^{-7}$ [a] ($M^{-1}\,s^{-1}$)	$k_r^{Q} \times 10^{-7}$ [a] ($M^{-1}\,s^{-1}$)	$k_q \times 10^{-7}$ ($M^{-1}\,s^{-1}$)	Method of determination of k_q [b]	Ref.
Alanine	—	—	0·2	B	147
			≤1	A	116
Histidine	0·1	0·7	5	A	116
			17	B	147
Tryptophan	200[d]	0·4	4	A	116
			9	B	147
Methionine	30	0·6	3	A	116
			3	B	147
Lysozyme	—	1·6	35; 41	C	148
			150; (80)[c]	B	147
Superoxide dismutase[e]	—	—	82	D	147
			260; (260)[c]	B	147
Aposuperoxide dismutase	—	—	110	D	147
			250; (260)[c]	B	147
Carbonic anhydrase	—	—	65	D	147
			80; (270)[c]	B	147
Guanosine monophosphate	100	—	—	A	116

[a] In H_2O–CH_3OH (1:1) solution.
[b] A: Pulse laser technique, 1,3-diphenylisobenzofuran as acceptor, methylene blue as sensitizer, H_2O–CH_3OH (1:1) solution. B: Direct generation of $^1O_2$ from $^3O_2$ by absorption of the 1·06-μm output of a Nd laser, bilirubin as acceptor, D_2O solutions. C: Acridine orange sensitized photo-inactivation of lysozyme, H_2O (pH 5·9) solution. D: Methylene blue sensitized photo-oxidation of bilirubin, D_2O solution.
[c] Calculated (see text).
[d] Values of k_q^{3S} (tryptophan) for thiazine dyes in H_2O solutions have recently been reported.[216]
[e] Quenching of methylene blue sensitized photo-oxidation of 1,3-diphenylisobenzofuran by superoxide dismutase (cuprein) has also been reported.[147]

Nothing is known about the mechanism of physical quenching of $^1O_2$ by **(23)**–**(25)**. A striking structural feature common to **(23)** and **(24)** (but not alanine) is the presence of a nitrogen atom in the heterocyclic ring. The quenching ability of methionine **(25)** may be attributed to the sulphide unit.

The fact that the k_q values for **(23)**–**(25)** are similar is apparently accidental. The mode of the 'reactive' quenching is better understood:[22, 149] The histidine **(23)** model compound, 4-methylimidazole, is photo-oxidized to acetylurea in the presence of methylene blue, whereas histidine itself gives rise to a variety of unidentified photo-oxidation products. In tryptophan **(24)**, the five-memberd nitrogen-containing ring is cleaved and *N*-formylkynurenine, a remarkable photodynamic sensitizer,[223] is formed. Photo-oxidation of methionine results in an almost quantitative conversion to the corresponding sulphoxide.

Table 6 also includes guanosine monophosphate, a very efficient inhibitor of DPBF photobleaching. Experiments indicate[116] that this is primarily due to the quenching of methylene blue triplets, rather than to the quenching of $^1O_2$. Reactions of nucleotides with the triplet state of this dye have been reported.[150]

Bilirubin and Biliverdin

The cause of jaundice in newborn infants (which may result in brain damage) is due to the insufficient activity of the hepatic system which effects the conversion of the lipid-soluble yellow bile pigment, bilirubin (**26**), to its water-soluble conjugate. This results in the deposition of excess of (**26**) in the skin and brain of the infant. The common treatment involves irradiating the infant with light in the wavelengths absorbed by bilirubin (ca. 450 nm). The bleaching of (**26**) is believed to be due to the self-sensitized photo-oxygenation of (**26**)[77] to give water-soluble degradation products which are excreted.[151]

M = $-CH_3$

V = $-CH{=}CH_2$

P = $-CH_2CH_2COOH$

(**26**) bilirubin (BR)
$k_q = 2.1\text{–}2.3 \times 10^9\ \text{M}^{-1}\ \text{s}^{-1}$

(**27**) biliverdin (BV)
$k_q = 3.3 \times 10^9\ \text{M}^{-1}\ \text{s}^{-1}$

In vitro studies supported the intermediary role of $^1O_2$ in the self-sensitized photo-oxygenation of (**26**).[152–154] Since the systems which generate $^1O_2$ have been shown to produce a variety of injuries, the discovery of physical $^1O_2$ quenching by (**26**)[62,151] and especially by its photo-oxidation product biliverdin (**27**)[153] is obviously of considerable interest.

Foote and Ching[62] studied kinetically the photo-oxidation of (**26**) in chloroform by several independent techniques. The rate of total consumption of $^1O_2$, $(k_q + k_r^Q)$, is $2{\cdot}5 \times 10^9\ \text{M}^{-1}\ \text{s}^{-1}$. The value for the physical quenching alone (k_q) is $2{\cdot}1 \times 10^9\ \text{M}^{-1}\ \text{s}^{-1}$, i.e. $^1O_2$ quenching by (**26**) accounts for about 83% of the total removal of $^1O_2$, with the reaction accounting for the remainder $(k_r = 0{\cdot}4 \times 10^9\ \text{M}^{-1}\ \text{s}^{-1})$. Thus, (**26**) is not only effective as a $^1O_2$ quencher but also as one of its most reactive known acceptors. This probably explains why the phototherapy of neonatal jaundice is so effective without severe photodynamic side effects: (**26**) produces $^1O_2$ on irradiation but it almost certainly is the most reactive local substrate in the lipid environment, in which it is localized.

Similar values of k_q $(2{\cdot}3 \pm 1{\cdot}0 \times 10^9\ \text{M}^{-1}\ \text{s}^{-1})$ and k_r^Q $(1{\cdot}7 \pm 0{\cdot}3 \times 10^8\ \text{M}^{-1}\ \text{s}^{-1})$ were found for bilirubin by Stevens and Small.[151] Moreover, they estimated values of $k_q = 3{\cdot}3 \times 10^9\ \text{M}^{-1}\ \text{s}^{-1}$ and $k_r^Q = 3 \times 10^6\ \text{M}^{-1}\ \text{s}^{-1}$ for biliverdin (**27**) using a competitive method with rubrene as the acceptor in CCl_4. Biliverdin is approximately as effective as β-carotene in inhibiting the photo-oxidation of bilirubin.[155]

The k_q constants for $^1O_2$ quenching by (**26**) and (**27**) are within an order of magnitude of diffusion-limited values estimated for the solvent used. For (**27**), with nine conjugated double bonds, the triplet state may be of sufficiently low energy to accommodate the energy-transfer quenching process which is virtually diffusion limited (see carotenoids). A similar energy transfer process is therefore unlikely for **(26)** (with five conjugated double bonds) and Stevens and Small[151] attributed physical quenching in this case to the reversible electron-transfer process[22]

$$BR + {}^1O_2 \longrightarrow (BR^+O_2^-) \longrightarrow BR + {}^3O_2 \qquad (22)$$

promoted by the high electron affinity (ca. 1·4 eV) of $^1O_2$, suggested also by Smith *et al.*[145] for $^1O_2$ quenching by azomethine dyes.

Sulphides

The singlet oxygen mechanism of oxidation of sulphides and disulphides is well established: oxidation of dibenzyl sulphide with $^1O_2$ generated by a microwave discharge can be completely suppressed by a $^1O_2$ quencher such as DABCO;[108] photo-oxidation of diethyl sulphide to diethyl sulphoxide can be competitively inhibited by diphenylanthracene ('reactive' quenching of $^1O_2$) or β-carotene;[73] disulphides give thiolsulphinates with $^1O_2$ generated either from $(PhO)_3PO_3$[156] or photochemically,[112,114,115,157,158] the latter reaction being effectively retarded by DABCO. Vinyl sulphides are oxygenated with $^1O_2$ to unstable dioxetanes.[159] Diphenyl sulphide and diphenyl disulphide have been reported to be virtually inert towards $^1O_2$.[9,112,160]

The $^1O_2$ quenching ability of dialkyl sulphides may remain obscured by their $^1O_2$ oxidation reaction. Foote *et al.*[73] found, however, that in the competitive oxidation of diphenylanthracene and diethyl sulphide in benzene the amount of diphenylanthracene $^1O_2$ oxygenation inhibited by diethyl sulphide is 20 times the amount of diethyl sulphoxide formed. This observation suggests that for every molecule of $(C_5H_5)_2S$ oxidized, 20 molecules of $^1O_2$ are quenched ($k_q/k_r^Q \approx 20$) and $k_q = 6 \times 10^6\ \mathrm{M^{-1}\ s^{-1}}$. In a direct methylene blue sensitized photo-oxidation of di-n-butyl sulphide in ethanol, a value of about 0·7 for k_q/k_r^Q was obtained.[161] Thus, in the protic solvent ethanol the oxidation to the sulphoxide is the predominant reaction.

From trapping experiments and solvent effects, Foote and Peters[160] collected much evidence which may best explain the results, assuming that both quenching and oxidation reactions originate from a common intermediate, i.e. the persulphoxide **(28)**:

$$[R_2S^+\text{–}OO^-] \xrightarrow[k_r^Q]{+R_2S} 2\,R_2SO \quad \text{trapping by second mole of } R_2S \qquad (23)$$

$$R_2S + {}^1O_2 \longrightarrow [R_2S^+\text{–}OO^-] \xrightarrow{k_q} R_2S + {}^3O_2 \quad \text{physical quenching} \qquad (24)$$

$$[R_2S^+\text{–}OO^-] \xrightarrow[k_{rear}]{} R_2SO_2 \quad \text{intramolecular rearrangement} \qquad (25)$$

The following experimental conditions are favoured: for reaction (23), 'high' (about room) temperatures, protic solvents; for reaction (24), 'high' (about room) temperatures, aprotic solvents; and for equation (25), low (−78 °C) temperatures, aprotic solvents and high dilution.

Protic solvents may act by decreasing the negative charge density on oxygen by hydrogen bonding, thus facilitating the nucleophilic attack by the second sulphide molecule. In aprotic solvents, intermolecular reaction is less efficient and (**28**) either decomposes giving $^3O_2$ (i.e. quenching, favoured at room temperature) or rearranges to a sulphone (at low temperatures). Thus, the mechanism of quenching by sulphides closely resembles the C-T mechanism [equation (17) or (18)]. However, an electron-transfer mechanism, similar to equation (22), cannot be ruled out.[160] A linear correlation between log k_q (sulphide) in the gas phase and ionization potentials (I_p) of the corresponding sulphides[162] suggest an increasing quenching ability with decreasing I_p. This behaviour is similar to that of amines (Figure 5) and may support both the above mechanisms of $^1O_2$ quenching by sulphides.

Interest was aroused by the possible $^1O_2$ quenching by α-lipoic acid (**29**) because the latter had been suggested[163] as being involved in photosynthesis. Furthermore, (**29**) exhibits a significant solubility in water and could find diagnostic use for involvement of $^1O_2$ in aqueous systems. Stevens *et al.*[164] found, however, that rubene-sensitized photo-oxidation leads to the removal of two molecules of (**29**) per $^1O_2$ molecule consumed ($k_r \approx 10^8$ M^{-1} s^{-1}); this fact indicates that processes (24) and (25) are relatively unimportant.

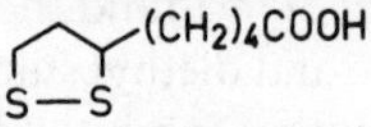

(CH$_2$)$_4$COOH
S—S

(29)

The importance of $^1O_2$ quenching in biopolymers by methionine, a natural amino acid containing a sulphide linkage, was stressed in the previous section. The known protective effect of sulphides, e.g. esters of thiodipropionic acid (especially as synergists in combination with phenolic antioxidants) in synthetic 'aprotic' polymers is of great practical importance.[165] What ever the predominant mechanism of protective action of sulphides may be, the physical and chemical quenching of $^1O_2$ during the outdoor exposure of polymers to heat, sunlight and oxygen can constitute a supplementary mode of their action.

Phenols

Phenols represent an important class of synthetic and biological antioxidants with ambivalent behaviour towards singlet oxygen: both oxidation and quenching reactions are known. The inhibiting effect of hydroquinone on photosensitized oxygenation of the acceptor (α-terpinene) was first mentioned by Schenck and Gollnick.[107] More than a decade later Foote *et al.*[73] confirmed the quenching ability of hydroquinone and measured the rate constant $k_q = 7 \times 10^7$ M^{-1} s^{-1}. Since then, several studies have shown that photo-oxidation of phenols can proceed by way of $^1O_2$ (although other free-radical mechanisms via $^3O_2$ can operate simultaneously).[166] However, the authors were unable to determine the quenching parameters from their experimental data.

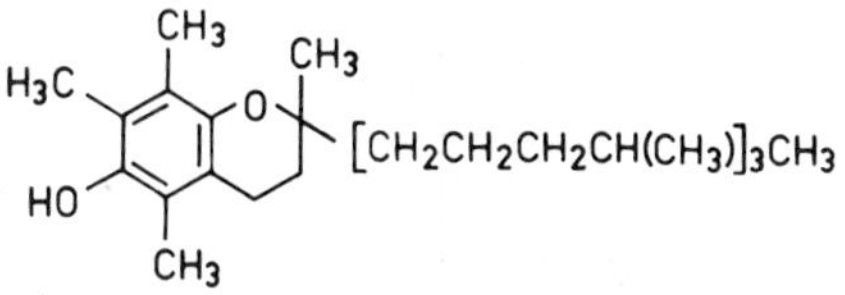

(30) (vitamin E)

The protective role of α-tocopherol (**30**, vitamin E), a biological protective phenolic agent inhibiting lipid peroxidation, was believed to involve 'reactive' quenching of $^1O_2$[166c–e] as well as the inhibition of free-radical reactions of $^3O_2$.[88] In 1974, three papers[167–169] appeared simultaneously dealing with the reaction rates and $^1O_2$ quenching of α-tocopherol (**30**). It was shown that (**30**) scavenges $^1O_2$ by a combination of chemical reaction and quenching (for rate constants, see Table 7). Since $k_q \gg k_r^Q$, the quenching process is almost entirely 'physical', that is (**30**) deactivates about 120 $^1O_2$ molecules before being destroyed.[167] It seems that physical $^1O_2$ quenching by (**30**), in addition to the generally accepted inhibition of autoxidation, may be an additional mechanism for the protective effect of (**30**) in photodynamic action.

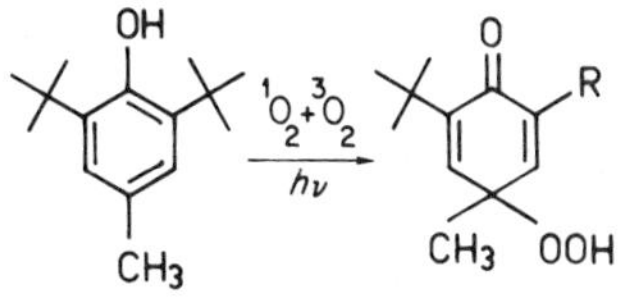

(31) (32) 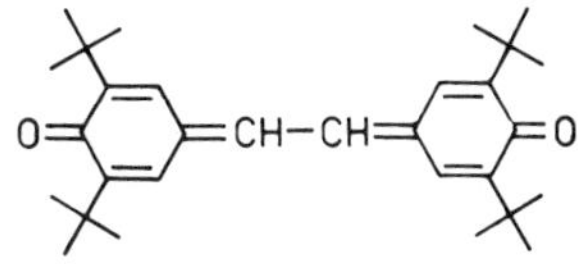 (33)

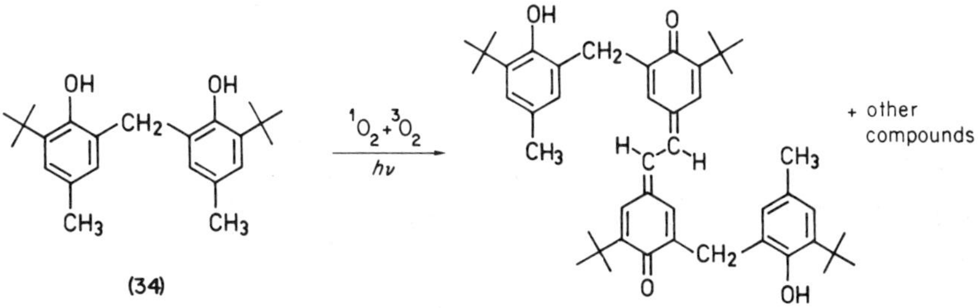

(34) (35)

Interesting observations were recently made by Taimr and Pospišil,[85] who studied the methylene blue sensitized photo-oxidations of antioxidants (**31**) and (**34**). The photo-oxidation of phenol (**31**) in CH_2Cl_2 leads to (**32**). The course of photo-oxidation of bisphenol (**34**) is different: the reaction is very slow and ceases after about 0·3 mol of oxygen per mole of (**34**) have been consumed.

Table 7 Phenols: rate constants of $^1O_2$ quenching (k_q) and oxidation reactions (k_r^Q)

Phenol	$k_q \times 10^{-7}$ ($M^{-1}\,s^{-1}$)	$k_r^Q \times 10^{-6}$ ($M^{-1}\,s^{-1}$)	Experimental conditions for determination of k_q and k_r^Q			Reference
			Solvent	Acceptor of $^1O_2$	Origin of $^1O_2$ [a]	
Hydroquinone	7·0	—	C_6H_5–CH_3OH (4:1)	2-Methyl-2-pentene	A-2	73
	6·9	—	Pyridine	Rubrene [f]	A-1	167
α-Tocopherol (**30**)	9	1·0 [e]	Cyclohexane	Rubrene	A-1	170
	17	—	Benzene	Rubrene	A-1	170
	25	2·1 [b]	Pyridine or isooctane	Rubrene	A-1	167
	62	46 [c]	Methanol	DPBF [d]	A-2	168; 166g
α-Tocopheryl acetate	⩽0·16	—	Pyridine	Rubrene	A-1	167
2,4,6-Triphenylphenol	2·24	— [g]	Benzene	2,5-Diphenylfuran	A-1	171b
	1·45	—	CH_3CN	2,5-Diphenylfuran	A-1	171b
	15·2	—	Methanol	2,5-Diphenylfuran	A-1	171b

[a] A: Photochemically generated $^1O_2$. Sensitizer: 1, rubrene; 2, methylene blue.
[b] From protoporphyrin-sensitized competitive oxidation of (**30**) and cholesterol.
[c] From methylene blue-sensitized oxidation of (**30**) and 2,5-diphenylfuran.
[d] 1,3-Diphenylisobenzofuran.
[e] From competitive photo-oxidation of (**30**) and 9,10-diphenylanthracene.
[f] (**30**) quenches neither triplet nor singlet states of rubrene.[167]
[g] 2,4,6-Triphenylphenoxy radical is formed during photo-oxidation as an intermediate, although no overall reaction with $^1O_2$ occurs.

Cyclohexadienone hydroperoxides were not detected; the reaction mixture contained, among other compounds, intensely brown-coloured oligomers having stilbenequinone structures, e.g. dimer (**35**). The above compounds are formed by transformation of the corresponding aryloxy radicals.[170] Since it is possible to infer that the photo-oxidation of (**34**) in CH_2Cl_2 is slowed down by forming products (**35**), the influence of (**35**) and also (**33**) on the $^1O_2$ photo-oxygenation of 2,3-dimethyl-2-butene was investigated. From Table 8 one can see that they exhibit a strong quenching effect which approaches that of β-carotene. The k_q value of the brown dimer (**35**) can be evaluated very approximately to be half of the k_q value of β-carotene, whereas the k_q value of (**33**) approaches a quarter of that value. The ratio $k_q(\beta\text{-carotene})/k_q(\mathbf{33}) \approx 4$ was confirmed in separate tests: (i) in methylene blue sensitized photo-oxygenation of 1,3-cyclohexadiene; (ii) in self-sensitized photo-oxygenation of rubrene; and (iii) in oxygenation of rubrene with $^1O_2$ generated from $(PhO)_3PO_3$.

Table 8 Methylene blue (10^{-3}M) sensitized photo-oxygenation of 2,3-dimethyl-2-butene (5×10^{-2}M) in CH_2Cl_2, inhibited by (**33**) and (**35**)

Quencher (M)	Initial rate (ml O_2/h)
None	58
(**33**) (0·02)	1·9
(**35**) (0·001)	9·0
β-Carotene (0·005)	1·4

In contrast to β-carotene, (**33**) is reported to be very stable towards $^1O_2$ oxidation. For this reason, the unique, stable stilbenequinones could generally be used as $^1O_2$ quenchers. Since (**33**) can easily be formed in the substrate protected by (**31**) by other oxidative routes,[85] (**33**) (and similar compounds, e.g. **35**) may exert their protective effect in the polymer by a dual mechanism: firstly, by $^1O_2$ quenching and secondly, by retarding the $^1O_2$ oxidation of phenol (**31**), thus saving it for other, probably more important, free-radical inhibitive processes.

No experiments were made to explain the quenching mechanism of (**33**) or (**35**). The extensive system of conjugated double bonds (including carbonyl groups) is a structural feature similar to that in the carotenoids and hence it would allow the operation of the energy-transfer mechanism.

In addition to the data in Table 7, 2,4,6-triphenylphenol was reported to quench $^1O_2$ rapidly with no appreciable accompanying chemical reaction with $^1O_2$.[171] In general, however, the question of the relationship between the structure and $k_q + k_r^Q$ values of phenols still remains open. For example, the known free-radical inhibitor 2,6-di-*tert*-butylphenol must possess considerably lower values of k_q and k_r^Q than α-tocopherol, because it has repeatedly been shown not to have any influence on $^1O_2$ oxygenations.[31,77,113,172,173] This effect is used even as an argument for exclusive $^1O_2$ involvement.[31,77,172] [However, Ouannès and Wilson[108] reported a small inhibitory effect of

2,6-di-*tert*-butylphenol at high concentration (0·3 M) on the photo-oxidation of rubrene.]

Few speculations concerning the mechanism of physical quenching by phenols have appeared. Fahrenholtz *et al.*[167] advocated the C-T mechanism which is consistent with the k_q values for α-tocopherol and hydroquinone, since the oxidation potential of the more highly alkylated phenol (**30**) would be expected to be the lower of the two. The very low value observed for α-tocopheryl acetate (high oxidation potential) is also consistent with the C-T mechanism. Foote *et al.*[168] suggested an electron-transfer mechanism, consistent with observed solvent effects displayed by (**30**). Furthermore, $k_q + k_r^Q$ values of a number of *para*-substituted 2,6-di-*tert*-butylphenols were determined by Thomas and Foote[171] and showed a linear correlation between the logarithm of the total rate of $^1O_2$ removal and their half-wave oxidation potential. The same correlation was given for certain phenol methyl ethers. A Hammett plot using σ^+ gave a rho-value of $-1{\cdot}72 \pm 0{\cdot}12$, indicating that some charge was developed in the quenching step.[171]

Nevertheless, very limited data are available for comparison. On account of the great importance of phenols as natural and synthetic anti-ageing agents for polymers, oils, etc., a more extensive examination of steric requirements *versus* photo-oxidative stability *versus* $^1O_2$ quenching ability remains a highly desirable task for future investigations in the $^1O_2$ quenching field.

Metal Chelates

Metal chelates are of particular interest because of their widespread presence in nature and also because of the considerable commercial potential of synthetic derivatives, such as light stabilizers and pigments. Naturally occurring metal chelates are better known as $^1O_2$ sensitizers (e.g. chlorophyll and protoporphyrin)[24] than $^1O_2$ quenchers. The only natural metal chelate known as a $^1O_2$ quencher—superoxide dismutase (erythrocuprein, 'singlet oxygen decontaminase'[174]) — contains protein-bound Cu(II) and shows a high $^1O_2$ quenching efficiency ($k_q = 1{\cdot}6 \times 10^9$ M^{-1} s^{-1}, see Table 6). However, the $^1O_2$ quenching rate is unchanged even when the copper is entirely removed to form the apoenzyme (Table 6).[147] Therefore, the Cu active site cannot be involved in the strong $^1O_2$ quenching activity of superoxide dismutase. Accordingly, a number of other Cu-containing proteins and low-molecular-weight Cu chelates such as $Cu(lys)_2$, $Cu(his)_2$, and $Cu(tyr)_2$ proved to be inefficient as $^1O_2$ quenchers.[174] Very recently, even the function of superoxide dismutase as a physical $^1O_2$ quencher has been questioned.[219]

Synthetic metal chelates have been known for a long time to be efficient quenchers of singlet and triplet excited states of ketones and aromatic compounds.[175] In 1967, Briggs and McKellar[176] found a good correlation between the triplet quenching efficiency of a series of nickel(II) chelates and their effectiveness as light stabilizers in polypropylene. They believe that stabilization by Ni(II) chelates is achieved by a mechanism of efficient energy transfer from the photoreactive macrocarbonyl groups or other UV-prodegradants to the chelate, where the energy is harmlessly dissipated. Subsequently, many papers appeared on this topic, some of them approving[177] and others disapproving[175j, n, 178] of the McKellar proposal because of the diffusion problems in polymers.

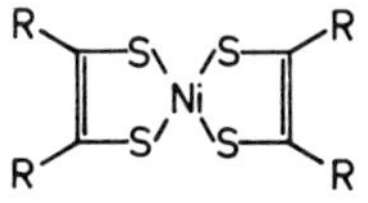

(36)

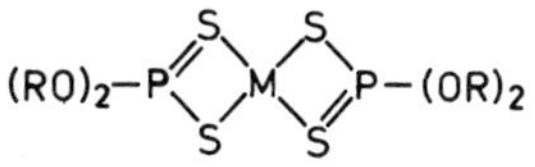

(37)

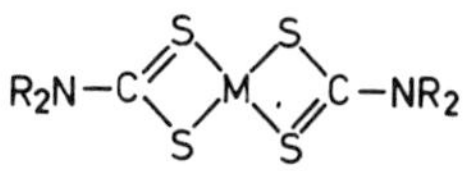

(38)

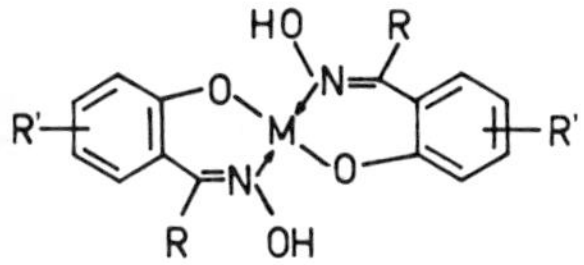

(39)

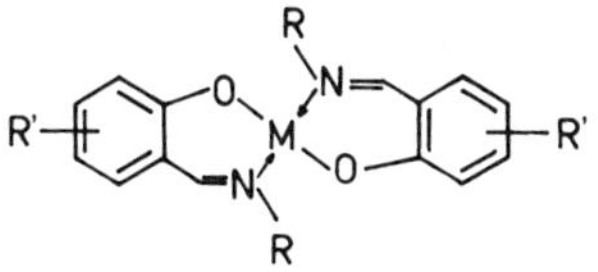

(40)

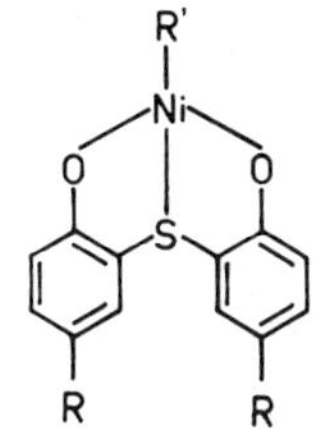

(41)

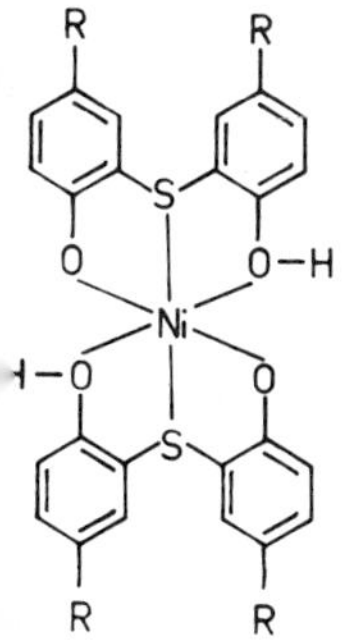

(42)

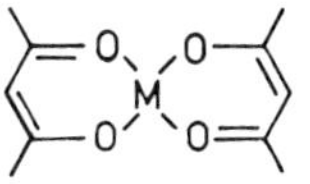

(43)

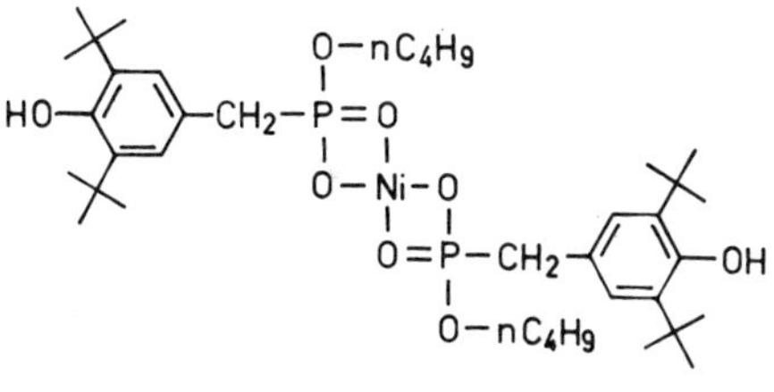

(44)

Low-molecular singlet oxygen can diffuse through appreciably longer distances in polymers than the metal chelates. With this in mind, Carlsson *et al.*[57] and Felder and Schumacher[127] investigated the $^1O_2$ quenching ability of some Ni(II) chelates and simple salts. They discovered that all of them are very powerful quenchers of $^1O_2$ in the liquid phase, better than e.g. DABCO. Since 1972, quenching rate constants (k_q) of numerous metal chelates of types **(36)**–**(44)** and some others were measured (Table 9). A paradoxical situation emerged: although the metal chelates are a class of $^1O_2$ quenchers for which the most k_q values are known, two principal problems still remain unresolved: firstly the relationship between the nature of metal as well as ligand structure and $^1O_2$ quenching efficiency, and secondly the mechanism by which the chelates quench $^1O_2$. Nevertheless, from results published in the literature, several conclusions can be drawn, as follows.

Table 9 Solution $^1O_2$ quenching rate constants of metal chelates

Compound	$k_q \times 10^{-8}$ ($M^{-1}\ s^{-1}$)	Method of determination [a]	Reference
Bisdithio-α-diketone chelates (**36**):			
R = methyl	280	A-4	58
R = phenyl	220	A-4	58
Bisdithiophosphate chelates (**37**):			
M = Ni(II), R = isopropyl	54	A-1; A-2	26, 97
M = Co(II), R = cyclohexyl	27	A-4	58
Bisdialkyldithiocarbamate chelates (**38**):			
M = Ni(II), R = *n*-butyl	70; 170	A-1	26, 57, 100
	80; 90	A-2	26, 100
	43	A-4	58
	40	E	179
	16	G	83
	> 10 [b]	B-1	26, 57, 100
	9; 10	C-1	26, 57, 100
	5·4	F	92
R = isopropyl	34	A-2	97
M = Fe(III), R = isopropyl	38; 43	A-2	26, 97, 100
	38	A-1	26
	12	C-1	26, 100
M = Co(II), R = isopropyl	19	A-2	26, 97, 100
	9	C-1	100
M = Zn(II), R = *n*-butyl	0·2	E	179
R = isopropyl	< 0·1	A-2	97, 100
Bis-2-hydroxyphenylaldoxime and -ketoxime chelates (**39**):			
M = Ni(II), R = R′ = H	59	A-4	58
R = CH_3, R′ = H	52	A-4	58
R = CH_3, R′ = 4-CH_3			94, 175 m
R = $C_{11}H_{23}$, R′ = 4-CH_3	27	D	94
R = *n*-$C_{12}H_{25}$, R′ = 4-CH_3	57	A-4	58
M = Pd(II), R = $C_{17}H_{35}$, R′ = 5-*tert*-C_4H_9	0·6	D	94

Bis-2-hydroxyphenylaldimine chelates (**40**):			
M = Ni(II), R′ = H, R = phenyl	78	A-4	58
R = NH_2	75	A-4	58
R = *n*-dodecyl	70	A-4	58
R = isopropyl	59	A-4	58
R = cyclohexyl	47	A-4	58
R = *sec*-butyl	40	A-4	58
R = 4′-aminobiphenyl	17	A-4	58
R—R = ethylene	53	A-4	58
R—R = *o*-phenylene	37	A-4	58
R—R = 1,8-naphthylene	120	A-4	58
R′ = 5-OCH_3, R = *n*-butyl	2; 4	C-1	26, 57, 100
	>10, 20	A-2	26, 57, 100
	24; 26; 40	A-1	26, 97, 100
	35	B-2	26, 57, 100
M = Co(II), R′ = H, R = *n*-dodecyl	24	A-4	58
R = phenyl	32	A-4	58
R—R = ethylene	100	A-4	58
M = Cu(II), R′ = H, R = *n*-dodecyl	0·5	A-4	58
R = phenyl	4	A-4	58
Bis-2-hydroxyphenylketimine chelates:			
Ni(II) 2,2′-[ethylenebis(nitroethylidyne)]diphenol	34	A-4	58
Ni(II) 2,2′-[ethylenebis(nitrilodecylidyne)]di-*p*-cresol	48	A-4	58
Co(II) 2,2′-[ethylenebis(nitrilomethylidyne)]diphenol	100	A-4	58
Thiobisphenolato Ni(II) chelates (**41**):			
R = *tert*-octyl, R′ = *n*-butylamine	0·8; 1	C-1	26, 57, 100
	1·1	A-3	98
	1·4; 1·8; 2	A-1	26, 57, 97, 100
	1·5	E	179
	2·7	F	92
	2·7	B-2	26, 57, 100
	2·8	C-2	100
	4	A-4	58
R′ = triethanolamine	1·1	A-4	58

Table 9—*continued*

Compound	$k_q \times 10^{-8}$ (M^{-1} s^{-1})	Method of determination[a]	Reference
$R' = H_2O$	1·4	A-4	58
R′ = cyclohexylamine	1·8	A-4	58
R′ = *n*-dodecylamine	2·6	A-4	58
R′ = *n*-propylamine	3	A-4	58
R′ = ethylamine	3·1	A-4	58
$R' = NH_3$	4·8	A-4	58
R = 3,4-dimethyl, $R' = H_2O$	1·2	A-4	58
R′ = aniline	1·9	A-4	58
R′ = *n*-butylamine	2·5	A-4	58
R′ = di-*n*-dodecylamine	3·2	A-4	58
R′ = cyclohexylamine	3·4	A-4	58
R′ = *n*-dodecylamine	3·9	A-4	58
R′ = ethylamine	4·1	A-4	58
Bis-thiobisphenolate Ni(II) chelates (**42**):			
R = *tert*-octyl	0·57	A-3	98
	0·8	F	92
	1; 1·3	A-1	57, 97, 100
	1·3	C-1	57, 100
	2	B-2	26, 57, 100
	2·7	A-4	58
	5	E	179
Bis-acetylacetonate chelates (**43**):			
M = Ni(II)	0·75	C-2	57, 97, 100
	0·82	A-3	98
	1·3	F	92
	1·5	A-2	100
	3	A-4	58
M = Cu(II)	0·04	A-3	98
	1	A-4	58
M = Zn(II)	0·11	A-3	98

M = Fe(III)	0·87	A-3	98
M = Co(II)	1	A-4	58
	1·46	A-3	98
M = Mn(III)	5	A-4	58
M = Co(III)	9·2	A-3	98
Bisphosphonate Ni(II) chelates (**44**):	0·09	A-3	98
	0·1	B-1	26, 57, 100
	0·14; 0·16	A-2	26, 57, 97, 100
	0·27	F	92
	0·34	C-2	57, 100
	2	A-4	58
Miscellaneous compounds:			
Ni(II) isopropylxanthate	54	A-1	26
Ni(II) 2-hydroxy-5-methylbenzophenone	39	A-4	58
Bis(dithioacetylacetonato(nickel(II)	4·8	F	92
Ni(II) 3,5-diisopropylsalicylate	0·5	A-4	58
Co(II) 3,5-diisopropylsalicylate	0·3	A-4	58
Ni(II) 4-*tert*-butylphenylsalicylate	0·15	A-3	98
Ferrocene	0·03	A-3	98
Miscellaneous salts:			
$NiCl_2 . 6H_2O$	3·1	C-2	57, 100
$CoCl_2 . 6H_2O$	0·48	C-2	57, 100
$MnCl_2 . 6H_2O$	<0·01	C-2	57, 100

[a] A: Quenching of rubrene self-sensitized photo-oxygenation: 1, in isooctane; 2, in CH_2Cl_2; 3, in CCl_4; 4, in toluene. B: Quenching of rubrene oxygenation with $^1O_2$ from $(PhO)_3PO_3$: 1, in CH_2Cl_2; 2, in isooctane. C: Quenching of rubrene oxygenation with $^1O_2$ from microwave discharge: 1, in hexadecane; 2, in 2-butoxyethanol. D: Quenching of oxygenation of 1,3-diphenylisobenzofuran (DPBF) by pulse laser technique, in benzene. E: Quenching of methylene blue-sensitized oxygenation of DPBF in flash-photolysed carbon disulphide solution. F: Calculated, using K(R) values from ref. 92 and assuming k_r (rubrene) = $7{\cdot}3 \times 10^7$ M^{-1} s^{-1} (see Table 1, first footnote, K) and k_d (bromobenzene) = $2{\cdot}94 \times 10^4$ s^{-1} (Table 1). Rubrene as oxygenated substrate, $^1O_2$ from microwave discharge, bromobenzene as solvent. G: Quenching of Rose Bengal-sensitized photo-oxidation of 2,2,6,6-tetramethyl-4-hydroxypiperidine in benzene–ethanol (8:1).

[b] Interference by chelate–ozonide reaction gave coloured products.

(a) $^1O_2$ quenching seems to be a property of the chelate as a whole, since free ligands (when stable) fail to quench $^1O_2$.[57,127,179]

(b) Considering the character of the central metal atom (and keeping the ligand constant), the $^1O_2$ quenching efficiency decreases in order Ni(II) > Co(II) ≫ other metals. Some anomalies in this respect have been reported[98] for acetylacetonates (**43**) (see Table 9). Among the dialkyldithiocarbamate chelates (**38**), Fe(III) ≈ Co(II).

(c) Among the same class of ligand, the k_q values of Ni(II) chelates are largely independent of the substitution pattern of the ligand. The following order of groups of Ni(II) chelates with decreasing average values of k_q may be established (average $k_q \times 10^{-8}$, $M^{-1}\ s^{-1}$; see also Table 9):

Bisdithio-α-diketone chelates (**36**)	250
Bisdithiophosphate (**37**) and bisalkylxanthate chelates	54
Bisdialkyldithiocarbamate chelates (**38**)	48
Bis-2-hydroxyphenylaldoxime and -ketoxime (**39**) chelates	45
Bis-2-hydroxyphenylaldimine and -ketimine (**40**) chelates	45
Thiobisphenolate chelates (**41**)	2.4
Bisthiobisphenolate chelates (**42**)	1.8
Bisacetylacetonate chelates (**43**)	1·5
Bisphosphonate chelates (**44**)	0·44

(d) Simple salts such as $NiCl_2·6H_2O$ and $CoCl_2·6H_2O$ can quench the rubrene $^1O_2$ oxygenation effectively (Table 9).

(e) The magnetism and coordination of the chelate appear to be unimportant factors for the $^1O_2$ quenching efficiency. Thus, in toluene solution the *N*-cyclohexyl- and *N*-dodecylformimidoylphenol Ni(II) chelates (**40**) (R=H) are reported to be diamagnetic square-planar complexes, while the more sterically hindered *N*-isopropyl- and *N*-*sec*-butyl-substituted derivatives (**40**) are tetrahedral and paramagnetic.[58] As can be seen from Table 9, these considerations do not in fact play any detectable role in influencing the quenching efficiency. Furthermore, the octahedral paramagnetic [e.g. (**43**), $NiCl_2·6H_2O$[180]], octahedral trimeric [e.g. Ni(II) bisacetylacetonate (**43**)[180]] and tetrameric [e.g. 2,2′-thiobis-4-*tert*-octylphenolate-*n*-butylamine nickel(II) (**41**)[92]] chelates all quench $^1O_2$. In this respect it is of interest that paramagnetic Ni(II) chelates were reported to quench triplet benzophenone with rate constants which were by an order of magnitude less than those of diamagnetic Ni(II) chelates.[175m,177]

(f) The mechanism of $^1O_2$ quenching by metal chelates is not yet known exactly. Carlsson *et al.*[57,100] found for the hydrated chlorides an increasing quenching efficiency in the sequence Mn(II) ≪ Co(II) < Ni(II) (see Table 9). Both Co(II) and Ni(II) chlorides have electronic absorptions corresponding to vacant energy levels at 8000–9000 cm^{-1}, whereas the Mn(II) salt with five unpaired d electrons has none.[180] This infers that energy transfer might be important for these chlorides[100] as well as other Ni(II) chelates,[94] because the oxygen $^1\Delta_g \rightarrow {}^3\Sigma_g^+$ transition occurs at ca. 8000 cm^{-1}. However, Zweig and

Henderson[58] failed to find any detectable relationship between experimental k_q values and those calculated according to equation (10), using extinction coefficients of a number of Ni(II) chelates at 7880 and 6280 cm^{-1}. Triplet levels reported for (**41**) ($E_T = 77{\cdot}1$ kcal mol^{-1}),[181] for (**42**) ($E_T = 76{\cdot}8$ kcal mol^{-1}),[182] and for (**44**) ($E_T = 70{\cdot}5$ kcal mol^{-1})[182] clearly do not allow a collisional energy transfer from $^1O_2$ to these Ni(II) chelates. On the other hand, it seems to be well confirmed that the low-lying energy levels of Ni(II) chelates are below the lowest triplet state of piperylene ($E_T = 56{\cdot}9$ kcal mol^{-1})[175] and anthracene $E_T = 42$ kcal mol^{-1}).[176] Moreover, there is a parallel between the ease of quenching of $^1O_2$ by metal chelates and their ability to quench triplet pentacene, a species which is roughly isoenergetic with singlet oxygen ($E_T = 23$ kcal mol^{-1}).[71] These results suggest that the same mechanism applies to the quenching of $^1O_2$ and triplet pentacene, i.e. that of energy transfer.[230]

(g) High resistence to $^1O_2$ attack has been reported for many Ni(II) chelates.[57, 94, 100] Some loss (3–10%) of (**38**) (R = *n*-butyl) occurred in bromobenzene solution after exposure to $^1O_2$ from a microwave discharge system.[175]

(h) On account of the obvious experimental difficulties involved a direct demonstration that $^1O_2$ quenching mechanism is operative and directly responsible for the excellent light-stabilizing efficiency of some Ni(II) chelates in solid polymers has not yet been proposed. However, an indirect evidence is available: (i) The introduction by homogeneous mixing of Ni(II) chelates (**37**) (R = isopropyl) and (**38**) (R = ethyl) into the polymer, *trans*-polyisoprene, significantly decreases the observed rate of oxygenation with $^1O_2$ from a microwave discharge system;[183] (ii) the photo-oxidation rates of polypropylene films, containing anthracene (a sensitizer, generating $^1O_2$ within the polymer) and some Ni(II) chelates is appreciably reduced, compared with polypropylene films containing anthracene only;[26, 184] (iii) self-sensitized $^1O_2$ oxygenation of tetracene in polystyrene films containing (**39**) was considerably inhibited; however, the quenching efficiency of (**39**) for the same tetracene $^1O_2$ oxygenation in CH_2Cl_2 solution was about three times higher;[93] (iv) moreover, the highly coloured Ni(II) chelates (**36**), which are comparable to β-carotene in $^1O_2$ quenching efficiency, k_q(**36**) $\approx 2{\cdot}5 \times 10^{10}$ M^{-1} s^{-1}, are described as excellent polyolefin stabilizers,[185] while the more than 100-times less efficient $^1O_2$ quenchers (**43**) [R=Ni(II)] are not effective photostabilizers for polyolefins.[58]

Inorganic Anions

The field of singlet oxygen quenching by inorganic anions is very recent, three of the four known values of rate constants, k_q being published in 1976. They span more than three orders of magnitude, from a low value for the bromide anion ($k_q \approx 1{\cdot}2 \times 10^6$ M^{-1} s^{-1}),[186] to a respectable value for the oxygen superoxide radical anion ($k_q = 7 \times 10^9$ M^{-1} s^{-1} in acetonitrile[187]) (Table 10, first column). A very pronounced solvent dependence seems to be involved. The azide anion ($k_q = 2{\cdot}2 \times 10^8$ M^{-1} s^{-1})[188] can quench $^1O_2$ in protic solvents such as methanol

Table 10 Inorganic anions: absolute rate constants for quenching of $^1O_2$ (k_q) and acetone triplet (k_Q^A) and Stern–Volmer constants for quenching of fluorescence of methylene blue (K_{SV}).

Anion	$k_q \times 10^{-7}$ (M^{-1} s^{-1})[a]		$k_Q^A \times 10^{-7}$ (M^{-1} s^{-1})[b]	K_{SV}[c]
$O_2^-\cdot$	700	(as $Me_4N^+O_2^-\cdot$ in dimethylsulphoxide)	—	—
	160	(as $Me_4N^+O_2^-\cdot$ in acetonitrile)	—	—
$NCSe^-$	—		630	—
I^-	28	[as crown-KI[d] in bromobenzene-acetone (2:1)]	710	15·4
	9·1	[as n-$Bu_4N^+I^-$ in bromobenzene-acetone (2:1)]		
	8·1	[as LiI in bromobenzene-acetone (2:1)]		
$S_2O_3^{2-}$	—		170	3·1
N_3^-	22	(as NaN_3 in methanol)	35	3·6
Br^-	0·12	[as crown-KBr[d] in bromobenzene-acetone (2:1)]	0·55	0·8
Cl^-	No detectable quenching		<0·001	0·2

[a] After ref. 187 for $O_2^-\cdot$, ref. 186 for I^-, Br^- and Cl^-, and ref. 188 for N_3^-.
[b] After ref. 196, measured in water.
[c] After ref. 72, calculated from $\Phi_0/\Phi = 1 + K_{SV}(Q)$ where Φ_0 and Φ = quantum yields of fluorescence in absence and in presence of quencher, respectively, which is present at concentration (Q) in methanol–water (20:80 v/v).
[d] Dicyclohexyl-18-crown-6.

and water but anions such as I^- and Br^- display the same $^1O_2$ quenching behaviour only in aprotic solvents.[186] In water I^- is efficiently oxidized with $^1O_2$ to I_3^-.[189]

Because of its efficiency in water, N_3^- became a popular diagnostic tool for the unambiguous involvement of $^1O_2$ in photo-oxidation of biological water-soluble substrates.[116,190–193,222,228] However, some care had to be taken in using N_3^- in tests for the participation of $^1O_2$ in dye-sensitized oxygenation reactions. Foote *et al.*[188] have shown that N_3^- can be consumed in $^1O_2$ oxygenations of olefinic substances by involving the dye rather than $^1O_2$. Hasty *et al.*[194] reached similar conclusions, because some observed $^1O_2$ oxygenation rates were retarded by factors of only 5–10, when reductions by factors of 150 up to 70 000 were expected on the basis of azide ion quenching of $^1O_2$. Furthermore, N_3^- can efficiently quench singlet and triplet excited states of sensitizers (see Table 10, second and third columns), if used in high concentrations (> 10^{-1} M). For example, with methylene blue (MB) as sensitizer at an azide concentration of 1·0 M, about 78% of the excited singlet state of MB is quenched.[72] As far as the triplet state of MB is concerned, oxygen is ca. 10^4 times more effective than the azide ion in quenching the monomer triplet state of MB, so that under 'normal' photo-oxidizing conditions (ca. 10^{-3} M $^3O_2$, 10^{-1} M N_3^-) most of the triplet state is quenched by oxygen rather than by N_3^-.[195]

It was found very recently[196] that quenching of carbonyl triplets by anions (see Table 10, column 2, for acetone) follows the pattern $Fe(CN)_6^{4-} > NCSe^- > I^- > S_2O_3^{2-} > N_3^- > Br^- > (Cl^-, NO_3^-)$. The mechanism is considered to involve the charge-transfer state ($M^{\overline{\cdot}}X\cdot$), where M^- is a negative carbonyl ion and X is an inorganic radical, e.g. I, N_3. The above order of anions resembles approximately the parallel order of $^1O_2$ quenching efficiency by anions (although the amount of data is limited): $I^- \geqslant N_3^- > Br^- > Cl^-$. Therefore, it is tempting to predict that efficient $^1O_2$ quenching by anions such as $Fe(CN)_6^{4-}$, $NCSe^-$ and $S_2O_3^{2-}$ should occur in aprotic solvents, and also to suggest that the quenching of this species by inorganic anions involves the formation of a charge-transfer complex between the electrophilic $^1O_2$ and nucleophilic anion, such as $(O_2^- \cdot X\cdot)$ or $(^1O_2^{\delta-} \cdots X^{-\delta+})$ [similar to equations (17) and (18)]. The latter assumption is strengthened by the observation that a similar parallel order of quenching efficiency of carbonyl triplets to that of $^1O_2$ also exists for amines, which quench $^1O_2$ by a C-T mechanism (compare, e.g., ref. 122 for quenching of the benzophenone triplet and Table 4 for $^1O_2$ quenching). Moreover, Rosenthal and Frimer[186] pointed out that among the halide ions, I^- possesses the highest polarizability, which means that its electron cloud can be most easily distorted. The electron affinity increases in order $I^-(70{\cdot}5) < N_3^-(70 \pm 3) < Br^-)77{\cdot}3) <$ $Cl^-(83{\cdot}5)$.[196] Consequently, I^- requires the lowest activation energy for the formation of the complex with electrophilic $^1O_2$. Unfortunately, no electron affinity value for the superoxide radical anion is known to explain its extraordinarily high $^1O_2$ quenching ability.

$$^1O_2 + O_2^- \cdot \longrightarrow O_2^- \cdot + {}^3O_2 + 22 \text{ kcal} \qquad (26)$$

Foote and Guiraud[187] assumed, that $O_2^- \cdot$ could quench $^1O_2$ by an electron transfer, according to equation (23), and that this reaction may be very rapid.

Other Compounds

The rate of quenching of $^1O_2$ by ground-state oxygen is $2{\cdot}7 \pm 0{\cdot}3 \times 10^3$ or $1{\cdot}7 \times 10^3$ M^{-1} s^{-1} in Freon-113[99,220] and $6 \pm 4 \times 10^4$ M^{-1} s^{-1} in benzene.[16] The quenching rate coefficient of $O_2(^1\Delta_g)$ by $O_2(^3\Sigma_g^-)$ was found to be $1{\cdot}0 \pm 0{\cdot}1 \times 10^{-18}$ cm^3 s^{-1} at 77 K.[221] Quenching by liquid nitrogen and argon was at least one order of magnitude slower.[221]

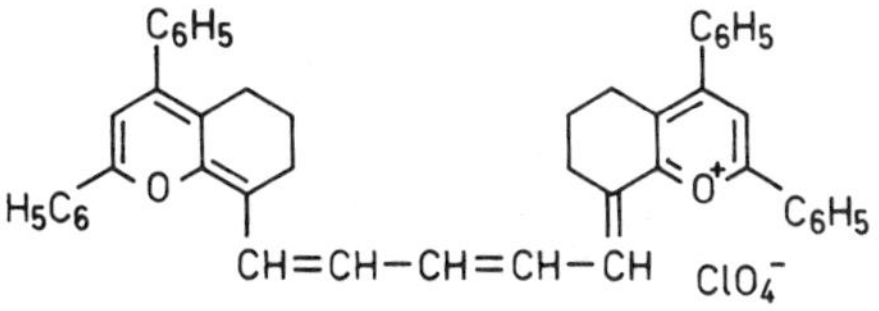

C6H5 C6H5 H5C6 O O+ C6H5 CH=CH−CH=CH−CH ClO4−

(45)

The polymethene pyrylium dye (**45**) is a very efficient quencher of $^1O_2$, probably owing to the extended system of conjugated double bonds. In a photooxygenation experiment, with methylene blue as sensitizer, a ruby laser as a pulse irradiation source, and 1,3-diphenylisobenzofuran as the acceptor, in acetonitrile, a value of $k_q = 3 \times 10$ M^{-1} s^{-1} was obtained for this dye.[49]

Another dye, leucomalachite green (**46**), had been used as an acceptor of $^1O_2$ in one of the first of Kautsky's experiments with singlet oxygen.[1b] However,

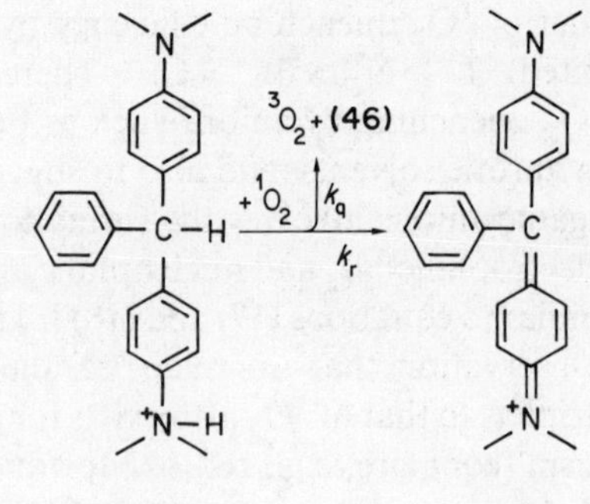

$^3O_2$ + (46)

$+^1O_2$ k_q k_r

(46) (47)

only very recently it was shown by Stevens[224] that (**46**) is also an efficient $^1O_2$ quencher, with $k_q = 1{\cdot}7 \pm 0{\cdot}7 \times 10^8\ \text{M}^{-1}\ \text{s}^{-1}$. For the rate constant of $^1O_2$ oxidation of (**46**) to malachite green (**47**), a value of $k_r = 1{\cdot}3 \pm 0{\cdot}6 \times 10^5\ \text{M}^{-1}\ \text{s}^{-1}$ was determined.[224]

REFERENCES

1. (a) H. Kautsky and H. de Bruin, *Naturwissenschaften*, **19**, 1043 (1931); (b) H. Kautsky, H. de Bruin, R. Neuwirth, and W. Baumeister, *Chem. Ber.*, **66**, 1588 (1933); (c) H. Kautsky, *Biochem. Z.*, **291**, 271 (1937); (d) H. Kautsky, *Trans. Faraday Soc.*, **35**, 216 (1939).
2. C. S. Foote and S. Wexler, *J. Amer. Chem. Soc.*, **86**, 3879 (1964); (b) C. S. Foote and S. Wexler, *J. Amer. Chem. Soc.*, **86**, 3880 (1964).
3. E. J. Corey and W. C. Taylor, *J. Amer. Chem. Soc.*, **86**, 3881 (1964).
4. (a) S. J. Arnold, E. A. Ogryzlo, and H. Witzke, *J. Chem. Phys.*, **40**, 1769 (1964); (b) R. J. Browne and E. A. Ogryzlo, *Proc. Chem. Soc.*, **1964**, 117; (c) J. S. Arnold, R. J. Browne, and E. A. Ogryzlo, *Photochem. Photobiol.*, **4**, 963 (1965); (d) R. J. Browne and E. A. Ogryzlo, *Can. J. Chem.*, **43**, 2915 (1965); (e) A. U. Khan and M. Kasha, *J. Amer. Chem. Soc.*, **88**, 1574 (1966); (f) S. J. Arnold, M. Kubo, and E. A. Ogryzlo, *Adv. Chem. Ser.*, **77**, 133 (1968).
5. S. N. Foner and R. L. Hudson, *J. Chem. Phys.*, **23**, 1974 (1955) and **25**, 601 (1956).
6. K. Gollnick and G. O. Schenck, in *1,4-Cycloaddition Reactions*, (Ed. J. Hammer), Academic Press, New York, 1967, p. 255.
7. C. S. Foote, *Accounts Chem. Res.*, **1**, 104 (1967).
8. A. Schönberg, *Preparative Organic Photochemistry*, Springer, New York, 1968.
9. K. Gollnick, *Adv. Photochem.*, **6**, 1 (1968).
10. J. Rigaudy, *Pure Appl. Chem.*, **16**, 169 (1968).
11. R. Higgins, C. S. Foote and H. Cheng, *Adv. Chem. Ser.*, **77**, 102 (1968).
12. G. O. Schenck, *Ann. N.Y. Acad. Sci.*, **171**, 67 (1970).
13. D. R. Kearns, *Chem. Rev.*, **71**, 395 (1971).
14. C. S. Foote, *Pure Appl. Chem.*, **27**, 635 (1971).
15. R. W. Denny and A. Nickon, *Org. Reactions*, **20**, 133 (1973).
16. B. Stevens, *Accounts Chem. Res.*, **6**, 90 (1973).
17. W. R. Adams, in *Methoden der Organischen Chemie*, Vol. 4, (Houben-Weyl), G. Thieme Verlag, Stuttgart, 1975, p. 1465.
18. W. Adam, *Chem. Ztg.*, **99**, 142 (1975).
19. G. Ohloff, *Pure Appl. Chem.*, **43**, 481 (1975).
20. P. D. Bartlett, *Chem. Soc. Rev.*, **5**, 149 (1976).
21. T. Wilson and J. W. Hastings, in *Photophysiology*, Vol. 5, (Ed. A. C. Giese), Academic Press, New York, 1970, p. 49.

22. J. D. Spikes and M. L. MacKnight, *Ann. N.Y. Acad. Sci.*, **171**, 149 (1970).
23. P. Douzou, *Res. Prog. Org.-Biol. Med. Chem.*, **3**, Pt. 1, 37 (1972).
24. J. Bland, *J. Chem. Educ.*, **53**, 274 (1976).
25. G. S. Egerton and A. G. Morgan, *J. Soc. Dyers Colourists*, **87**, 268 (1971).
26. D. J. Carlsson and D. M. Wiles, *Rubb. Chem. Technol.*, **47**, 991 (1974).
27. B. Rånby and J. F. Rabek, *Photodegradation, Photo-oxidation and Photostabilization of Polymers*, Wiley, New York, 1975, p. 254.
28. J. F. Rabek and B. Rånby, *Polym. Eng. Sci.*, **15**, 40 (1975).
29. J. F. Rabek, in *Degradation of Polymers*, (Ed. C. H. Bamford and C. F. H. Tipper), Elsevier, Amsterdam, 1975, p. 485.
30. S. P. Pappas and R. M. Fischer, *Pigment Resin Technol.*, **1975**, 3.
31. J. Griffiths and C. Hawkins, *J. Soc. Dyers Colourists*, **89**, 173 (1973).
32. R. P. Wayne, *Adv. Photochem.*, **7**, 311 (1969).
33. J. N. Pitts, Jr., *Ann. N.Y. Acad. Sci.*, **171**, 239 (1970).
34. A. P. Altshuller and J. J. Bufalini, *Envir. Sci. Technol.*, **5**, 39 (1971).
35. R. H. Kummler and M. H. Bortner, *Amer. Chem. Soc. Div. Pet. Chem. Prepr.*, **16**, A44 (1971).
36. M. Kasha and A. U. Khan, *Ann. N.Y. Acad. Sci.*, **171**, 5 (1970).
37. V. Y. Shlyapintokh and V. B. Ivanov, *Usp. Khim.*, **45**, 202 (1976).
38. G. Herzberg, *Spectra of Diatomic Molecules*, 2nd Edition, Van Nostrand, Princeton, N.J., 1950.
39. K. Kawaoka, A. U. Khan, and D. R. Kearns, *J. Chem. Phys.*, **46**, 1842 (1967).
40. A. U. Khan and D. R. Kearns, *J. Chem. Phys.*, **48**, 3272 (1968).
41. R. A. Ackermann, J. N. Pitts, Jr., and R. P. Steer, *J. Chem. Phys.*, **51**, 843 (1969).
42. R. H. Kummler and M. H. Bortner, *Ann. N.Y. Acad. Sci.*, **171**, 273 (1970).
43. R. P. Wayne, *Ann. N.Y. Acad. Sci.*, **171**, 199 (1970).
44. R.-D. Penzhorn, H. Güsten, U. Schurat, and K. H. Becker, *Envir. Sci. Technol.*, **8**, 907 (1974).
45. M. J. E. Gauthier and D. R. Snelling, *J. Photochem.*, **4**, 27 (1975).
46. R. A. Ackerman, J. N. Pitts, Jr., and R. P. Steer, *J. Chem. Phys.*, **52**, 1603 (1970).
47. D. R. Kearns, *Amer. Chem. Soc. Div. Pet. Chem. Prepr.*, **16**, A9 (1971).
48. C. K. Ducan and D. R. Kearns, *Chem. Phys. Lett.*, **12**, 306 (1971).
49. P. B. Merkel and D. R. Kearns, *J. Amer. Chem. Soc.*, **94**, 7244 (1972).
50. H. E. A. Kramer and A. Maute, *Photochem. Photobiol.*, **17**, 413 (1973).
51. R. H. Young, R. Martin, K. Wehrly, and D. Feriozi, *Amer. Chem. Soc. Div. Pet. Chem. Prepr.*, **16**, A89 (1971).
52. P. B. Merkel, R. Nilsson, and D. R. Kearns, *J. Amer. Chem. Soc.*, **94**, 1030 (1972).
53. R. O. Loufty and A. C. Sommersall, *Can. J. Chem.*, **54**, 760 (1976).
54. K. Furukawa, E. W. Gray, and E. A. Ogryzlo, *Ann. N.Y. Acad. Sci.*, **171**, 175 (1970).
55. K. Gollnick, T. Franken, G. Schade, and G. Dörhöfer, *Ann. N.Y. Acad. Sci.*, **171** 89 (1970).
56. D. Belluš, H. Lind and J. F. Wyatt, *Chem. Commun.*, **1972**, 1199.
57. D. J. Carlsson, G. D. Mendenhall, T. Suprunchuk, and D. M. Wiles, *J. Amer. Chem. Soc.*, **94**, 8960 (1972).
58. A. Zweig and W. A. Henderson, *J. Polym. Sci., Polym. Chem. Ed.*, **13**, 717 (1975).
59. W. F. Smith, Jr., *J. Amer. Chem. Soc.*, **94**, 186 (1972).
60. R. H. Young, R. L. Martin, D. Feriozi, D. Brewer, and R. Kayser, *Photochem. Photobiol.*, **17**, 233 (1973).
61. R. S. Davidson and K. R. Trethewey, *Chem. Commun.*, **1975**, 674.
62. S. C. Foote and T.-Y. Ching, *J. Amer. Chem. Soc.*, **97**, 6209 (1975).
63. O. Raab, *Z. Biol.*, **39**, 524 (1900).
64. H. V. Tappeiner and A. Yodlbauer, *Dtsch. Arch. Klin. Med.*, **80**, 427 (1904).
65. B. C. L. Weedon, in *Carotenoids*, (Ed. O. Isler), Birkhäuser Verlag, Basle, 1971, p. 29.
66. N. I. Krinsky, in *Carotenoids*, (Ed. O. Isler), Birkhäuser Verlag, Basle, 1971, p. 669.

67. C. S. Foote and R. W. Denny, *J. Amer. Chem. Soc.*, **90**, 6233 (1968).
68. R. Becker and M. Kasha, *J. Amer. Chem. Soc.*, **77**, 3669 (1955).
69. M. Chessin, R. Livingston, and T. G. Truscott, *Trans. Faraday Soc.*, **62**, 1519 (1966).
70. P. S. Vincett, E. M. Voight, and K. E. Rieckhoff, *J. Chem. Phys.*, **55**, 4130 (1971).
71. W. G. Herkstroeter, *J. Amer. Chem. Soc.*, **97**, 4161 (1975).
72. R. S. Davidson and K. R. Trethewey, *J. Amer. Chem. Soc.*, **98**, 4008 (1976).
73. C. S. Foote, R. W. Denny, L. Weaver, Y. Chang, and J. Peters, *Ann. N.Y. Acad. Sci.*, **171**, 139 (1970).
74. E. J. Land, A. Sykes, and T. G. Truscott, *Chem. Commun.*, **1970**, 332.
75. C. S. Foote, Y. C. Chang, and R. W. Denny, *J. Amer. Chem. Soc.*, **92**, 5216 (1970).
76. (a) H. Claes, *Biochem. Biophys. Res. Commun.*, **3**, 585 (1960); (b) H. Claes and T. O. M. Nakayama, *Z. Naturforsch. B*, **14**, 746 (1959).
77. A. F. McDonagh, *Biochem. Biophys. Res. Commun.*, **44**, 1306 (1971).
78. N. A. Evans and I. H. Leaver, *Aust. J. Chem.*, **27**, 1797 (1974).
79. J. F. Rabek and B. Rånby, *J. Polym. Sci. A1*, **12**, 278 (1974).
80. J. F. Rabek and B. Rånby, *J. Polym. Sci. A1*, **12**, 295 (1974).
81. H. H. Wasserman and J. E. VanVerth, *J. Amer. Chem. Soc.*, **96**, 585 (1974).
82. G. Gellerstedt and E.-L. Pettersson, *Acta Chem. Scand.*, *Ser. B*, **29**, 1005 (1975).
83. V. B. Ivanov, V. Y. Shlyapintokh, O. M. Kvostach, A. B. Shapiro, and E. G. Rosantsev, *J. Photochem.*, **4**, 313 (1975).
84. C. W. Jefford, A. F. Boschung, T. A. B. M. Bolsman, R. M. Moriarty, and B. Melnick, *J. Amer. Chem. Soc.*, **98**, 1017 (1976).
85. L. Taimr and J. Pospišil, *Angew. Macromol. Chem.*, **52**, 31 (1976).
86. L. A. Sternson and R. A. Wiley, *Chem.-Biol. Interaction*, **5**, 717 (1972).
87. R. E. Boyer, C. G. Lindstrom, B. Darby, and M. Hylarides, *Tetrahedron Lett.*, **1975**, 4111.
88. S. M. Anderson and N. I. Krinsky, *Photochem. Photobiol.*, **18**, 403 (1973).
89. S. M. Anderson, N. I. Krinsky, M. J. Stone, and D. C. Clagett, *Photochem. Photobiol.*, **20**, 65 (1974).
90. C. S. Foote, E. R. Peterson, and K.-W. Lee, *J. Amer. Chem. Soc.*, **94**, 1032 (1972).
91. P. R. Bolduc and G. L. Goe, *J. Org. Chem.*, **39**, 3178 (1974).
92. J. P. Guillory and C. F. Cook, *J. Polym. Sci. A1*, **11**, 1927 (1973).
93. E. V. Bystritskaya and O. N. Karpuchin, *Dokl. Akad. Nauk SSSR*, **221**, 1100 (1975).
94. A. Farmilo and F. Wilkinson, *Photochem. Photobiol.*, **18**, 447 (1973).
95. M. M. Mathews-Roth, T. Wilson, E. Fujimori, and N. I. Krinsky, *Photochem. Photobiol.*, **19**, 217 (1974).
96. I. B. C. Matheson and J. Lee, *Chem. Phys. Lett.*, **14**, 350 (1972).
97. D. J. Carlsson, T. Suprunchuk, and D. M. Wiles, *J. Polym. Sci. B*, **11**, 61 (1973).
98. P. Hrdlovič, J. Daneček, M. Karvaš, and J. Durmis, *Chem. Zvesti*, **28**, 792 (1974).
99. I. B. C. Matheson, J. Lee, B. S. Yamanashi, and M. L. Wolbarsht, *J. Amer. Chem. Soc.*, **96**, 3343 (1974).
100. D. J. Carlsson, T. Suprunchuk, and D. M. Wiles, *Can. J. Chem.*, **52**, 3728 (1974).
101. I. Sachihiko, B. H. Suong, and S. Takeo, *Tetrahedron Lett.*, **1969**, 279.
102. G. O. Schenck and G. Schade, *Chimia*, **24**, 13 (1970).
103. G. R. Seely and T. H. Meyer, *Photochem. Photobiol.*, **13**, 27 (1971), and references cited therein.
104. M. M. Mathews-Roth, M. A. Pathak, T. B. Fitzpatrick, L. C. Harber, and E. H. Kass, *New Engl. J. Med.*, **282**, 1231 (1970).
105. E. A. Ogryzlo and C. W. Tang, *J. Amer. Chem. Soc.*, **92**, 5034 (1970).
106. K. Furukawa and E. A. Ogryzlo, *Amer. Chem. Soc. Div. Pet. Chem. Prepr.*, **16**, A37 (1971).
107. G. O. Schenck and K. Gollnick, *J. Chim. Phys.*, **55**, 892 (1958).
108. C. Ouannès and T. Wilson, *J. Amer. Chem. Soc.*, **90**, 6527 (1968).
109. R. H. Young and R. L. Martin, *J. Amer. Chem. Soc.*, **94**, 5183 (1972).

110. W. S. Glesson, A. D. Broadbent, E. Whittle, and J. N. Pitts, Jr., *J. Amer. Chem. Soc.*, **92**, 2068 (1970).
111. M. H. Fisch, J. C. Gramain, and J. A. Olesen, *Chem. Commun.*, **1971**, 663.
112. R. W. Murray and S. L. Jindal, *Amer. Chem. Soc. Div. Pet. Chem. Prepr.*, **16**, A72 (1971).
113. J. Griffiths and C. Hawkins, *Chem. Commun.*, **1972**, 463.
114. R. W. Murray and S. L. Jindal, *J. Org. Chem.*, **37**, 3516 (1972).
115. R. W. Murray and S. L. Jindal, *Photochem. Photobiol.*, **16**, 147 (1972).
116. R. Nilsson, P. B. Merkel, and D. R. Kearns, *Photochem. Photobiol.*, **16**, 117 (1972).
117. J. W. Peters, P. J. Bekowies, A. M. Winer and J. N. Pitts, Jr., *J. Amer. Chem. Soc.*, **97**, 3299 (1975).
118. A. P. Schaap, A. L. Thayer, E. C. Blossey and D. C. Neckers, *J. Amer. Chem. Soc.*, **97**, 3741 (1975).
119. N. Shimizu and P. D. Barlett, *J. Amer. Chem. Soc.*, **98**, 4193 (1976).
120. C. F. Deneke and N. I. Krinsky, *J. Amer. Chem. Soc.*, **98**, 3041 (1976).
121. K. Gollnick and J. H. E. Lindner, *Tetrahedron Lett.*, **1973**, 1903.
122. S. G. Cohen, A. Parola, and G. H. Pearson, Jr., *Chem. Rev.*, **73**, 141 (1973).
123. R. Keese, *Angew. Chem.*, **87**, 568 (1975).
124. R. F. Bartholomew and R. S. Davidson, *J. Chem. Soc. (C)*, **1971**, 2347.
125. R. S. Atkinson, D. R. G. Brimage, R. S. Davidson, and E. Gray, *J. Chem. Soc. Perkin Trans. I*, **1973**, 960.
126. H. J. Heller and H. R. Blattmann, *Pure Appl. Chem.*, **36**, 141 (1973).
127. B. Felder and R. Schumacher, *Angew. Makromol. Chem.*, **31**, 35 (1973).
128. J. H. E. Lindner, H. J. Kuhn and K. Gollnick, *Tetrahedron Lett.*, **1972**, 1705.
129. D. Herlem, Y. Hubert-Brierre, and F. Khuong-Huu, *Tetrahedron Lett.*, **1973**, 4173.
130. F. Khuong-Huu, D. Herlem, and Y. Hubert-Brierre, *Tetrahedron Lett.*, **1975**, 359.
171. V. B. Ivanov, V. Y. Shlyapintokh, O. M. Kvostach, A. B. Shapiro, and E. G. Rosantsev, *Izv. Akad. Nauk SSSR, Ser. Khim.*, **1974**, 1916.
172. I. Morishima, K. Yoshikava, T. Yonezawa, and H. Matsumoto, *Chem. Phys. Lett.*, **16**, 336 (1972).
133. G. W. Kinka and L. R. Faulkner, *J. Amer. Chem. Soc.*, **98**, 3897 (1976), and references cited therein.
134. G. J. Hoytink, *Accounts Chem. Res.*, **2**, 114 (1969).
135. D. Belluš, unpublished results.
136. R. A. Caldwell and R. E. Schwerzel, *J. Amer. Chem. Soc.*, **94**, 1035 (1972).
137. R. E. Schwerzel and R. A. Caldwell, *J. Amer. Chem. Soc.*, **95**, 1382 (1973).
138. J. J. McCullough, B. R. Ramachadran, F. F. Snyder, and G. N. Taylor, *J. Amer. Chem. Soc.*, **97**, 6767 (1975).
139. V. Y. Shlyapintokh, V. B. Ivanov, O. M. Kvostach, A. B. Shapiro, and E. G. Rosantsev, *Dokl. Akad. Nauk SSSR*, **225**, 1132 (1975).
140. T.-Y. Ching and C. S. Foote, *Tetrahedron Lett.*, **1975**, 3771.
141. E. F. Ullman and P. Singh, *J. Amer. Chem. Soc.*, **94**, 5077 (1972).
142. P. Singh and E. F. Ullmann, *J. Amer. Chem. Soc.*, **98**, 3018 (1976).
143. E. G. Janzen, *Accounts Chem. Res.*, **4**, 31 (1970).
144. F. D. Greene and K. E. Gilbert, *J. Org. Chem.*, **40**, 1409 (1975).
145. W. F. Smith, Jr., W. G. Herkstroeter and K. L. Eddy, *J. Amer. Chem. Soc.*, **97**, 2764 (1975).
146. B. W. Glad and J. D. Spikes, *Radiat. Res.*, **27**, 237 (1966).
147. I. B. C. Matheson, R. D. Etheridge, N. R. Kratowich, and J. Lee, *Photochem. Photobiol.*, **21**, 165 (1975).
148. (a) H. Schmidt and P. Rosenkranz, *Z. Naturforsch.*, **27b**, 1436 (1972); (b) P. Rosenkranz, A. Al-Ibrahim, and H. Schmidt, this book, Chapter 19.
149. J. R. Fischer, G. R. Julian, and S. J. Rogers, *Physiol. Chem. Phys.*, **6**, 179 (1974).
150. A. Knowles, *Photochem. Photobiol.*, **13**, 473 (1971).

151. B. Stevens and R. D. Small, *Photochem. Photobiol.*, **23**, 33 (1976), and references cited therein.
152. (a) R. Bonnett and H. C. M. Stewart, *Biochem. J.*, **1972**, 130; (b) D. A. Lightner and G. B. Quistad, *Nature New. Biol.*, **236**, 203 (1972); *Science, N.Y.*, **175**, 324 (1972) and *FEBS Lett.*, **25**, 94 (1972); (c) R. Bonnett and J. C. M. Stewart, *Arch. Int. Physiol. Biochem.*, **80**, 951 (1972) and *Chem. Commun.*, **1972**, 596.
153. A. F. McDonagh, *Biochem. Biophys. Res. Commun.*, **48**, 408 (1972).
154. I. B. C. Matheson, N. U. Curry, and J. Lee, *J. Amer. Chem. Soc.*, **96**, 3348 (1974).
155. A. F. McDonagh, *Biochem. Biophys. Res. Commun.*, **44**, 1306 (1971).
156. R. W. Murray, R. D. Smetana, and E. Block, *Tetrahedron Lett.*, **1971**, 299.
157. E. Block and J. O'Connor, *J. Amer. Chem. Soc.*, **96**, 3921 (1974).
158. F. E. Stary, S. L. Jindal, and R. W. Murray, *J. Org. Chem.*, **40**, 58 (1975).
159. (a) W. Adam and J.-C. Liu, *J. Amer. Chem. Soc.*, **94**, 1206 (1972); (b) W. Ando, K. Watanabe, J. Suzuki, and T. Migita, *J. Amer. Chem. Soc.*, **96**, 6766 (1974); (c) W. Ando, K. Watanabe, and T. Migita, *Chem. Commun.*, **1975**, 961.
160. C. S. Foote and J. W. Peters, *J. Amer. Chem. Soc.*, **93**, 3795 (1971).
161. M. Casagrande, G. Gennari, and G. Cauzzo, *Gazz. Chim. Ital.*, **104**, 1251 (1974).
162. R. A. Ackerman, I. Rosenthal, and J. N. Pitts, Jr., *J. Chem. Phys.*, **54**, 4960 (1971).
163. J. A. Barltrop, P. M. Hayes, and M. Calvin, *J. Amer. Chem. Soc.*, **76**, 4348 (1954).
164. B. Stevens, S. R. Perez, and R. D. Small, *Photochem. Photobiol.*, **19**, 315 (1974).
165. (a) P. N. Neureiter and D. E. Bown, *Ind. Eng. Chem. Prod. Res. Dev.*, **1**, 236 (1962). (b) J. C. W. Chien and C. R. Boss, *J. Polym. Sci. A1*, **10**, 1579 (1972); (c) C. R. H. I. De Jonge, H. J. Hageman, W. G. B. Huysmans, and W. J. Mijs, *J. Chem. Soc. Perkin Trans. II*, **1973**, 1276; (d) P. Studt, *Erdöl Kohle*, **27**, 195 (1974).
166. (a) T. Matsuura, N. Yoshimura, A. Nishinaga, and I. Saito, *Tetrahedron Lett.*, **1969**, 1669; (b) I. Saito, S. Kato, and T. Matsuura, *Tetrahedron Lett.*, **1970**, 239; (c) G. W. Grams, *Tetrahedron Lett.*, **1971**, 4832; (d) G. W. Grams and K. Eskins, *Biochemistry*, **11**, 606 (1972); (e) G. W. Grams, K. Eskins, and G. E. Inglett, *J. Amer. Chem. Soc.*, **94**, 866 (1972); (f) T. Matsuura, N. Yoshimura, A. Nishinaga, and I. Saito, *Tetrahedron*, **28**, 4933 (1972); (g) C. S. Foote, M. Thomas, and T.-Y. Ching, *J. Photochem.*, **5**, 172 (1976).
167. S. R. Fahrenholtz, F. H. Doleiden, A. M. Trozzolo, and A. A. Lamola, *Photochem. Photobiol.*, **20**, 505 (1974).
168. C. S. Foote, T.-Y. Ching, and G. G. Geller, *Photochem. Photobiol.*, **20**, 511 (1974).
169. B. Stevens, R. D. Small, Jr., and S. R. Perez, *Photochem. Photobiol.*, **20**, 515 (1974).
170. L. Taimr and J. Pospišil, *Angew. Makromol. Chem.*, **28**, 13 (1973).
171. (a) M. J. Thomas, *Dissertation*, University College of Los Angeles, 1975, cited in ref. 168; (b) C. S. Foote, personal communication, paper submitted for publication; (c) C. S. Foote, M. Thomas, and T.-Y. Ching, *J. Photochem.*, **5**, 172 (1976).
172. C. S. Foote, S. Wexler and W. Ando, *Tetrahedron Lett.*, **1965**, 4111.
173. J. P. Dale, R. Magous, and M. Mousseron-Canet, *Photochem. Photobiol.*, **15**, 411 (1972).
174. U. Weser, W. Paschen, and M. Younes, *Biochem. Biol. Res. Commun.*, **66**, 769 (1975).
175. (a) G. Porter and M. P. Wright, *J. Chim. Phys.*, **55**, 705 (1958); (b) H. Linschitz and K. Sarkanen, *J. Amer. Chem. Soc.*, **80**, 4826 (1958); (c) G. Porter and M. P. Wright, *Discuss. Faraday Soc.*, **27**, 18 (1959); (d) H. Linschitz and L. Pekkarieneu, *J. Amer. Chem. Soc.*, **82**, 2411 (1960); (e) W. M. Moore, G. S. Hammond, and R. P. Foss, *J. Chem. Phys.*, **32**, 1594 (1960); (f) J. A. Bell and H. Linschitz, *J. Amer. Chem. Soc.*, **85**, 528 (1963); (g) G. S. Hammond and R. P. Foss, *J. Phys. Chem.*, **68**, 3799 (1964); (h) R. P. Foss, D. O. Cowan, and G. S. Hammond, *J. Phys. Chem.*, **68**, 3747 (1964); (i) J. C. W. Chien and W. P. Conner, *J. Amer. Chem. Soc.*, **90**, 1001 (1968); (j) D. J. Carlsson, D. E. Sproude, and D. M. Wiles, *Macromolecules*, **5**, 659 (1972). (k) P. Hrdlovič, I. Lukáč, and Z. Maňásek, *Chem. Zvesti*, **26**, 433 (1972); (l) J. P.

Guillory and C. F. Cook, *J. Amer. Chem. Soc.*, **95**, 4885 (1973); (m) A. Adamczyk and F. Wilkinson, *J. Appl. Polym. Sci.*, **18**, 1225 (1974); (n) J. Flood, K. E. Russell, D. J. Carlsson, and D. M. Wiles, *Can. J. Chem.*, **52**, 688 (1974).
176. (a) P. J. Briggs and J. F. McKellar, *Chem. Ind.*, **1967**, 622; (b) P. J. Briggs and J. F. McKellar, *J. Appl. Polym. Sci.*, **12**, 1825 (1968).
177. D. J. Harper and J. F. McKellar, *J. Appl. Polym. Sci.*, **18**, 1233 (1974).
178. J. P. Guillory and R. S. Becker, *J. Polym. Sci. A1*, **12**, 993 (1974).
179. J. Flood, K. E. Russel, and J. K. S. Wan, *Macromolecules*, **6**, 669 (1973).
180. F. A. Cotton and G. Wilkinson, *Advanced Inorganic Chemistry*, 2nd Edition, Interscience, New York, 1966, Ch. 28.
181. G. A. George, *J. Appl. Polym. Sci.*, **18**, 117 (1974).
182. S. W. Beavan and D. Phillips, *J. Photochem.*, **3**, 349 (1974/5).
183. A. K. Breck, C. L. Taylor, K. E. Russell, and J. K. S. Wan, *J. Polym. Sci. A1*, **12**, 1505 (1974).
184. D. J. Carlsson and D. M. Wiles, *J. Polym. Sci. B*, **11**, 759 (1973).
185. N. Uri, *Israel. J. Chem.*, **8**, 125 (1970).
186. I. Rosenthal and A. Frimer, *Photochem. Photobiol.*, **23**, 209 (1976).
187. C. S. Foote and H. J. Guiraud, *J. Amer. Chem. Soc.*, **98**, 1984 (1976).
188. C. S. Foote, T. T. Fujimoto, and Y. C. Chang, *Tetrahedron Lett.*, **1972**, 45.
189. A. G. Kepka and L. I. Grossweiner, *Photochem. Photobiol.*, **18**, 49 (1973).
190. R. Nilsson and D. R. Kearns, *Photochem. Photobiol.*, **17**, 65 (1973).
191. P. Walrant and R. Santus, *Photochem. Photobiol.*, **19**, 411 (1974).
192. I. Saito, K. Inoue, and T. Matsuura, *Photochem. Photobiol.*, **21**, 27 (1975).
193. K. Kobayashi and T. Ito, *Photochem. Photobiol.*, **23**, 21 (1976).
194. N. Hasty, P. B. Merkel, P. Radlick, and D. R. Kearns, *Tetrahedron Lett.*, **1972**, 49.
195. R. Nilsson, P. B. Merkel, and D. R. Kearns, *Photochem. Photobiol.*, **16**, 109 (1972).
196. A. Treinin and E. Hayon, *J. Amer. Chem. Soc.*, **98**, 3884 (1976).
197. R. H. Young, D. Brewer, and R. A. Keller, *J. Amer. Chem. Soc.*, **95**, 375 (1973).
198. D. R. Adams and F. Wilkinson, *J. Chem. Soc. Faraday Trans. II*, **68**, 586 (1972).
199. R. H. Young, K. Wehrly, and R. L. Martin, *J. Amer. Chem. Soc.*, **93**, 5774 (1971).
200. B. Stevens and S. R. Perez, *Mol. Photochem.*, **6**, 1 (1974).
201. R. Nilsson and D. R. Kearns, *J. Phys. Chem.*, **78**, 1681 (1974).
202. H.-D. Brauer and H. Wagener, *Ber. Bunsenges. Phys. Chem.*, **79**, 597 (1975).
203. A. C. Long and D. R. Kearns, *J. Amer. Chem. Soc.*, **97**, 2018 (1975).
204. P. B. Merkel and D. R. Kearns, *J. Amer. Chem. Soc.*, **97**, 462 (1975).
205. Y. Usui, *Chem. Lett.*, **1973**, 747.
206. O. Straub, in *Carotenoids*, (Ed. O. Isler), Birkhäuser Verlag, Basle, 1971, p. 771.
207. W. R. Ware, *J. Phys. Chem.*, **66**, 455 (1962).
208. D. F. Evans, *J. Chem. Soc.*, **1960**, 1735.
209. P. Mathis and J. Kleo, *Photochem. Photobiol.*, **18**, 343 (1973).
210. I. B. C. Matheson and J. Lee, *J. Amer. Chem. Soc.*, **94**, 3310 (1972).
211. S. D. Worley and M. J. S. Dewar, *J. Chem. Phys.*, **50**, 654 (1969).
212. V. I. Vedeneyev, L. V. Gurvich, V. N. Kondratyev, V. A. Medvedev, and Y. L. Frankevich, *Bond Energies, Ionization Potentials and Electron Affinities*, Arnold, London, 1966.
213. P. Bischof, J. A. Hashmall, E. Heilbronner, and V. Hornung, *Tetrahedron Lett.*, **1969**, 4025.
214. C. S. Foote, S. Wexler, W. Ando, and R. Higgins, *J. Amer. Chem. Soc.*, **90**, 975 (1968).
215. J. B. Guttenplan and S. G. Cohen, *Tetrahedron Lett.*, **1972**, 2163.
216. R. Pottier, O. Bagno, and J. Joussot-Dubien, *Photochem. Photobiol.*, **21**, 159 (1975).
217. K. Kikuchi, H. Kokobun, and M. Kikuchi, *Bull. Chem. Soc. Japan*, **48**, 1378 (1975).
218. B. M. Monroe, *172nd Amer. Chem. Soc. Nat. Meeting, San Francisco, 1976, Div. Org. Chem., Prepr. No. 15.*

219. E. A. Mayeda and A. J. Bard, *J. Amer. Chem. Soc.*, **96**, 4023 (1976).
220. I. B. C. Matheson, J. Lee, B. S. Yamanashi, and M. L. Wolbarsht, *Chem. Phys. Lett.*, **27**, 355 (1974).
221. D. L. Huestis, G. Black, S. A. Edelstein, and R. L. Sharpless, *J. Chem. Phys.*, **60**, 4471 (1974).
222. J. O. Grunewald, J. C. Walker, and E. R. Strope, *Photochem. Photobiol.*, **24**, 29 (1976).
223. P. Walrant, R. Santus, and M. Charlier, *Photochem. Photobiol.*, **24**, 13 (1976).
224. B. Stevens, this book, Chapter 8.
225. R. H. Young and D. R. Brewer, this book, Chapter 6.
226. M. Nowakowska, this book, Chapter 25.
227. G. Brunow, I. Forsskåhl, A. C. Grönlund, G. Lindström, and K. Nyberg, this book, Chapter 32.
228. C. S. Foote, this book, Chapter 11.
229. E. A. Ogryzlo, this book, Chapter 2.
230. H. Furue and K. E. Russell, this book, Chapter 33.
231. A. Garner and F. Wilkinson, this book, Chapter 7.

10

Mechanism and Kinetics of Chemical Reactions of Singlet Oxygen with Organic Compounds

K. GOLLNICK

Institut für Organische Chemie der Universität München, Karlstrasse 23, D-8000 München 2, West Germany

Singlet (molecular) oxygen in either its $^1\Sigma_g^+$ or $^1\Delta_g$ state with energies of 37·5 and 22·6 kcal mol^{-1}, respectively, above that of the corresponding triplet ground state may be generated from triplet ground-state (molecular) oxygen, $^3\Sigma_g^- O_2$, by a microwave discharge [1] or by photosensitization.[2–4] Microwave discharge of other simple gases such as CO_2,[5] NO,[6] N_2O [6] and NO_2 [6,7] may produce singlet oxygen, and the photochemical or thermal decomposition of molecules (or short-lived intermediates) which contain peroxy groups may also give rise to singlet oxygen: for example, by photolysis of ozone,[8] from H_2O_2/NaOCl,[9] from triphenyl phosphite–ozone adduct,[10] from the 9,10-endoperoxide of 9,10-diphenylanthracene,[11] from primary and secondary peroxy radicals,[12] and from potassium tetraperoxochromate.[13]

Of the two electronically excited singlet oxygen molecules, so far only the one lower in energy, $^1\Delta_g O_2$, appears to be involved in reactions with substrates which are presently considered as typical 'singlet oxygen acceptors', although $^1\Sigma_g^+ O_2$ may be generated in the gas phase [14,15] and may be produced primarily by photosensitization via energy transfer from sensitizers whose triplet energies exceed that of about 38 kcal mol^{-1}.[16] However, if $^1\Sigma_g^+ O_2$ is transformed collisionally into $^1\Delta_g O_2$ rather than into $^3\Sigma_g^- O_2$[15] at rates that exceed the corresponding quenching of $^1\Delta_g O_2$ to $^3\Sigma_g^- O_2$ by factors of the order of 10^4–10^5 (Table 1), only reactions of $^1\Delta_g O_2$ should be observable in the gas phase, with solids (i.e. on the surface of or within the solid if the latter can be penetrated by oxygen) as well as even in rather concentrated solutions of potential acceptors of singlet oxygen. Using sensitizers of triplet energies ranging from about 34 to 70 kcal mol^{-1}, no variation in products, product distributions and relative reactivities have been encountered.[17] Therefore, the term 'singlet oxygen' and the symbol $^1O_2$ will be used synonymously for $^1\Delta_g O_2$ in this paper.

In the gas phase, singlet oxygen has an intrinsic lifetime, τ_i, of about 65 min;[8] the gas-phase lifetime may be shortened considerably by energy pooling [equation (2)], by dimole formation [equation (3)], by collision with 'inert'

Table 1 Rate constants for the deactivation of $^1\Sigma_g^+ O_2$ (ref. 15) and of $^1\Delta_g O_2$ (ref. 8) by several quenchers in the gas phase at room temperature

Quencher M	$^1\Sigma_g^+ O_2+M$ k_4 (l mol^{-1} s^{-1})	$^1\Delta_g O_2+M$ k_4 l mol^{-1} s^{-1}
O_2	6×10^4	$1{\cdot}4 \times 10^3$
N_2	$1{\cdot}4 \times 10^6$	4×10^1
CH_4	8×10^6	
CO_2	$2{\cdot}8 \times 10^7$	$2{\cdot}4 \times 10^3$
NH_3	$1{\cdot}6 \times 10^8$	
CH_3OH	$2{\cdot}6 \times 10^8$	
H_2O	$6{\cdot}0 \times 10^8$	$8{\cdot}4 \times 10^3$

molecules M (=quencher) [equation (4)] (see Table 1), and by reaction with suitable substrates A [equation (5)] (see Table 2) which may form more or less stable primary products:

$$^1O_2 \longrightarrow {}^3O_2+h\nu(\lambda=12\,700\ \text{Å}),\quad \tau_i=64{\cdot}5\ \text{min} \tag{1}$$

$$^1O_2+{}^1O_2 \longrightarrow O_2\,(^1\Sigma_g^+)+O_2\,(^3\Sigma_g^-),\ k_2=10^3\text{–}10^4\ \text{l mol}^{-1}\ \text{s}^{-1} \tag{2}$$

$$^1O_2+{}^1O_2 \longrightarrow 1[O_2\text{—}O_2]^* \longrightarrow 2\,{}^3O_2+h\nu(\lambda=6340\ \text{Å})\quad k_3=0{\cdot}28\ \text{l mol}^{-1}\ \text{s}^{-1} \tag{3}$$

$$^1O_2+M \longrightarrow {}^3O_2+M \tag{4}$$

$$^1O_2+A \longrightarrow AO_2\ \text{(primary products)} \tag{5}$$

In solid matrices and in solution, the lifetime of singlet oxygen in the absence of substrates A will depend upon the nature of M according to

$$\tau_{^1O_2}=\frac{1}{k_4[M]} \tag{6}$$

or

$$\tau_{^1O_2}=\frac{1}{k_4'} \tag{7}$$

where M = matrix molecules or solvent molecules, respectively, since reactions (2) and (3) do not seem to play a significant role in either of these phases.

No lifetimes have been measured so far for $^1O_2$ in solid matrices. However, for solid matrices in which the sensitizer–acceptor pairs such as methylene blue—rubrene or eosin—diphenylanthracene were applied as monolayers separated by oxygen permeable layers of barium or cadmium stearate, Schnuriger and Bourdon[19] showed that half of the singlet oxygen molecules were deactivated after a diffusion path of about 115 Å, and that singlet oxygen reactions were traceable up to diffusion paths of 500 Å.

Table 2 Rate constants for the reaction of $^1\Delta_g\ O_2$ with several olefinic substrates in the gas phase at room temperature[18]

Substrate A	$^1\Delta_g\ O_2 + A$ k_5 ($1\ mol^{-1}\ s^{-1}$)	k_5/k_{TME}
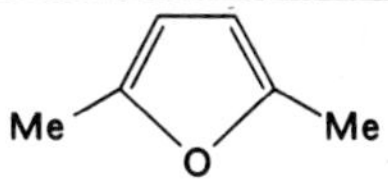 Me O Me	$1{\cdot}6 \times 10^7$	16
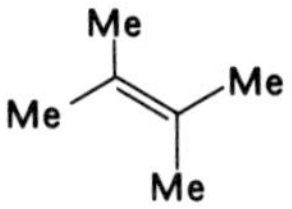 Me Me Me Me	$1{\cdot}0 \times 10^6$	$\equiv 1{\cdot}00$
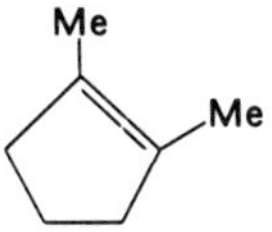 Me Me	4×10^5	0·40
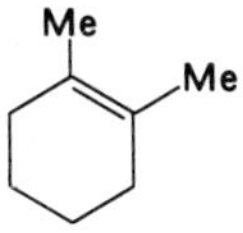 Me Me	4×10^5	0·40
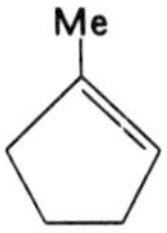 Me	$1{\cdot}5 \times 10^4$	0·015
	9×10^4	0·09

Recently, several workers have determined the lifetime of singlet oxygen in various solvents. Table 3 gives those determined by Kearns and co-workers.[20–22] The k_5 values of reaction (5) given in Table 4 were calculated from reported β-values [see equation (8)] using the $^1O_2$ lifetimes given in Table 3.

In solution, the lifetime of singlet oxygen varies over a range of three powers of magnitude, from about 10^{-6}s in water to about 10^{-3}s in perhalogenated solvents.

The shortest lifetimes of singlet oxygen are found in solvents which contain hydroxyl groups, owing to the successful transfer of electronic energy to the vibrational energy levels of the O—H moiety.[20–22] Deuteration should therefore result in relatively large changes in $\tau_{^1O_2}$, whereas k_5 should remain relatively unchanged. Consequently, β, the 'half-value acceptor concentration',[24,25]

Table 3 Lifetimes of singlet oxygen in Solution[a] at room temperature[20–22]

Solvent M	$10^6\tau_{^1O_2}$ (s)	Solvent M	$10^6\tau_{^1O_2}$ (s)
H_2O	2	D_2O	20
H_2O–CH_3OH (1:1)	3·5	D_2O–CH_3OH (1:1)	11
CH_3OH	7	D_2O–CD_3OD (1:1)	35
C_2H_5OH	12	*tert*-C_4H_9OH	34[b]
Cyclohexane	17	Pyridine	33[b]
Benzene	24	CS_2	200
Acetone	26	CCl_4	700
Acetone-d_6	26	$CDCl_3$	300
Acetonitrile	30	C_6F_6	600
Chloroform	60	CF_3Cl (Freon-11)	1000

[a] $\tau_{^1O_2} = 1/k'_4$, equation (7).
[b] From ref. 23.

should be drastically decreased by O-deuteration of alcoholic and aqueous alcoholic solutions. This effect has been suggested as a means of establishing the participation of singlet oxygen in oxygenation reactions.[20]

$$\beta(\text{mol l}^{-1}) = \frac{k_4[\text{M}]}{k_5} = \frac{k'_4}{k_5} = \frac{1}{\tau_{^1O_2}k_5} \tag{8}$$

If the relative rates, k_5/k_{TME}, in Tables 2 and 4 are compared, fairly good agreement between the corresponding values determined in the gas phase and in solution is found. However, the absolute rates of singlet oxygen reactions in solutions appear to be about thirty times larger than those of the corresponding gas-phase reactions. This result may indicate that, in the gas phase, the energies of activation are about 2 kcal mol^{-1} l higher than those in solution.

It has been shown[29] that, in the gas phase, collisional deactivation of $^1O_2$ by dimethylfuran (DMF) and tetramethylethylene (TME; 2,3-dimethyl-2-butene) according to equation (4) was negligibly slow compared with the chemical reaction (5).

Similarly, we found that, in the liquid phase, 1,3-cyclohexadiene and olefins such as 2-methyl-2-butene, *cis*- and *trans*-2-butene, limonene, α- and β-pinene and Δ^3-carene do not quench the rates of singlet oxygen reactions with DMF and TME.[30] This result indicates that, in solution, simple 1,3-dienes and olefins either do not act as quenchers of $^1O_2$ according to equation (4) or quench singlet oxygen with efficiencies that are similar to or less than those obtained with solvents, i.e. k_4 (simple olefins) $\leqslant k_4$ (solvents).

However, if we proceed from monoolefins and 1,3-dienes to conjugated polyenes, the rates of deactivation of $^1O_2$ may pass and finally far exceed the rates of chemical reaction. Thus, β-carotene, containing eleven conjugated double bonds, quenches singlet oxygen to $^3O_2$ in a diffusion-controlled process,[31,32] i.e. at a rate which exceeds that of the chemical reaction between

Table 4 Rate constants for the reaction of $^1\Delta_g\,O_2$ with several olefinic substrates in methanol at room temperature

Substrate A	$^1\Delta_g\,O_2 + A$ k_5 ($l\ mol^{-1}\ s^{-1}$)	k_5/k_{TME}	Calc. from ref.
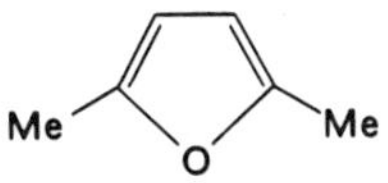	$4{\cdot}0 \times 10^8$ $1{\cdot}0 \times 10^{8\,a}$	8·51	21 26
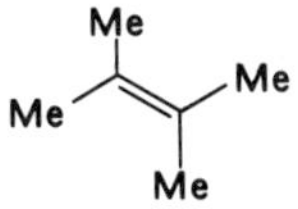	$4{\cdot}7 \times 10^7$ $4{\cdot}0 \times 10^7$	≡1·00	27 21
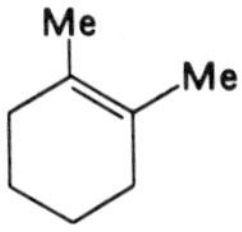	$1{\cdot}0 \times 10^7$ $4{\cdot}7 \times 10^6$	0·21 0·10	21 27
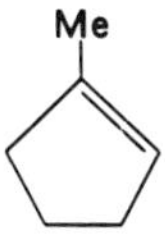	$7{\cdot}3 \times 10^5$	0·015	28
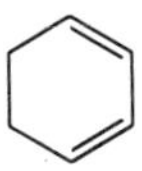	$3{\cdot}8 \times 10^{6\,a}$	0·08	26

[a] In methanol–*tert*-butanol (1 : 1).

β-carotene and $^1O_2$ by a factor of at least 10^3 (see Table 7). Singlet oxygen quenching by β-carotene is due to energy transfer from $^1O_2$ to a lower lying triplet state of the hydrocarbon,[33] a process in which the total spin of the system is conserved. Whenever a collisional, spin-allowed, energetically 'down-hill' energy transfer is possible, diffusion-limited rates may be expected if the energy difference between the 'quenchee' (energy donor) and the quencher (energy acceptor) amounts to at least 2 kcal mol^{-1}.[34]

With regard to products of singlet oxygen reactions with hydrocarbons that contain one or more C=C double bonds, there appear to be only three types of *primary* products which are formed in three types of reactions:

I. Endoperoxide synthesis by 1,4-addition of singlet oxygen to *cis*-1,3-diene systems:

(9)

II. Allyl hydroperoxide synthesis by an ene reaction of singlet oxygen with a C=C double bond system which contains at least one allylic hydrogen atom:

(10)

III. Dioxetane synthesis by 1,2-addition of singlet oxygen to an activated C=C double bond:

(11)

D = electron donor substituent

In all three types of reactions, singlet oxygen may attack the unsaturated hydrocarbons in a concerted fashion. Whereas this view appears to be undisputed for case I (endoperoxide synthesis), cases II (allyl hydroperoxide synthesis) and III (dioxetane synthesis) have not yet been completely settled. Depending on the nature of the substituents attached to the reaction centres and on the bulkiness of more distant parts of the molecule, stereoelectronic and steric effects determine the regiospecificity, the regioselectivity, the stereospecificity, and the stereoselectivity of the reactions and thus determine the product distributions observed with real systems. About 800 hydrocarbons, belonging to various classes of organic compounds as listed in Table 5, have been submitted to singlet oxygen oxygenation reactions (for some recent reviews, see refs. 4 and 35–47).

Since triplet ground-state oxygen, $^3O_2$, does not seem to produce cycloaddition reactions,† the production of endoperoxides, e.g. from anthracenes,[37] ergosterol,[35,49] or α-terpinene,[4,50] may establish the presence of singlet oxygen in an oxidizing system.

† Recently, however, Barton[48] found that 1,3-dienes, e.g. ergosterol, may react with $^3O_2$ to give endoperoxides if strong Lewis acids such as tris(*p*-bromophenyl)ammoniumyl hexachloroantimonate are present.

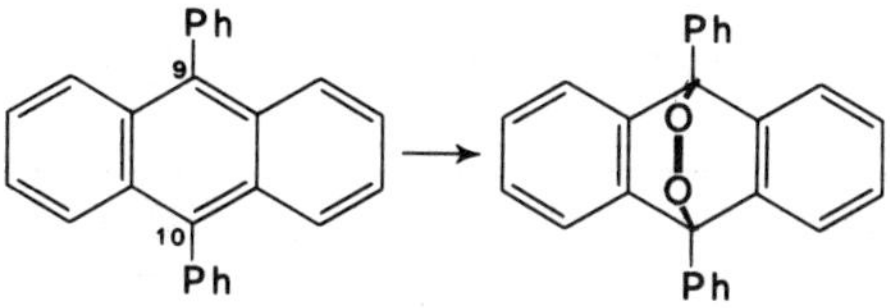

(12)

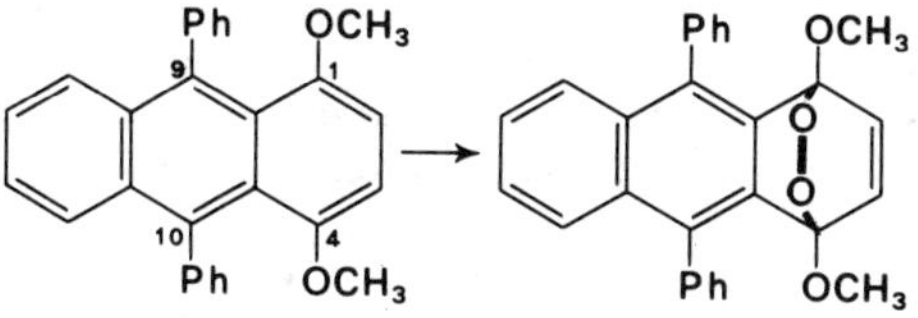

(13)

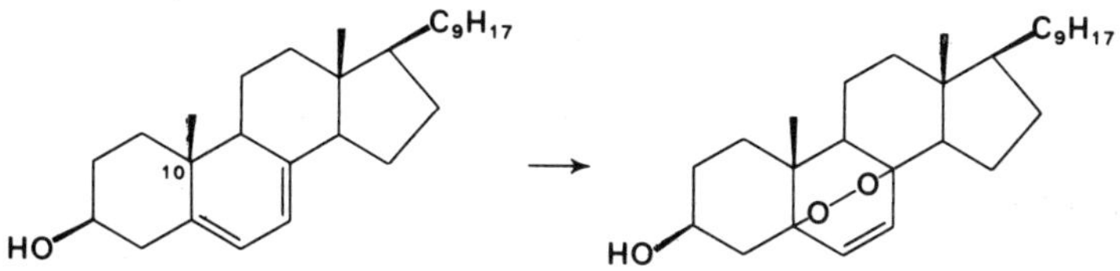

(14)

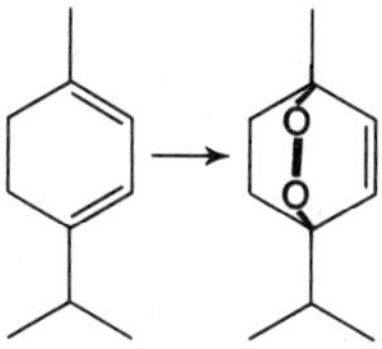

(15)

Generally, cycloaddition of the electrophilic singlet oxygen to anthracenes occurs regiospecifically at the *meso*-positions 9 and 10. Powerful electron-donating substituents attached to positions 1 and 4, however, may direct the $^1O_2$ attack regiospecifically towards these positions. Ergosterol is regio-specifically oxygenated at the conjugated diene system since the disubstituted C=C double bond of the side-chain will be attacked by singlet oxygen about 1000 times less efficiently than the 1,3-cyclohexadiene moiety (compare the k_5 values of 1,4-cycloadditions, Table 6, with those of the ene-reaction of disubstituted olefins, Table 7). Furthermore, this cycloaddition occurs stereo-specifically on the α-side of the molecule since the methyl group attached to position C-10 on the β-side exerts a powerful steric hindrance towards a singlet oxygen attack on the β-side of the diene system.

Methoxy and methyl groups attached to the ends of a 1,3-diene system enhance the reactivity toward attack by singlet oxygen. Therefore, benzene and naphthalene, which do not react with $^1O_2$, are converted into singlet oxygen substrates by methoxy and methyl substitution, as was shown, for example,

Table 5 Classes of hydrocarbons submitted to singlet oxygen oxygenation (cases I, II, and III)

Class of compounds	Examples
(1) Simple olefins and 1,3-dienes	Open-chain and cyclic alkyl- and aryl-substituted ethylenes and 1,3-dienes
(2) Aromatic and hetero-aromatic compounds	Substituted benzenes and phenols, naphthalenes, anthracenes, furans, thiophenes, pyrroles, imidazoles, oxazoles, thiazoles
Natural products:	
(3) Terpenes	
(a) Mono-	Limonene, carenes, pinenes, α-terpinene
(b) C_{13} compounds	α- and β-ionone
(c) Sesqui-	Caryophyllenes, thujopsene, farnesene, α-gurjunene
(d) Di-	Gibberelins, resin acids, vitamin A
(e) Tri-	Squalene
(f) Tetra-	β-carotene
(4) Steroids	Cholesterol, ergosterol, vitamin D_2
(5) Fatty acids	Oleic acid, linoleic acid
(6) Flavones	3-hydroxyflavones
(7) Tetracyclines	Terramycin, aureomycin
(8) Vitamins	A, B group (folic acid), C (ascorbic acid), D_2, E (α-tocopherol), K_1 and Q (ubiquinone)
(9) Amino acids	Tryptophan, histidine, tyrosine (peptides, proteins)
(10) Nucleic acids	Purines (guanine), pyrimidines
(11) Blood and bile pigments	Porphyrins, bilirubin, biliverdin

Table 6 Rate constants for the reaction of $^1\Delta_g\ O_2$ with cyclic 1,3-dienes, aromatic and heteroaromatic compounds in solution at room temperature

Substrate A	Solvent	$^1\Delta_g\ O_2 + A$ $k_5\ (l\ mol^{-1}\ s^{-1})$	From ref. or calc. from ref.
1,3-Cyclopentadiene	MeOH–*tert*-BuOH (1:1)	$5{\cdot}6 \times 10^7$	26
1,3-Cyclohexadiene	MeOH–*tert*-BuOH (1:1)	$3{\cdot}8 \times 10^6$	26
α-Terpinene	MeOH–H_2O (2:3)	$3{\cdot}2 \times 10^8$	30
Ergosterol	Pyridine	$6{\cdot}7 \times 10^6$	51
Pentacene	Benzene	$4{\cdot}2 \times 10^9$	52
Tetracene	Benzene	$4{\cdot}2 \times 10^7$	52
Anthracene	Benzene	$1{\cdot}5 \times 10^5$	52
9-Methoxyanthracene	Benzene	$2{\cdot}5 \times 10^6$	52
9,10-Dimethoxy-anthracene	Benzene	$1{\cdot}4 \times 10^7$	52
α-Tocopherol	MeOH	$4{\cdot}7 \times 10^7$ $k_4 = 6{\cdot}7 \times 10^8$	53, 54
Bilirubin (dianion):			
pH > 8	D_2O	$3{\cdot}0 \times 10^9$	55
pH < 7	D_2O	$1{\cdot}5 \times 10^7$	55
Diphenylisobenzofuran	Benzene	7×10^8	52
	MeOH	8×10^8	21
	Freon-113	4×10^7 (?)	56
2,5-Dimethylfuran (DMF)	MeOH	4×10^8	21

with 1,2,4,5-tetramethoxybenzene,[57,58] and octamethylnaphthalene.[59] Similarly, vinyl-substituted benzenes such as benzhydrylidenecyclobutane,[60] α-vinylnaphthalene,[61] and indene[62] as well as 2-vinylthiophenes[63] undergo cycloaddition reactions with singlet oxygen. The creation of a new cisoid 1,3-diene system in the case of benzhydrylidenecyclobutane and indene causes a second molecule of singlet oxygen to cycloadd, sometimes only after the primary endoperoxide has rearranged.

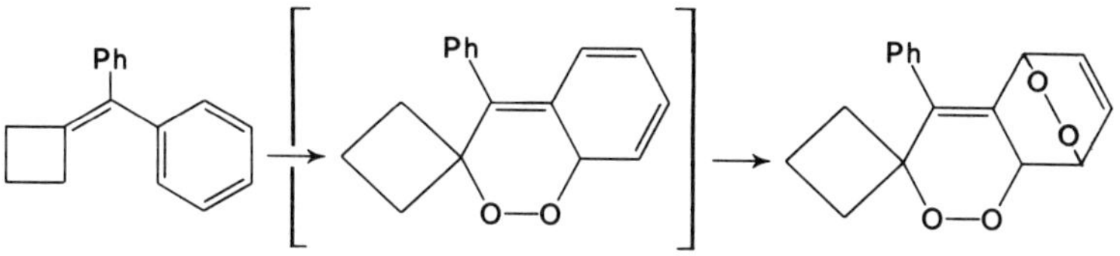

(16)

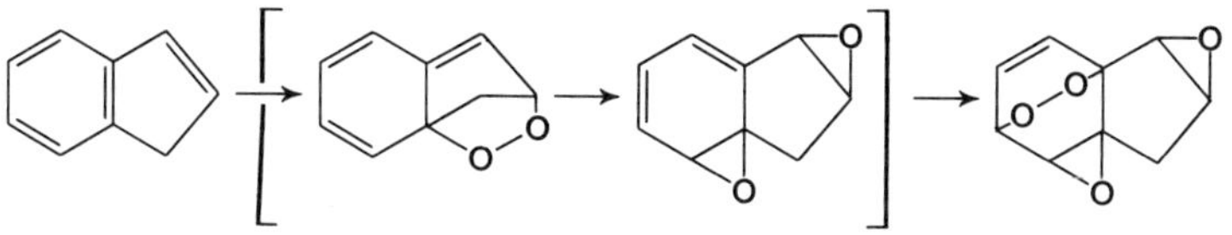

(17)

Whereas endoperoxides derived from carbocyclic 1,3-dienes are generally rather stable, those derived from heteroxyclic 1,3-dienes are generally rather unstable at room temperature and their formation as primary oxygenation products is frequently inferred only from the structures of the isolated products. Scheme 1 shows the various pathways of thermal and/or photochemical transformations that have been observed.

Transformations of anthracene endoperoxides in steps 1 (which forms singlet oxygen[64] and may be accompanied by a bright luminescence[65,66]), 2, 3, 4, and 6 have been reviewed.[35,37] Endoperoxides of cyclopentadienes and cyclohexadienes may undergo steps 3, 4, and 6,[35] while those of fulvenes are reported to rearrange to 1,2-dioxetanes (step 5)[67] or via steps 7, 7a, and 7b to enol lactones.[68,69] Transformations of furan endoperoxides (ozonides) occur easily according to steps 8, 9, and 10,[35,47] whereas the endoperoxides of oxazoles (ozonides), of imidazoles, and of purines rearrange via step 8 and those of thiophenes and thiazoles via step 9.[47] Thiophene endoperoxides seem to be able to rearrange also according to step 7.[70] Pyrrole endoperoxides, on the other hand, rearrange in steps 2, 3, and 11,[47] and recently their transformations to 1,2-dioxetanes (step 5)[71] as well as to pyrroles and singlet oxygen (step 1)[72] were reported.

1,2-Cycloaddition of singlet oxygen to activated C=C double bonds was reported for enamines,[73,74] *N*-heterocyclic compounds,[75] betaines,[76] vinyl ethers,[77,78] benzo[*b*]furans, [79,80] and vinyl thioethers.[81] Stereospecific *cis*-cycloaddition was demonstrated with *cis*- and *trans*-1,2-diethoxyethylenes.[78]

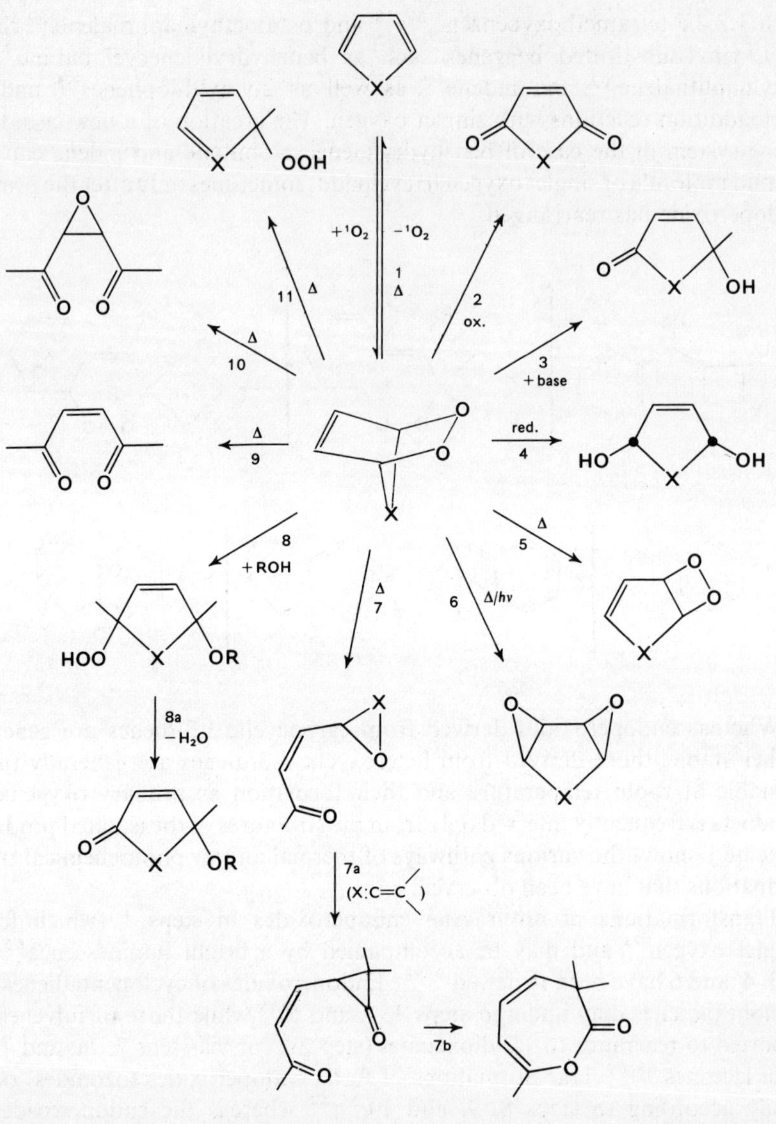

Scheme 1 Thermal and/or photochemical transformations of endoperoxides

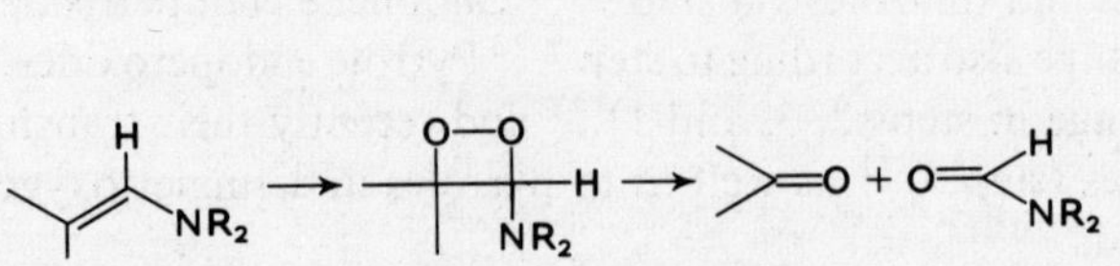

(18)

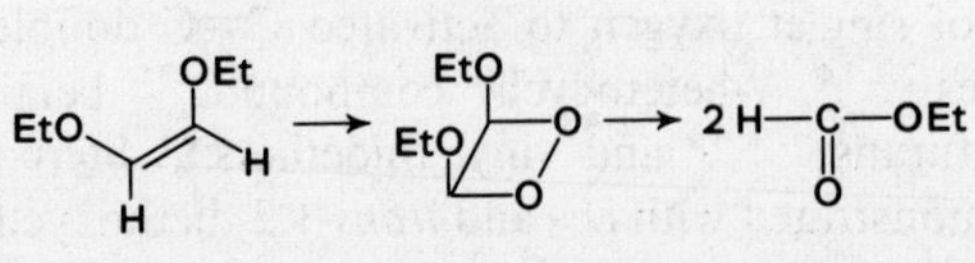

(19)

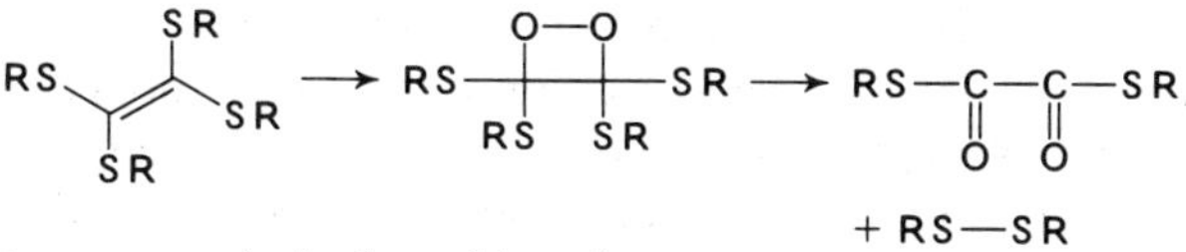

RS—C(SR)=C(SR)—SR → RS—C(SR)(O—O)C(SR)—SR → RS—C(=O)—C(=O)—SR + RS—SR (20)

1,2-Dioxetanes are relatively stable at low temperatures; at room temperature they usually decompose by C—C cleavage[77–80] and/or by C—Héteroatom cleavage (C—D/C—R cleavage)[81–85]

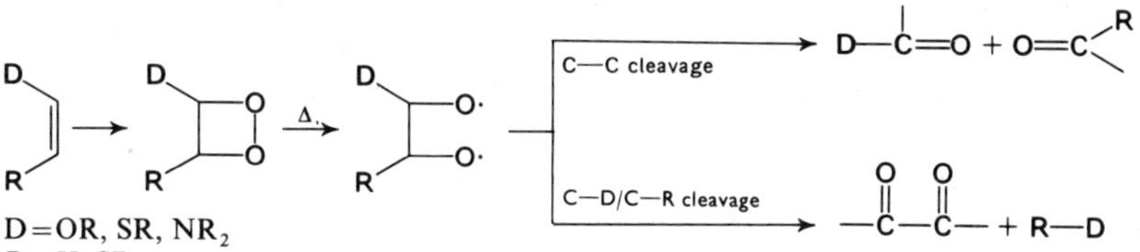

C—C cleavage → D—C=O + O=C<R (21)

C—D/C—R cleavage → —C(=O)—C(=O)— + R—D (22)

D = OR, SR, NR_2
R = H, SR

Monoolefins which do not possess allylic hydrogen atoms, such as ethylene,[86] styrene, and stilbenes,[87] or whose allylic hydrogen atoms are not available for the ene reaction, as in cyclopropenes,[88] adamantylideneadamantane,[89,90] and 7,7′-binorbornylidene,[91] may also undergo 1,2-cycloaddition reactions with singlet oxygen. The 1,2-dioxetanes of the latter two olefins are unusually stable; when heated they decompose via C—C cleavage to give the corresponding ketones. Since the 1,2-dioxetane synthesis is accompanied by epoxide formation, the intermediacy of a perepoxide was postulated,[90] which was assumed either to rearrange to the 1,2-dioxetane or to react further in a complicated way to give finally the epoxide. However, the solvent dependence of the ratio of 1,2-dioxetane synthesis to epoxide formation does not support this hypothesis.[91]

The ene mechanism requires that the double bond of the olefin is inevitably shifted into the allylic position during allylic hydroperoxide synthesis with singlet oxygen. This requirement distinguishes the allylic hydroperoxide synthesis by singlet oxygen from that by triplet ground-state oxygen, which proceeds via the autoxidation mechanism. In the latter reaction, dehydrogenation in the allylic position of the olefin occurs, which is followed by $^3O_2$ addition to the allylic radical, thus leading to allylic hydroperoxides with shifted and non-shifted double bonds.

Ene reaction:

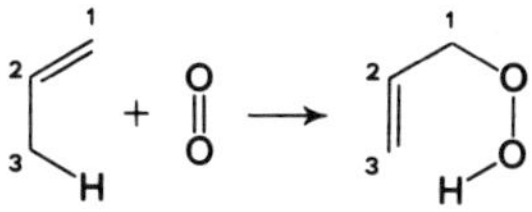

(23)

utoxidation:

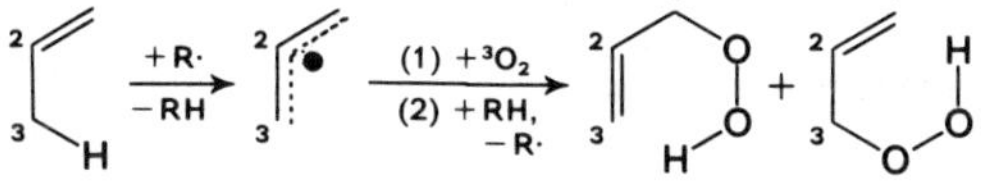

+R· / −RH ; (1) $+^3O_2$ (2) +RH, −R· (24)

In some cases, singlet-oxygen products are indistinguishable from those of autoxidation. For example, the ene reaction with TME affords necessarily only the tertiary hydroperoxide, 2,3-dimethyl-3-hydroperoxy-1-butene. This tertiary hydroperoxide seems also to be the only product of TME autoxidation although one would expect that, in addition, a primary hydroperoxide, 1-hydroperoxy-2,3-dimethyl-2-butene, should be formed. Thus, TME should not be used as a probe for the occurrence of singlet oxygen in oxidizing systems.

Many olefins, however, do give distinct product distributions in the two oxygenation reactions. The optically active (+)-limonene,[4,17,24,92] α-pinene,[24,93] and cholesterol[51,94,95] are appropriate olefins to serve as such probes.

The stereoelectronic effects of alkyl groups attached to the carbon atoms of the olefinic double bond, the requirement of 'axially positioned' allylic hydrogen atoms, and the steric effects executed by bulky groups towards an attack of singlet oxygen as well as in the expected product (e.g. 1,3-diaxial interaction in rigid cyclohexenes) have recently been discussed extensively by ourselves[4,24,25,27,96] and by others.[28,36,39,40,42,44]

From reported β-values and from reported relative rates, we have calculated absolute rate constants, k_5, for the basic open-chain and cyclic olefins listed in Table 7 by using the singlet oxygen lifetimes listed in Table 3.

Table 7 Rate constants for the reaction of $^1\Delta_g\ O_2$ with olefins in methanol at room temperature

Substrate A		$^1\Delta_g\ O_2 + A$ k_5 ($l\ mol^{-1}\ s^{-1}$)	Calc. from ref.
Polyenes:			
β-Carotene		$\leqslant 10^7$	
		$k_4 = 1{\cdot}4 \times 10^{10}$	31
Tetra-substituted olefins:			
$Me_2C{=}CMe_2$ (TME)		$4{\cdot}7 \times 10^7$	27
		$4{\cdot}0 \times 10^7$	21
		$1{\cdot}5 \times 10^7$ [a]	30
Me—C=CMe₂ with Ph-X (*m*- and *p*-) on C	e.g. X = *p*-NMe₂	$1{\cdot}6 \times 10^7$ [b]	
	= H	$4{\cdot}0 \times 10^6$ [b]	97
$\rho = -0{\cdot}93$	= *p*-CN	$1{\cdot}0 \times 10^6$ [b]	
Me—C(Ph-X)=C(Me)—Ph—X	e.g. X = *p*-OMe	$6{\cdot}8 \times 10^5$	
	= H	$2{\cdot}9 \times 10^5$	98
	= *p*-NO₂	$1{\cdot}9 \times 10^4$	
$\rho = -1{\cdot}35$			
1,2-Dimethylcyclohexene		$1{\cdot}0 \times 10^7$	21
		$4{\cdot}7 \times 10^6$	27

Table 7 *continued*

Substrate A	$^1\Delta_g\ O_2 + A$ k_5 (l mol^{-1} s^{-1})	Calc. from ref.
Tri-substituted olefins:		
$Me_2C{=}CH{-}Me$	$2{\cdot}5 \times 10^6$	27
	$2{\cdot}4 \times 10^6$	26
$Me{-}C(Et){=}CH{-}Me$ *trans-*	$1{\cdot}9 \times 10^6$	26
cis-	$1{\cdot}4 \times 10^6$	26
$Me_2C{=}CH{-}Et$	$7{\cdot}7 \times 10^5$	27
	$8{\cdot}8 \times 10^5$	99
in MeOH–Benzene (1:4)	$7{\cdot}7 \times 10^5$	
tert-BuOH	$5{\cdot}8 \times 10^5$	
MeOH-*tert*-BuOH (1:1)	$3{\cdot}7 \times 10^5$	
Pyridine	$6{\cdot}2 \times 10^5$	99
CS_2	$6{\cdot}3 \times 10^5$	
Acetone	$4{\cdot}8 \times 10^5$	
Benzene	$4{\cdot}2 \times 10^5$	
Average value:	$(6{\cdot}1 \pm 1{\cdot}7) \times 10^5$	
$Me_2C{=}CH$-*i*-Pr	$1{\cdot}1 \times 10^5$	27
$Me_2C{=}CH$-*tert*-Bu	$3{\cdot}3 \times 10^4$	27
1-Methylcyclopentene	$7{\cdot}3 \times 10^5$	28
1-Methylcyclohexene	$1{\cdot}2 \times 10^5$	27, 28
1-Methylcycloheptene	$1{\cdot}1 \times 10^6$	28
1-Methylcyclooctene	$1{\cdot}8 \times 10^5$	28
Cholesterol	$3{\cdot}4 \times 10^4$ [d]	51
Δ^3-Carene	$3{\cdot}5 \times 10^5$	27
α-Pinene	$1{\cdot}7 \times 10^4$	27
2-Methylnorbornene	$4{\cdot}6 \times 10^4$ [e]	100
2,7,7-Trimethyl-2-norbornene	$1{\cdot}6 \times 10^3$ [e]	101
Di-substituted olefins:		
Methyl oleate *cis*-$Me{-}(CH_2)_7{-}CH{=}CH{-}(CH_2)_7{-}CO_2CH_3$	$7{\cdot}4 \times 10^4$ [d]	102
$Me{-}CH{=}CH{-}iPr$ *cis-*	$1{\cdot}2 \times 10^4$ [c]	26
trans-	$2{\cdot}2 \times 10^3$ [c]	26
Cyclopentene	$8{\cdot}7 \times 10^4$	103
Cyclohexene	$5{\cdot}4 \times 10^3$	27
	$2{\cdot}2 \times 10^3$ [c]	26
Cycloheptene	$5{\cdot}5 \times 10^3$	103
Cyclooctene	$7{\cdot}6 \times 10^2$	103
$\Delta^{4(10)}$-Carene	$7{\cdot}0 \times 10^4$	4
Methylenecyclopentane	$4{\cdot}8 \times 10^4$ [e]	100
2-Methylenenorbornane	$2{\cdot}1 \times 10^4$ [e]	100
7,7-Dimethyl-2-methylenenorbornane	$5{\cdot}4 \times 10^3$ [e]	101
β-Pinene	$3{\cdot}7 \times 10^3$	25

[a] In MeOH–H_2O (2:3).
[b] In MeOH, containing 2% of pyridine.
[c] In MeOH–*tert*-BuOH (1:1).
[d] In pyridine.
[e] In acetonitrile.

The electrophilic nature of singlet oxygen is unequivocally shown by the negative ρ-value determined from Hammett plots of the ene reactions with *m*- and *p*-substituted α,β,β-trimethylstyrenes[97] and with *p,p'*-disubstituted *trans*-α,α'-dimethylstilbenes.[98] Methyl substitution of a hydrogen atom, attached to a carbon atom of the C=C double bond, increases the rate remarkably (by a factor of about 20–30 per methyl group), whereas substitution by a phenyl group increases the reactivity of the olefin towards singlet oxygen attack only slightly (by a factor of about 2 per phenyl group).

Almost no regioselectivity is observed with simple olefins such as 2-methyl-2-butene, 2-methyl-2-pentene, and 3-methyl-2-pentenes. Slightly more complicated olefins, however, such as 2,4,4-trimethyl-2-pentene and 2,4-dimethyl-2-pentene, yield only secondary hydroperoxides owing to a regiospecific attack of $^1O_2$ on the C-3 position of the olefins. In 2,4,4-trimethyl-2-pentene, the *tert*-butyl group does not offer an allylic hydrogen atom so that the ene reaction must occur regiospecifically. In 2,4-dimethyl-2-pentene, however, the isopropyl group offers an allylic hydrogen atom so that the formation of a tertiary hydroperoxide (by $^1O_2$ attack on the C-2 position) would be expected.

$$\mathrm{Me{-}\overset{2}{C}(Me){=}\overset{3}{C}H\text{-}\mathit{tert}\text{-}Bu \longrightarrow Me{-}C({=}CH_2){-}CH(OOH)\text{-}\mathit{tert}\text{-}Bu} \quad (25)$$

$$\mathrm{Me{-}\overset{2}{C}(Me){=}\overset{3}{C}H\text{-}i\text{-}Pr \longrightarrow Me{-}C({=}CH_2){-}CH(OOH){-}CH(Me)_2} \quad (26)$$

$$\mathrm{Me{-}\overset{2}{C}(Me){=}\overset{3}{C}H\text{-}i\text{-}Pr \not\longrightarrow Me_2C(OOH){-}CH{=}C(Me)_2} \quad (27)$$

Since the ene reaction is appreciably exothermic as well as associated with rather small activation energies (about 2–4 kcal mol^{-1} in solution[104–106]), the geometry of the transition state should resemble that of the starting material (Hammond principle). Thus, the unavailability of the allylic hydrogen atom of the isopropyl group has been made responsible for the non-appearance of the tertiary hydroperoxide because, from models, free rotation of the isopropyl group about the C-3/isopropyl bond is highly restricted, with the allylic hydrogen atom located in the plane of the double bond ('equatorially positioned').

The decrease of the rate constant from $2{\cdot}4 \times 10^6$ l mol^{-1} s^{-1} for 2-methyl-2-butene to $3{\cdot}3 \times 10^4$ l mol^{-1} s^{-1} for 2,4,4-trimethyl-2-pentene, i.e. by about two orders of magnitude, however, is due mainly to steric hindrance exerted by the *tert*-butyl group on the $^1O_2$ attack on the C-3 position of the 2,4,4-trimethyl-2-pentene molecule.

Similar arguments have been used in order to explain relative rates, regio- and stereoselective (and regio- and stereospecific) ene reactions with other olefins

Competition between ene reaction and 1,2-dioxetane formation (followed by C—C cleavage) as a function of the solvent polarity was reported by Hasty and Kearns[107] for the singlet-oxygen oxygenation of 2,5-dimethyl-2,4-hexadiene (see Table 8).

Table 8 Solvent effects on product distribution, relative rates, and partial rate constants of singlet-oxygen oxygenation of 2,5-dimethyl-2,4-hexadiene at room temperature

Solvent	Exptl. values after ref. 107		Calculated values[b]		
	Product ratio, (**3**):(**2**)	$k(\mathbf{1})/k_5$[a]	$k(\mathbf{1})$ ($l\ mol^{-1}\ s^{-1}$)	$k(\mathbf{2})$ ($l/mol^{-1}\ s^{-1}$)	$k(\mathbf{3})$ ($l\ mol^{-1}\ s^{-1}$)
Acetonitrile	0·01	6·3	$7{\cdot}6 \times 10^5$	$7{\cdot}6 \times 10^5$	$0{\cdot}08 \times 10^5$
Methylene chloride	0·1	5·0	$6{\cdot}0 \times 10^5$	$5{\cdot}4 \times 10^5$	$0{\cdot}6 \times 10^5$
Acetone	0·2	3·2	$3{\cdot}8 \times 10^5$	$3{\cdot}2 \times 10^5$	$0{\cdot}6 \times 10^5$
Aqueous acetone (25% H_2O)	1·5	13·0	$1{\cdot}6 \times 10^6$	$6{\cdot}4 \times 10^5$	$9{\cdot}6 \times 10^5$
Methanol	2·6	28·0	$3{\cdot}4 \times 10^6$	$9{\cdot}4 \times 10^5$	$24{\cdot}6 \times 10^5$
Aqueous methanol (30% H_2O)	5·5	29·0	$3{\cdot}5 \times 10^6$	$5{\cdot}4 \times 10^5$	$29{\cdot}6 \times 10^5$

[a] k_5 of 1-methylcyclohexene.
[b] With $k_5 = 1{\cdot}2 \times 10^5\ l\ mol^{-1}\ s^{-1}$.

$$(\mathbf{1}) \xrightarrow[k(\mathbf{1})]{+{}^1O_2} (\mathbf{2}) + (\mathbf{3}) \qquad (28)$$

Since the rate constant of the ene reaction of 2-methyl-2-pentene is virtually independent of the solvent (Table 7), it would be expected that the ene reaction of 1-methylcyclohexene [used to determine the relative rates of reaction of (**1**) in various solvents] would also occur with a solvent-independent rate. Using a k_5 value of $1{\cdot}2 \times 10^5\ l\ mol^{-1}\ s^{-1}$ for 1-methylcyclohexene (Table 7), we calculated the partial rate constants $k(\mathbf{2})$ and $k(\mathbf{3})$ of Table 8 and found that, whereas the partial rate constant $k(\mathbf{2})$ for the formation of the ene product (**2**) varies by a factor of less than 3, the partial rate constant $k(\mathbf{3})$ for the formation of the dioxetane (**3**) increases by a factor of about 400 by changing the solvent from acetonitrile to aqueous methanol.

From solvent incorporation products, Hasty and Kearns[107] have suggested that the formation of the dioxetane (**3**) is preceded by a perepoxide intermediate, whereas the ene product (**2**) is formed in a concerted reaction.† The invariance

† Earlier, a perepoxide had been postulated as an intermediate in the ene reaction.[108–110] However, experimental evidence disfavoured it as an intermediate.[111–113] For more details, see ref. 96.

of $k(\mathbf{2})$ with solvent and the appreciable dependence of $k(\mathbf{3})$ upon the solvent polarity support their suggestion.

Allylic hydroperoxides are reduced to the corresponding allylic alcohols with retention of configuration. Occasionally, a 1,3-rearrangement which shifts the double bond back into the original position and thus makes the product indistinguishable from an autoxidation product,[114,115] water elimination, affording an α,β-unsaturated carbonyl compound, and Hock cleavage[4] may occur. In the last case, two carbonyl compounds are formed which are indistinguishable from those produced via the dioxetane pathway. However, if the mechanism of the Hock cleavage prevails, H/D exchange should occur, whereas the dioxetane pathway should not produce an α-deuterated carbonyl compound provided that the carbonyl compound itself does not enolize under the reaction conditions.

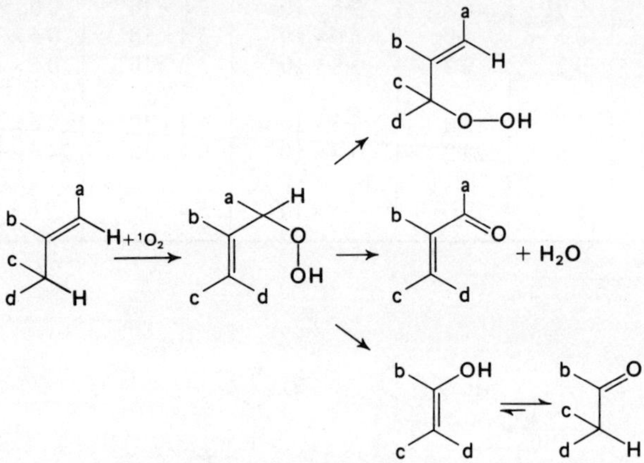

(29)

(30)

(31)

Ene reactions leading to allenes (in addition to the usual products) have been observed in the special case of 1,3,3-trimethyl-2-vinylcyclohexenes.[116–118] Ene reactions may also occur with allylic systems in which one or more carbon atoms are replaced by nitrogen, as was reported for phenyl hydrazones[119] and a Δ^1-pyrroline-N-oxide.[120] In two other special cases, ene-like reactions have been postulated to proceed with the enol form of p-hydroxyphenylpyruvic acid[121] and with a bis(trimethylsilyl) ketene acetal.[122]

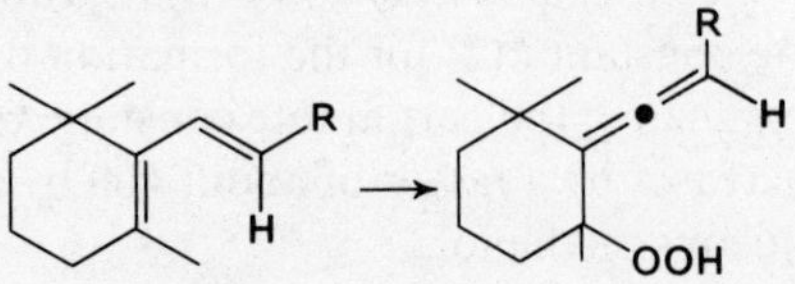

(32)

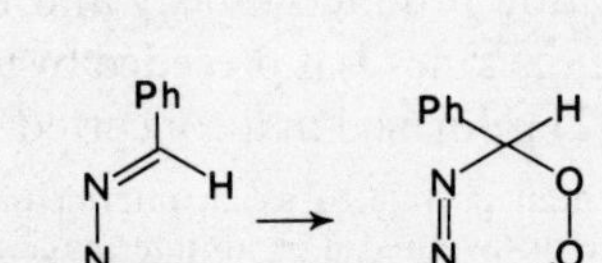

(33)

$$\text{Me}_3\text{SiO}(\text{Me}_3\text{SiO})\text{C}{=}\text{CH}(tert\text{-Bu}) \longrightarrow \text{Me}_3\text{SiO–C(=O)–CH}(tert\text{-Bu})\text{–O–O–SiMe}_3 \qquad (34)$$

Owing to its electrophilic nature, singlet oxygen reacts with neutral nucleophiles such as sulphides and amines as well as with anions such as iodide and azide.

Sulphides are oxidized to sulphoxides,[4, 123–125] which may be further oxidized by singlet oxygen to sulphones, although at a reduced rate,[126] and disulphides afford thiolsulphinates.[127, 128] In the gas phase, quenching of $^1O_2$ by sulphides has been observed, which increases with decreasing ionization potentials of the sulphides.[125]

$$2\ \text{Et—S—Et} + {}^1\text{O}_2 \longrightarrow 2\ \text{Et—SO—Et} \qquad (35)$$

Alkylamines such as 2-butylamine, benzylamine and *p*-substituted benzylamines,[129] *N*,2,2,5,5-pentamethylpiperidine,[130] and triethylamine,[131] as well as alkaloids such as nicotine,[132] pseudopelletierine, *N*-methylgranatanine and tropinone,[133, 134] 20α-dimethylaminopregnane,[135] and codeine,[136] undergo singlet oxygen reactions. Occasionally, an oxidation reaction with triplet ground-state oxygen via free-radical intermediates may be involved to a certain extent.[137]

In alcoholic solutions, demethylation from the $\overset{|}{\text{—N—CH}_3}$ moiety of codeine, pentamethylpiperidine, and dimethylaminopregnane is observed. In aprotic solvents such as benzene, the methyl group of the $\overset{|}{\text{—N—CH}_3}$ moiety of tropinone, pseudopelletierine, and *N*-methylgranatanine is oxidized to the formyl group, thus giving *NN*-disubstituted formamides.

$$\text{codeine (N—CH}_3) \xrightarrow{+{}^1\text{O}_2\ (\text{in ROH H}_2\text{O})} \text{norcodeine (N—H)} \quad (+\text{H—CHO}) \quad (+\text{H}_2\text{O}_2) \qquad (36)$$

$$\text{tropinone (N—CH}_3) \xrightarrow[\text{in benzene}]{+{}^1\text{O}_2} \text{N—C(=O)—H} \quad (+\text{H}_2\text{O}) \qquad (37)$$

Nicotine was the first example for which singlet-oxygen quenching was shown to occur.[132] Since then, many amines have been investigated as singlet-oxygen quenchers in the gas phase[138] and in the liquid phase,[139, 140] and good correlations between the quenching rates and the ionization potentials of the amines have been established for both phases.

Table 9 shows some recently obtained rate constants for the interaction of singlet oxygen with sulphides, amino acids, and anions, and Table 10 reports our unpublished data on amine interactions with singlet oxygen.

Table 9 Rate constants of singlet oxygen interactions with sulphides, amino acids, and anions in solution at room temperature

Substrate A	Solvent	Reaction: k_5 $(\mathrm{l\,mol^{-1}\,s^{-1}})$	Quenching: k_4 $(\mathrm{l\,mol^{-1}\,s^{-1}})$	Ref.
Et—S—Et	MeOH	3×10^5	$5{\cdot}7\times10^6$	124
$Me—S—CH_2—CH_2—CH(NH_2)—CO_2H$ (Methionine)	$MeOH–H_2O$ (1:1)	5×10^6	$2{\cdot}5\times10^7$	141
Tryptophan	$MeOH–H_2O$ (1:1)	4×10^6	$3{\cdot}6\times10^7$	141
Histidine	$MeOH–H_2O$ (1:1)	7×10^6	$4{\cdot}3\times10^7$	141
Iodide ion	Bromobenzene–acetone (2:1)		$8{\cdot}1\times10^7$	142
Azide ion	MeOH		$2{\cdot}2\times10^8$	113
	$MeOH–H_2O$ (2:3)	$1{\cdot}6\times10^7$	$4{\cdot}0\times10^8$	30

Table 10 Rate constants of singlet oxygen interactions with amines in *tert*-butanol–water (5:1)[a] at room temperature[129]

Amine	Ionization potential (eV)[b]	Reaction: k_5 $(\mathrm{l\,mol^{-1}\,s^{-1}})$	Quenching: k_4 $(\mathrm{l\,mol^{-1}\,s^{-1}})$
Ammonia	10·19	—	$<10^4$
Pyridine	9·26	—	$<10^4$
tert-Butylamine	8·83	—	$1{\cdot}2\times10^5$
2-Butylamine	8·70	$1{\cdot}7\times10^4$	$2{\cdot}5\times10^5$
Benzylamine	8·64	$2{\cdot}2\times10^4$	$1{\cdot}4\times10^4$
p-Methoxybenzylamine		$1{\cdot}7\times10^4$	$4{\cdot}2\times10^4$
p-Methylbenzylamine		$2{\cdot}5\times10^4$	$0{\cdot}9\times10^4$
p-Chlorobenzylamine		$2{\cdot}1\times10^4$	$3{\cdot}5\times10^4$
m-Chlorobenzylamine		$2{\cdot}1\times10^4$	$2{\cdot}4\times10^4$
Hexamethylenetetramine	8·26[c]	—	$4{\cdot}8\times10^4$
Nicotine[d]		$1{\cdot}4\times10^6$	$5{\cdot}1\times10^6$
1,4-Diazabicyclo[2.2.2]octane (DABCO)	7·52	—	$2{\cdot}4\times10^6$

[a] With $\tau_{^1O_2}=29\times10^{-6}$ s (by linear interpolation).
[b] From ref. 143.
[c] From ref. 144.
[d] In methanol; calculated from ref. 132.

A zwitterion $\rangle \overset{+}{S}—O—O^-$ that may be preceded by a charge-transfer complex was proposed as the oxidizing intermediate in the singlet-oxygen oxidations of sulphides.[124] Since protic solvents are needed for the oxidation reaction to occur, however, we have suggested[45] that the zwitterion is protonated to $\rangle \overset{+}{S}—OOH$, which is then easily attacked by a nucleophilic sulphide molecule:

$$\text{Et—}\ddot{\text{S}}\text{—Et} + {}^1O_2 \rightleftharpoons \text{C—T complex} \longrightarrow Et_2\overset{+}{S}\text{—O—O}^- \overset{H^+}{\rightleftharpoons} Et_2\overset{+}{S}\text{—O—OH} \xrightarrow{+\ddot{S}Et_2} 2\,\text{Et—SO—Et} + H^+ \qquad (38)$$

$$\text{C—T complex} \longrightarrow Et_2S + {}^3O_2$$

A similar mechanism was recently proposed by us for the interaction of amines with singlet oxygen:[45, 145]

$$\text{H—}\overset{|}{\underset{|}{C}}\text{—}\ddot{N}\text{—} + {}^1O_2 \rightleftharpoons \text{C—T complex} \longrightarrow \text{H—C—}\overset{+}{N}\text{—O—O}^- \longrightarrow \text{—}\overset{H}{C}\text{=}\overset{+}{N}\text{—} + HO_2^- \qquad (39)$$

$$\text{C—T complex} \longrightarrow \text{amine} + {}^3O_2$$

$$\text{(aqueous solvent) } (+H_2O): \quad \text{—}\overset{H}{C}\text{=}\overset{+}{N}\text{—} \longrightarrow \rangle C\text{=O} + HN\langle + H_2O_2$$

$$\text{aprotic solvent: } \text{base:} \; \text{H—}\overset{|}{\underset{O\text{—OH}}{C}}\text{—N—} \longrightarrow \text{—}\underset{\|}{\underset{O}{C}}\text{—N—} + H_2O$$

This mechanism accounts for the fact that amines such as DABCO, which cannot form a C=N double bond (Bredt rule), and *tert*-butylamine, which has no α-CH group, only quench $^1O_2$ but do not react with singlet oxygen. It is evident, then, that ammonium compounds should neither react with $^1O_2$ nor quench singlet oxygen.

Anions such as primary and secondary alcoholates[6, 45, 146, 147] which afford carbonyl compounds, dithiocarbamates,[148] dimedone anion,[149] and certain α-ketocarboxylic acids and their anions[150] should interact with singlet oxygen in an analogous manner:

$$\mathrm{RR'C(H){-}O^-} + {}^1\mathrm{O_2} \longrightarrow \mathrm{RR'C(H){-}O{-}O{-}O^-} \longrightarrow \mathrm{RR'C{=}O} + \mathrm{HO_2^-} \quad (40)$$

R = Alkyl, Aryl; R′, H, Alkyl, Aryl

$$\mathrm{Alk{-}CO{-}CO_2^-} + {}^1\mathrm{O_2} \longrightarrow \mathrm{Alk{-}CO(O{-}O^-){-}C(=O)O} \longrightarrow \mathrm{Alk{-}C(=O)OO^-} + \mathrm{CO_2} \quad (41)$$

The inorganic anions, iodide and azide ions, are rather powerful singlet-oxygen quenchers (see Table 9). Quenching by azide ions has been recommended as a means of establishing the presence of singlet oxygen in oxidizing systems.[141] In the presence of olefins, however, quenching of $^1O_2$ by N_3^- is accompanied by a reaction which yields azido hydroperoxides, in addition to the usual allylic hydroperoxides.[108–113]

Careful analysis of the products formed with α-terpinene in the presence of $^1O_2/N_3^-$ as well as with $^3O_2/N_3$ in protic solvents showed that about 96% of the interaction of $^1O_2$ with N_3^- leads to quenching, and only about 4% of the interactions yield an addition product which, after protonation, reacts with the olefins to give azido hydroperoxides.[6,45]

$$N_3^- + {}^1O_2 \longrightarrow N_3O_2^- \longrightarrow N_3^- + {}^3O_2 \quad (42)$$

$$N_3O_2^- \overset{+H^+}{\rightleftharpoons} N_3O_2H \quad (43)$$

$$N_3\cdot + {}^3O_2 \rightleftharpoons N_3O_2\cdot \xrightarrow[-R\cdot]{+RH} N_3O_2H \quad (44)$$

$$N_3{-}OOH + \text{α-terpinene} \longrightarrow \text{(allylic cation, OOH, H)} \xrightarrow{+N_3^-} \text{(azido hydroperoxide: OOH, H, }N_3) \quad (45)$$

+ other addition products

ACKNOWLEDGEMENTS

The author expresses his gratitude to Dr. H. J. Kuhn, Institut für Strahlenchemie im Max-Planck-Institut für Kohlenforschung, Mülheim-Ruhr, for his steady efforts to keep our file of the tremendously growing singlet-oxygen literature up to date, and to the Fonds der Chemischen Industrie, Frankfurt am Main, for the continuous support of his work.

REFERENCES

1. E. J. Corey and W. C. Taylor, *J. Amer. Chem. Soc.*, **86**, 3881 (1964).
2. H. Kautsky, *Biochem. Z.*, **291**, 271 (1937).
3. C. S. Foote and S. Wexler, *J. Amer. Chem. Soc.*, **86**, 3879 (1964).
4. K. Gollnick, *Adv. Photochem.*, **6**, 1 (1968).
5. K. Gollnick and G. Schade, *Tetrahedron Lett.*, **1973**, 857.
6. K. Gollnick and D. Haisch, unpublished results; D. Haisch, *Dissertation*, Univ. München, 1976.
7. T. J. Cook and T. A. Miller, *Chem. Phys. Lett.*, **25**, 396 (1974).
8. R. P. Wayne, *Adv. Photochem.*, **7**, 311 (1969).
9. C. S. Foote and S. Wexler, *J. Amer. Chem. Soc.*, **86**, 3880 (1964).
10. R. W. Murray and M. L. Kaplan, *J. Amer. Chem. Soc.*, **90**, 537 (1968).
11. H. H. Wasserman and J. R. Scheffer, *J. Amer. Chem. Soc.*, **89**, 3073 (1967).
12. J. A. Howard and K. U. Ingold, *J. Amer. Chem. Soc.*, **90**, 1056 (1968); R. E. Kellog, *J. Amer. Chem. Soc.*, **91**, 5433 (1969).
13. J. W. Peters, J. N. Pitts, Jr., I. Rosenthal, and H. Fuhr, *J. Amer. Chem. Soc.*, **94**, 4348 (1972).
14. A. U. Khan and M. Kasha, *J. Amer. Chem. Soc.*, **88**, 1574 (1966).
15. S. J. Arnold, M. Kubo, and E. A. Ogryzlo, *Adv. Chem. Ser.*, No. 77, 133 (1968).
16. D. R. Kearns and A. U. Khan, *Photochem. Photobiol.*, **10**, 193 (1969).
17. K. Gollnick, T. Franken, G. Schade, and G. Dörhöfer, *Ann. N.Y. Acad. Sci.*, **171**, 89 (1970).
18. J. T. Herron and R. E. Huic, *J. Chem. Phys.*, **51**, 4164 (1969).
19. B. Schnuriger and J. Bourdon, *Photochem. Photobiol.*, **8**, 361 (1968).
20. P. B. Merkel, R. Nilsson, and D. R. Kearns, *J. Amer. Chem. Soc.*, **94**, 1030 (1972).
21. P. B. Merkel and D. R. Kearns, *J. Amer. Chem. Soc.*, **94**, 7244 (1972).
22. C. A. Long and D. R. Kearns, *J. Amer. Chem. Soc.*, **97**, 2018 (1975).
23. R. H. Young, D. Brewer, and R. A. Keller, *J. Amer. Chem. Soc.*, **95**, 375 (1973).
24. K. Gollnick and G. O. Schenck, *Pure Appl. Chem.*, **9**, 507 (1964).
25. G. O. Schenck and K. Gollnick, *Forschungsber. Landes Nordrhein-Westfalen*, No. 1256, Westdeutscher Verlag, Opladen and Köln, 1963.
26. R. Higgins, C. S. Foote, and H. Cheng, *Adv. Chem. Ser.*, No. 77, 102 (1968).
27. K. Gollnick, *Adv. Chem. Ser.*, No. 77, 78 (1968).
28. C. S. Foote, *Pure Appl. Chem.*, **27**, 635 (1971).
29. G. A. Hollinden and R. B. Timmons, *J. Amer. Chem. Soc.*, **92**, 4181 (1970).
30. K. Gollnick and G. Schade, unpublished results.
31. C. S. Foote and R. W. Denny, *J. Amer. Chem. Soc.*, **90**, 6233 (1968).
32. G. R. Seely and T. H. Meyer, *Photochem. Photobiol.*, **13**, 27 (1971).
33. W. G. Herkstroeter, *J. Amer. Chem. Soc.*, **97**, 4161 (1975).
34. K. Sandros, *Acta Chem. Scand.*, **18**, 2355 (1964).
35. K. Gollnick and G. O. Schenck, in *1, 4-Cycloaddition Reactions*, (Ed. J. Hamer), Academic Press, New York, 1967, Ch. 10, p. 255.
36. C. S. Foote, *Accounts Chem. Res.*, **1**, 104 (1968).
37. J. Rigaudy, *Pure Appl. Chem.*, **16**, 1 (1968).
38. T. Wilson and J. W. Hastings, in *Photophysiology, Current Topics in Photophysiology and Photochemistry*, Vol. 5 (Ed. A. C. Giese), Academic Press, New York, 1970, Ch. 3, p. 50.
39. D. R. Kearns, *Chem. Rev.*, **71**, 395 (1971).
40. W. R. Adams, in *Oxidation*, Vol. 2 (Ed. R. L. Augustine and D. J. Trecker), Marcel Dekker, New York, 1971, Ch. 2, p. 65.
41. I. R. Politzer, G. W. Griffin, and J. L. Laseter, *Chem.-Biol. Interactions*, **3**, 73 (1973).
42. R. W. Denny and A. Nickon, *Org. Reactions*, **20**, 133 (1973).

43. G. Ohloff, *Pure Appl. Chem.*, **43**, 481 (1975).
44. W. R. Adams, *Houben-Weyl*, Vol. IV/5b, Thieme Verlag, Stuttgart, 1975, p. 1465.
45. K. Gollnick, in *Radiation Research; Biomedical, Chemical and Physical Perspectives*, (Ed. O. F. Nygaard, H. I. Adler, and W. K. Sinclair), Academic Press, New York, 1975, p. 590.
46. C. S. Foote, in *Free Radicals in Biology*, Vol. 3 (Ed. W. A. Pryor), Academic Press, New York, 1976, p. 85.
47. T. Matsuura and I. Saito, in *Photochemistry of Heterocyclic Compounds* (Ed. O. Buchardt), Wiley, New York, 1976, Ch. 7, p. 456.
48. D. H. R. Barton, *J. Chem. Soc., Perkin Trans. I*, **1975**, 2055.
49. A. Windaus and J. Brunken, *Justus Liebigs Ann. Chem.*, **460**, 225 (1928).
50. G. O. Schenck and K. Ziegler, *Naturwissenschaften*, **32**, 157 (1944).
51. G. O. Schenck, K. Gollnick, and O. A. Neumüller, *Justus Liebigs Ann. Chem.*, **603**, 46 (1957).
52. B. Stevens, S. R. Perez, and J. A. Ors, *J. Amer. Chem. Soc.*, **96**, 6846 (1974).
53. B. Stevens, R. D. Small, Jr., and S. R. Perez, *Photochem. Photobiol.*, **20**, 515 (1974).
54. C. S. Foote, T. Y. Ching, and G. G. Geller, *Photochem. Photobiol.*, **20**, 511 (1974).
55. I. B. C. Matheson, N. U. Curry, and J. Lee, *J. Amer. Chem. Soc.*, **96**, 3348 (1974).
56. I. B. C. Matheson and J. Lee, *Chem. Phys. Lett.*, **14**, 350 (1972).
57. I. Saito, M. Imuta, and T. Matsuura, *Tetrahedron*, **28**, 5307 (1972).
58. I. Saito, M. Imuta, and T. Matsuura, *Tetrahedron*, **28**, 5313 (1972).
59. H. Hart and A. Oku, *Chem. Commun.*, **1972**, 254.
60. G. Rio, D. Bricout, and L. Lecombe, *Tetrahedron Lett.*, **1972**, 3583.
61. M. Matsumoto and K. Kondo, *Tetrahedron Lett.*, **1975**, 3935.
62. P. A. Burns, C. S. Foote, and S. Mazur, *J. Org. Chem.*, **41**, 899 (1976).
63. M. Matsumoto, S. Dobashi, and K. Kondo, *Tetrahedron Lett.*, **1975**, 4471.
64. H. H. Wasserman and J. R. Scheffer, *J. Amer. Chem. Soc.*, **89**, 3073 (1967).
65. C. Dufraisse, L. Velluz, and L. Velluz, *C.R. Hebd. Séanc. Acad. Sci., Paris*, **208**, 1822 (1939).
66. C. Dufraisse, J. Rigaudy, I. I. Basselier, and N. K. Cuong, *C.R. Hebd. Séanc. Acad. Sci., Paris*, **260**, 5031 (1965).
67. J. P. LeRoux and C. Goasdoue, *Tetrahedron*, **31**, 2761 (1975).
68. W. Skorianetz, K. H. Schulte-Elte, and G. Ohloff, *Helv. Chim. Acta*, **54**, 1913 (1971).
69. A. Kawamoto, H. Kosugi, and H. Uda, *Chem. Lett.*, **1972**, 807.
70. C. N. Skold and R. H. Schlessinger, *Tetrahedron Lett.*, **1970**, 791.
71. D. A. Lightner and C. S. Pak, *J. Org. Chem.*, **40**, 2724 (1975).
72. D. A. Lightner, G. S. Bisacchi, and R. D. Norris, *J. Amer. Chem. Soc.*, **98**, 802 (1976).
73. C. S. Foote and J. W.-P. Lin, *Tetrahedron Lett.*, **1968**, 3267.
74. J. E. Huber, *Tetrahedron Lett.*, **1968**, 3271.
75. D. A. Lightner and L. K. Low, *J. Heterocycl. Chem.*, **12**, 793 (1975).
76. H. Takeshita, A. Mori, and S. Ohta, *Bull. Soc. Chem. Japan*, **47**, 2437 (1974).
77. S. Mazur and C. S. Foote, *J. Amer. Chem. Soc.*, **92**, 3225 (1970).
78. P. D. Bartlett and A. P. Schaap, *J. Amer. Chem. Soc.*, **92**, 3223 (1970).
79. G. Rio and J. Berthelot, *Bull. Soc. Chim. Fr.*, **1971**, 1705.
80. J. J. Basselier, J. C. Cherton and J. Caille, *C.R. Hebd. Séanc. Acad. Sci., Paris*, **273C**, 514 (1971).
81. W. Adam and J. C. Liu, *J. Amer. Chem. Soc.*, **94**, 1206 (1972).
82. W. Ando, T. Saiki, and T. Migita, *J. Amer. Chem. Soc.*, **97**, 5028 (1975).
83. H. H. Wasserman and S. Terao, *Tetrahedron Lett.*, **1975**, 1735.
84. W. Ando and K. Watanabe, *Chem. Commun.*, **1975**, 961.
85. W. Ando and K. Watanabe, *Tetrahedron Lett.*, **1975**, 4127.
86. D. J. Bogan, R. S. Sheinson, and F. W. Williams, *J. Amer. Chem. Soc.*, **98**, 1034 (1976).

87. G. Rio and J. Berthelot, *Bull. Soc. Chim. Fr.*, **1969**, 3609.
88. I. R. Politzer and G. W. Griffin, *Tetrahedron Lett.*, **1973**, 4775.
89. J. H. Wieringa, J. Strating, H. Wynberg, and W. Adam, *Tetrahedron Lett.*, **1972**, 169.
90. A. P. Schaap and G. R. Faler, *J. Amer. Chem. Soc.*, **95**, 3381 (1973).
91. P. D. Bartlett and M. S. Ho, *J. Amer. Chem. Soc.*, **96**, 627 (1974).
92. G. O. Schenck, K. Gollnick, G. Buchwald, S. Schroeter, and G. Ohloff, *Justus Liebigs Ann. Chem.*, **674**, 93 (1964).
93. G. O. Schenck, H. Eggert, and W. Denk, *Justus Liebigs Ann. Chem.*, **584**, 177 (1953).
94. A. Nickon and F. Bagli, *J. Amer. Chem. Soc.*, **83**, 1498, (1961).
95. M. J. Kulig and L. L. Smith, *J. Org. Chem.*, **38**, 3639 (1973).
96. K. Gollnick and H. J. Kuhn, in *Singlet Oxygen* (Ed. H. H. Wasserman and R. W. Murray), Academic Press, New York, to be published.
97. C. S. Foote and R. W. Denny, *J. Amer. Chem. Soc.*, **93**, 5162 (1971).
98. K. Gollnick and F. Kotsonis, unpublished results; F. Kotsonis, *M.S. Thesis*, Univ. Arizona, 1969.
99. C. S. Foote and R. W. Denny, *J. Amer. Chem. Soc.*, **93**, 5168 (1971).
100. C. W. Jefford, M. H. Laffer, and A. F. Boschung, *J. Amer. Chem. Soc.*, **94**, 8905 (1972).
101. C. W. Jefford and A. F. Boschung, *Helv. Chim. Acta*, **57**, 2242 (1974).
102. F. H. Doleiden, S. R. Fahrenholtz, A. A. Lamola, and A. M. Trozzolo, *Photochem. Photobiol.*, **20**, 519 (1974).
103. T. Matsuura, A. Horinaka, and R. Nakashima, *Chem. Lett.*, **1973**, 887.
104. G. O. Schenck and E. Koch, *Z. Elektrochem.*, **64**, 170 (1960).
105. E. Koch, *Tetrahedron*, **24**, 6295 (1968).
106. G. A. Hollinden and R. B. Timmons, *J. Amer. Chem. Soc.*, **92**, 4181 (1970).
107. N. M. Hasty and D. R. Kearns, *J. Amer. Chem. Soc.*, **95**, 3380 (1973).
108. W. Fenical, D. R. Kearns, and P. Radlick, *J. Amer. Chem. Soc.*, **91**, 3396 (1969).
109. W. Fenical, D. R. Kearns, and P. Radlick, *J. Amer. Chem. Soc.*, **91**, 7771 (1969).
110. D. R. Kearns, W. Fenical, and P. Radlick, *Ann. N.Y. Acad. Sci.*, **171**, 34 (1970).
111. K. Gollnick, D. Haisch, and G. Schade, *J. Amer. Chem. Soc.*, **94**, 1747 (1972).
112. C. S. Foote, T. T. Fujimoto, and Y. C. Chang, *Tetrahedron Lett.*, **1972**, 45.
113. N. M. Hasty, P. B. Merkel, P. Radlick, and D. R. Kearns, *Tetrahedron Lett.*, **1972**, 49.
114. G. O. Schenck, O. A. Neumüller, and W. Eisfeld, *Justus Liebigs Ann. Chem.*, **618**, 202 (1958).
115. B. Lythgoe and S. Trippett, *J. Chem. Soc.*, **1959**, 471.
116. S. Isoe, S. Be Hyeon, H. Ichikawa, S. Katsumura, and T. Sakan, *Tetrahedron Lett.*, **1968**, 5561.
117. M. Mousseron-Canet, J. P. Dalle, and J. C. Mani, *Tetrahedron Lett.*, **1968**, 6037.
118. C. S. Foote and M. Brenner, *Tetrahedron Lett.*, **1968**, 6041.
119. G. O. Schenck and H. Wirth, *Naturwissenschaften*, **40**, 141 (1953).
120. T. Y. Ching and C. S. Foote, *Tetrahedron Lett.*, **1975**, 3771.
121. I. Saito, Y. Chujo, H. Shimazu, M. Yamane, T. Matsuura, and H. J. Cahnmann, *J. Amer. Chem. Soc.*, **97**, 5272 (1975).
122. W. Adam and J. C. Liu, *J. Amer. Chem. Soc.*, **94**, 2894 (1972).
123. G. O. Schenck and C. H. Krauch, *Angew. Chem.*, **74**, 510 (1962).
124. C. S. Foote, L. Weaver, Y. Chang, R. W. Denny, and J. Peters, *Ann. N.Y. Acad. Sci.*, **171**, 139 (1970).
125. R. A. Ackerman, I. Rosenthal, and J. N. Pitts, Jr., *J. Chem. Phys.*, **54**, 4960 (1971).
126. G. O. Schenck and C. H. Krauch, *Chem. Ber.*, **96**, 517 (1963).
127. R. W. Murray and S. L. Jindal, *J. Org. Chem.*, **37**, 3516 (1972).
128. B. Stevens, S. R. Perez, and R. D. Small, *Photochem. Photobiol.*, **19**, 315 (1974).
129. K. Gollnick and J. H. E. Lindner, unpublished results.

130. D. Bellus, H. Lind, and J. F. Wyatt, *Chem. Commun.*, **1972**, 1199.
131. W. F. Smith, *J. Amer. Chem. Soc.*, **94**, 186 (1972).
132. G. O. Schenck and K. Gollnick, *J. Chim. Phys.*, **55**, 892 (1958).
133. M. H. Fisch, J. C. Gramain, and J. A. Olesen, *Chem. Commun.*, **1970**, 13.
134. M. H. Fisch, J. C. Gramain, and J. A. Olesen, *Chem. Commun.*, **1971**, 663.
135. F. Khuong-Huu-Laine and D. Herlem-Gaullier, *Tetrahedron Lett.*, **1970**, 3649.
136. J. H. E. Lindner, H. J. Kuhn, and K. Gollnick, *Tetrahedron Lett.*, **1972**, 1705.
137. R. S. Davidson and K. R. Trethewey, *Chem. Commun.*, **1975**, 674.
138. E. A. Ogryzlo and C. W. Tang, *J. Amer. Chem. Soc.*, **92**, 5034 (1970).
139. C. Ouannes and T. Wilson, *J. Amer. Chem. Soc.*, **90**, 6527 (1968).
140. R. H. Young, R. L. Martin, D. Feriozi, D. Brewer, and R. Kayser, *Photochem. Photobiol.*, **17**, 233 (1973).
141. R. Nilsson, P. B. Merkel, and D. R. Kearns, *Photochem. Photobiol.*, **16**, 117 (1972).
142. I. Rosenthal and A. Frimer, *Photochem. Photobiol.*, **23**, 209 (1976).
143. L. J. Franklin, J. G. Dillard, H. M. Rosenstock, J. T. Herron, K. Droxy, and F. H. Field, in *Ionization Potentials and Heats of Formation of Gaseous Positive Ions*, NSRDS-NBS, No. 26, U.S. Government, Washington, D.C., 1962.
144. S. D. Worley and M. J. S. Dewar, *J. Chem. Phys.*, **50**, 654 (1969).
145. K. Gollnick and J. H. E. Lindner, *Tetrahedron Lett.*, **1973**, 1903.
146. G. O. Schenck and C. H. Krauch, unpublished results; C. H. Krauch, *Dissertation*, Univ. Göttingen, 1960.
147. H. H. Wasserman and J. E. VanVerth, *J. Amer. Chem. Soc.*, **96**, 585 (1974).
148. T. Yamase, H. Kokado, and E. Inuoe, *Bull. Chem. Soc. Japan*, **45**, 726 (1972).
149. R. H. Young, *Chem. Commun.*, **1970**, 704.
150. C. W. Jefford, A. F. Boschung, T. A. B. M. Bolsman, R. M. Moriarty, and B. Melnick, *J. Amer. Chem. Soc.*, **98**, 1017 (1976).

11

Mechanisms of Photo-oxidation

C. S. FOOTE

Department of Chemistry, University of California, Los Angeles, Calif. 90024, USA

Organic molecules are converted into reactive electronic excited states on absorption of light. An electron is excited to a higher orbital without change of spin; thus the first state formed is the singlet, in which there are no unpaired spins. The singlet in many cases undergoes a spin inversion to give the triplet state (which has two unpaired electrons). Both states involve electrons promoted to higher orbitals; these orbitals bind the electrons less strongly than those of the ground state; electrons in these orbitals are therefore more readily removed by oxidizing agents than are those in the ground state. In a similar manner, the 'holes' left by the promoted electron are in orbitals which bind electrons comparatively strongly; the excited molecule is thus also more readily reduced than the ground state (Figure 1). In addition to direct oxidation and reduction

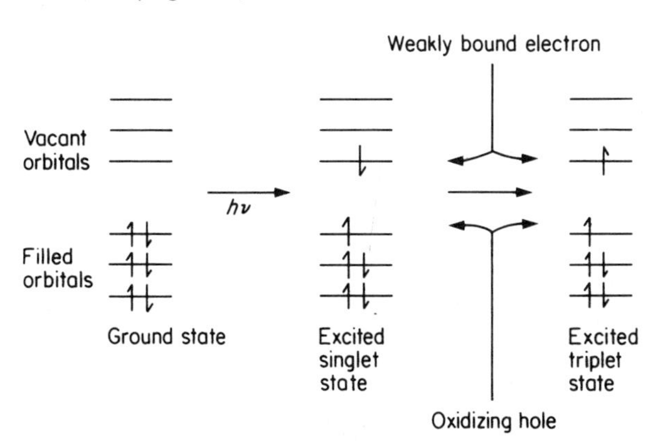

Figure 1 Orbital occupancy in ground-state and excited-state molecules

reactions, excited molecules are also capable of participating in photosensitized oxidations, in which a molecule other than the primary absorber is oxidized.[1]

The interest of chemists was attracted to these reactions by the fact that biologists had observed that the combination of sensitizing dyes, light, and oxygen are capable of damaging virtually all classes of organism.[2] The chemical basis of these effects has been traced to damage to many different cell constituents, including lipids (which are peroxidized), certain enzymes and peptides, and nucleic acids.[2] Recent studies have led to a rapid increase in the understanding of photo-oxidation mechanisms; it is now possible to recognize several distinct and different mechanistic pathways and to begin to predict which mechanism will occur in a given case.[3]

MECHANISMS

With few exceptions, sensitized oxidations proceed via the sensitizer triplet state, at least in part because the lifetime of the triplet is much longer than that of the singlet.[1–3] The most effective sensitizers are those which give a long-lived triplet state with high quantum yield. Many dyes, such as methylene blue and eosin, pigments (chlorophyll, haematoporphyrin and flavins) and aromatic hydrocarbons (rubrene and anthracene) are effective sensitizers. Most of these compounds absorb visible or near-UV light, which is effective with these sensitizers.

There are two major classes of reaction open to the sensitizer triplet.[4] In the first, the 'sensitizer' interacts with another molecule directly, usually with hydrogen-atom or electron transfer. The radicals thus formed react further with oxygen or other organic molecules. This reaction has been classified 'Type I' by Gollnick.[1a] In the second class of reaction, the sensitizer triplet interacts with oxygen; this class is called Type II. The most common Type II interaction is energy transfer to give singlet molecular oxygen, which reacts further with various acceptors.[5] Less efficiently, electron transfer to oxygen occurs, with the formation of the superoxide ion ($O_2^{-}\cdot$); this reaction occurs on less than 1% of the deactivating collisions of oxygen with the eosin triplet.[6] These reactions are summarized below:

$$\text{H or electron transfer} \xleftarrow[\text{Substrate}]{\text{Type I}} {}^3\text{Sens} \xrightarrow[O_2]{\text{Type II}} {}^1O_2(\gg O_2^{-}\cdot)$$

$$\downarrow \qquad k_I \qquad k_{II} \approx 1\text{–}3 \times 10^9\ \text{M}^{-1}\ \text{s}^{-1} \qquad \searrow$$

Reactions (Type I) — Reactions (Type II)

RATES

The rates of reaction, k_I and k_{II}, of the two processes are now well enough known that many of the factors which govern them can be stated definitively. The rate of the Type I process depends on the sensitizer and the substrate and varies over a wide range. Table 1 shows how this rate varies for a few selected cases.

Benzophenone is a stronger hydrogen abstractor by a factor of 10^4 than is eosin towards ethanol; however, both benzophenone and eosin react very

Table 1 Rates ($M^{-1}\,s^{-1}$) of Type I process for sample cases[a]

Substrate	Sensitizer	
	$(C_6H_5)_2C{=}O$	Eosin
Ethanol	$\sim 10^6$	~ 100
Dimethylaniline	$2{\cdot}7 \times 10^9$	2×10^9

[a] Data taken from ref. 3(b).

rapidly with the much stronger reductant dimethylaniline. As might be expected, structures which favour Type I (substrate–sensitizer) chemistry are those which are readily oxidized (phenols, amines, etc.) or readily reduced (quinones, etc.). Compounds which are not so readily oxidized or reduced (olefins, dienes, and aromatic compounds) more often favour Type II reactions; however, Type II reactions of amines,[7] phenols,[8] and other substrates are also known. The rates for the Type II process depend mainly on the oxygen concentration in solution, since the rate constant (k_{II}) falls in the range $1\text{–}3 \times 10^9\ M^{-1}\,s^{-1}$ with a few exceptions.[1(a), 9] Thus, for example, oxygen is less soluble in water than in organic solvents, so that in air-saturated water, the product $k_2[O_2]$ is much smaller than in organic solvents saturated with oxygen.

The competition which determines whether a Type I or Type II reaction occurs is thus between substrate and oxygen for a triplet sensitizer. Table 2

Table 2 Rates (s^{-1}) of Type I and Type II reactions in C_2H_5OH under O_2[a]

Sensitizer	$k_I[S]$	$k_{II}[O_2]$
$(C_6H_5)_2C{=}O$	$\sim 10^7$	$\sim 2 \times 10^7$
Eosin	$\sim 10^3$	$\sim 2 \times 10^7$

[a] Calculated from data in ref. 3(b).

shows that for benzophenone in oxygen-saturated ethanol, the Type I process (with ethanol as solvent) as substrate competes effectively with the Type II process; with eosin under the same conditions, however, the Type II process predominates, and would also predominate even at very low oxygen concentration. Thus, depending on the sensitizer, substrate, and concentrations of substrate and oxygen, the mechanism of the photo-oxidation may change from Type I to Type II. It is also important to remember that binding of a dye to macromolecular substrates is likely to favour Type I mechanisms.[10]

PRODUCTS

Type I chemistry usually involves the production of free radicals or radical ions. These radicals have a very wide variety of possible reactions, such as reaction with or electron transfer to oxygen, electron or hydrogen abstraction from other substrates, initiation of chain autoxidation, or recombination. Two examples of Type I reactions are shown below: the oxidation of alcohols by benzophenone, which can lead either to a ketone or hydroxyhydroperoxide, depending on the conditions,[11] and the oxidation of amines, which may proceed by either hydrogen or electron transfer.[12]

$$(CH_3)(H)C(C_2H_5)(OH) \xrightarrow{^3(C_6H_5)_2C=O} (CH_3)\dot{C}(C_2H_5)(OH) \xrightarrow{O_2} (CH_3)(OO\cdot)C(C_2H_5)(OH) \xrightarrow{RH}$$

$$(CH_3)(OOH)C(C_2H_5)(OH) + (C_6H_6)\dot{C}(C_6H_6)(OH) \longrightarrow \text{dimerization}$$

$$R_2NCH_3 \xrightarrow[\text{Sens}]{h\nu} R_2NCH_2^{\cdot} \xrightarrow{O_2} \longrightarrow R_2NCHO \longrightarrow R_2NH$$

The type II reaction produces singlet oxygen as the primary reactive species. Its nature and lifetime are adequately described in several of the other papers in this book, and will not be further discussed here. There is a second competition at this stage, between the rate of decay of singlet oxygen (k_d) and its reaction with substrate A given by the product $k_A[A]$; if $k_A[A] \ll k_d$, the major result of a Type II reaction is simply quenching of triplet sensitizer, and again one may observe no reaction. Rates for some typical substrates are given by Gollnick's in Chapter 10,[13] as are the various classes of reaction open to singlet oxygen.

$$^3O_2 \xrightarrow[k_d]{} {}^1O_2 \xrightarrow[k_A]{A} AO_2 \text{ (product)}$$

DISTINGUISHING TYPE I AND TYPE II REACTIONS

As a general problem, it is frequently necessary to distinguish between Type I and Type II reactions, and several methods are available. Simple kinetic studies may be misleading since both Type I and Type II reactions can give Stern–Volmer plots under the appropriate conditions. One of the most powerful techniques for demonstrating the intermediacy of singlet oxygen is by competitive inhibition with a known singlet oxygen acceptor which is not a good Type I substrate.[3,14] However, to be most useful, such results should be obtained under conditions which demonstrate that quenching is competitive, and should yield numerical values which can be compared with values determined from other studies. 'One-point' experiments with quenchers are of very limited value, since most singlet oxygen quenchers are also capable of quenching sensitizer excited states or reacting with free radicals.

Kearns and co-workers[15] have developed a technique for demonstrating singlet oxygen intermediacy based on the fact that the lifetime of singlet oxygen is much longer in D_2O than in H_2O; this fact can result in a higher rate of conversion of substrate in D_2O than in water. However, there are several assumptions which must be made before this technique can be used. The assumption that there is no change in efficiency of the Type I reaction in D_2O compared with H_2O has not yet been tested. Secondly, this technique will yield positive results only if the singlet oxygen reaction is in the first-order range, that is, where $k_A[A] < k_d$; if the reverse is true, there will be no change in efficiency of the singlet oxygen reaction in D_2O compared with water.

Another effective technique for determining whether the reaction is of Type I or Type II is to determine whether substrate and oxygen are competitive or not; in reactions in which the Type I reaction is much slower than Type II, there is no dependence on oxygen pressure even at very low oxygen concentrations.[1] In a Type I reaction, there is often a dependence on oxygen pressure; oxygen may actually inhibit the reaction although this is not the only possible behaviour.

Guanosine and GMP

As an example of techniques which can be used to differentiate Type I and Type II reactions, we studied the photo-oxidation of guanosine and guanosine-5-monophosphate, sensitized by Rose Bengal.[16] Several different techniques were used in the study. In the scheme shown below, k_d is the rate of decay of singlet oxygen, k_g is the rate of reaction of guanine derivative (G) with $^1O_2$, and k_x is the rate of reaction of additive X with singlet oxygen. In general, if G reacts only with $^1O_2$, the value of k_g determined will not depend on the medium (air or oxygen, which would change the proportion of Type I reaction, or D_2O or H_2O, which would change k_d[15]). The results are listed in Table 3. References for the techniques used can be found in ref. 17.

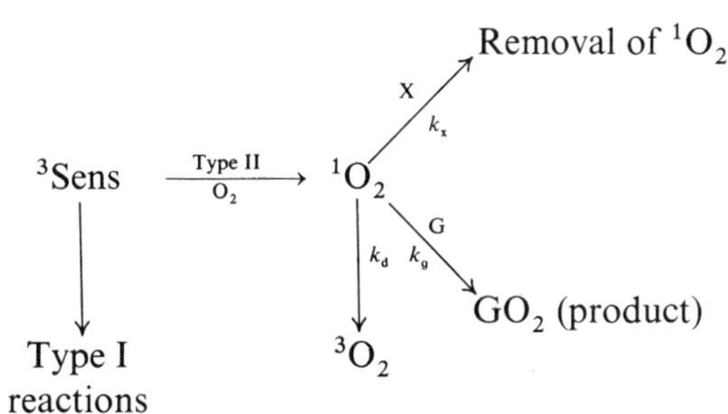

A study of the concentration dependence of the rate of disappearance of guanosine in air and oxygen-saturated water as a function of [G] was carried out by monitoring the change in guanosine concentration by high-performance liquid chromatography. Excellent kinetics were observed but, because guanosine is so insoluble, the intercept values are uncertain; hence the values listed in Table 3 for this technique are not very reliable. The guanosine-inhibited photo-

Table 3 Kinetic parameters and rates of reaction of guanosine and guanosine monophosphate, pH 8

Acceptor	Technique	Conditions	Parameter[a]	Value	k_g (M^{-1} s^{-1} $\times 10^{-7}$)[b]
Guanosine	Disappearance	H_2O/air	k_d/k_g	$1{\cdot}0 \times 10^{-3}$ M	~50[c]
		H_2O/O_2		$1{\cdot}8 \times 10^{-3}$ M	~20[c]
Guanosine	Young's	D_2O/air	k_d/k_g	$3{\cdot}0 \times 10^{-3}$ M	1·65
		D_2O/O_2		$3{\cdot}5 \times 10^{-3}$ M	1·43
Guanosine	Histidine competition	H_2O/O_2	k_h/k_g	5·8	0·86
		H_2O/O_2		4·2	1·2
Guanosine	NaN_3 competition	H_2O/O_2	k_{N_3}/k_g	35	0·82
		H_2O/O_2		26	1·07
Guanosine	DABCO competition	H_2O/O_2	k_{DAB}/k_g	0.187	3·9[d]
GMP	Young's	H_2O/air	k_d/k_g	$6{\cdot}7 \times 10^{-2}$ M	0·75
		H_2O/O_2		$7{\cdot}6 \times 10^{-2}$ M	0·66
		D_2O/air		$4{\cdot}6 \times 10^{-3}$ M	1·08
		D_2O/O_2		$4{\cdot}5 \times 10^{-3}$ M	1·14

[a] k_d = decay rate of $^1O_2$; k_g, k_h, k_{N_3}, k_{DAB} = rates of reaction of guanine derivative, histidine, azide, and DABCO, respectively, assuming $^1O_2$ is reactive species.
[b] Believed to be accurate to within ±20% except as noted.
[c] Inaccurate because of poor intercept caused by solubility limitations; probably accurate only to within an order of magnitude.
[d] Based on k_{DAB} in CH_3OH, which is probably higher than in H_2O.

oxidation of diphenylfuran ('Young's technique'[18]) was studied in D_2O under air and oxygen; results could not readily be obtained in water because of solubility limitations. The inhibition of guanosine photo-oxidation by histidine,[15(a)] DABCO,[18] and sodium azide[19] were studied; the value for DABCO listed in Table 3 is relative to the rate of quenching of $^1O_2$ by DABCO in methanol,[18] which is probably higher than that in water, so that the rate constant calculated is probably high. Inhibition with all three quenchers was competitive, indicating a singlet oxygen mechanism.

Guanosine-5′-monophosphate (which is more soluble than guanosine, and thus easier to study) was used to inhibit the photo-oxidation of diphenylfuran in D_2O and in H_2O, under air and oxygen (Young's technique). The values of k_g (rate of reaction of guanine with singlet oxygen) calculated from the data are independent of the medium, consistent with expectation for a singlet oxygen mechanism; the data are given in Table 3.

The results show that the guanosine reaction is competitively inhibited by singlet oxygen acceptors, and that guanosine and GMP inhibit the photo-oxidation of diphenylfuran. The effect of oxygen pressure is negligible, and the value of k_d/k_g is about 13 times as large in H_2O as in D_2O, consistent with the larger k_d values in H_2O. The results are thus consistent with the intermediacy of singlet oxygen in the reaction. The results are surprising, since this case is expected to be a borderline Type I–Type II reaction; indeed, evidence for some Type I reaction has been presented.[20] It may be that the aberrant values of k_d/k_g determined from the direct disappearance studies reflect the participation of some Type I reaction, but the results of this series can be trusted only to within an order of magnitude because of the uncertainty of the intercept. The other values in Table 3 are probably accurate to within about ±20%.

ELECTRON-TRANSFER TO SINGLET OXYGEN

Reactions of the four major classes of singlet oxygen substrate (dienes, olefins, electron-rich olefins and sulphides) have been shown rigorously by careful mechanistic studies to proceed by way of singlet oxygen.

Several of the more electron-rich substrates react with singlet oxygen in a fashion which is not characteristic of less electron-rich compounds. Electron-rich olefins, (enamines[21] and vinyl ethers[22–25]) react to give dioxetanes, phenols react to give hydroperoxy dienones,[8] amines react to give a variety of products,[26] and sulphides give sulphoxides;[27] quenching of singlet oxygen has also been shown with the last three classes under some conditions.

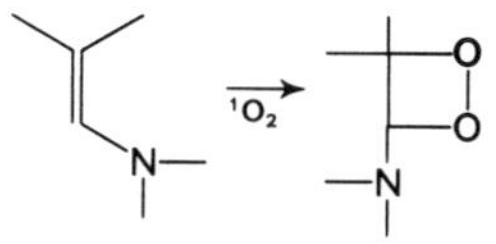

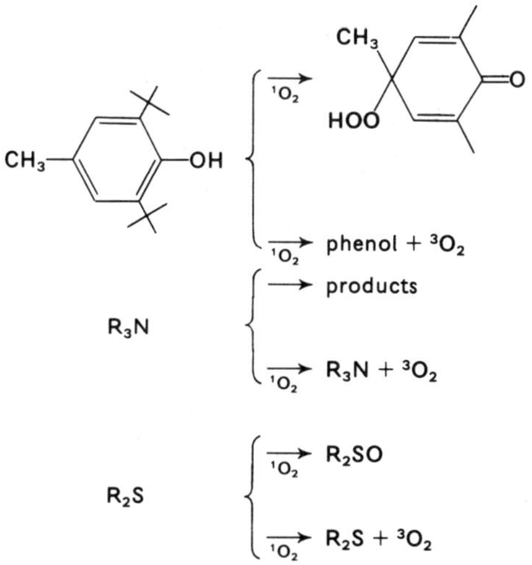

$^1O_2$

CH$_3$ OH $\xrightarrow{^1O_2}$ CH$_3$ HOO O

$\xrightarrow{^1O_2}$ phenol + $^3O_2$

$$R_3N \begin{cases} \longrightarrow \text{products} \\ \xrightarrow{^1O_2} R_3N + {}^3O_2 \end{cases}$$

$$R_2S \begin{cases} \xrightarrow{^1O_2} R_2SO \\ \xrightarrow{^1O_2} R_2S + {}^3O_2 \end{cases}$$

We have preliminary indications that all of these reactions may proceed by electron transfer from the electron-rich compound (D) to singlet oxygen to give a radical cation–superoxide ion pair or charge-transfer complex ($D^{+\cdot}O_2^{-}\cdot$)[28] (from the argument at the beginning of this paper, it will be seen that singlet oxygen is 1 V more oxidizing than ground-state oxygen). Recombination of the ion pairs would give the product (DO_2); back-transfer of an electron would give quenching of singlet oxygen.

$$D + {}^1O_2 \longrightarrow D^{+\cdot} + O_2^{-}\cdot \longrightarrow DO_2$$
$$\searrow D + {}^3O_2$$

Some of the evidence (at present only suggestive) which bears on this mechanism is the following.

1. Bartlett *et al.*[24] found that *p*-dioxene gives products derived from both an ene reaction and a 2+2 reaction, but that the 2+2 product is favoured in polar solvents, suggesting that the transition state for the 2+2 reaction is more polar than that of the ene reaction. The 2+2 reaction is Woodward–Hoffman forbidden unless it goes 2s+2a, and the electron-transfer mechanism may circumvent this difficulty.

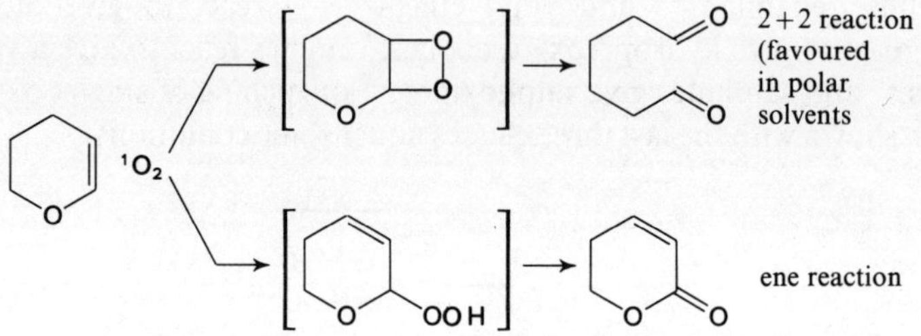

$^1O_2$

2+2 reaction (favoured in polar solvents

OOH

ene reaction

2. Mazur[29] noticed that the compounds he studied seemed to fall into two groups: those with low enough oxidation potentials for the electron-transfer reaction to be exothermic gave the 2+2 product, whereas those of higher potential which had alternate pathways open gave the alternate chemistry. This dichotomy is illustrated by the pair of reactions below;[21(b), 30] the first reaction is estimated to be endothermic for electron transfer to singlet oxygen; the second is probably exothermic.[29]

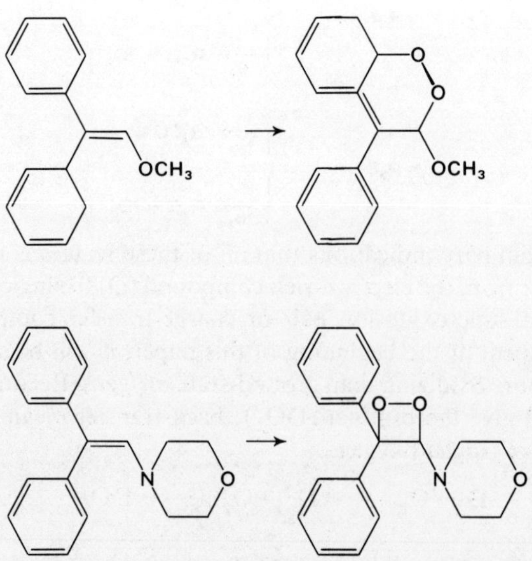

OCH_3 → OCH_3

N O → O–O N O

3. Thomas and co-workers[8(c, d)] showed that phenols reacted with singlet oxygen at a rate the logarithm of which is linearly correlated with their half-wave oxidation potential; the same rate was given by phenols and phenol methyl ethers of the same oxidation potential; this is strong evidence for electron transfer, rather than hydrogen abstraction. Thomas and co-workers were also able to obtain direct spectroscopic evidence for the intermediacy of phenoxy radicals in the reaction of triphenylphenol with singlet oxygen. The mechanism consistent with these findings appears to be that shown below. The quenching of singlet oxygen by amines such as DABCO probably also goes through a similar electron or charge-transfer mechanism.[31]

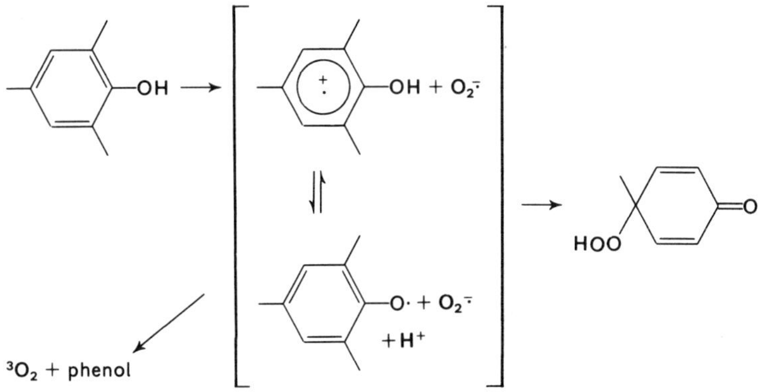

4. In an attempt to study this type of reaction directly, we carried out the reaction of certain radical cations and donor–$SbCl_5$ complexes with superoxide ion.[28] These studies were frustrating because of our inability to prepare radical cations of appropriate molecules, but preliminary results indicate the formation of products from sulphides and from electron-rich olefins which suggest a chemistry similar to that of intermediates in photo-oxidation.

We have recently found a new photo-oxidation reaction, which, although it does not proceed by way of singlet oxygen, may proceed via the same substrate cation–oxygen anion pair as the above reaction.[28] This technique rests on the fact that an excited molecule is a more powerful oxidizing agent than the corresponding ground-state molecule[32, 33] (as mentioned earlier).

Our rationale for this reaction was based on the report by Shigemitsu and Arnold[34] of the use of cyano aromatics as photosensitizers for the addition of solvent (SH) to electron-rich olefins, and that of Roth and Manion[35] for their dimerization. These workers demonstrated an electron-transfer pathway for these (anaerobic) reactions.

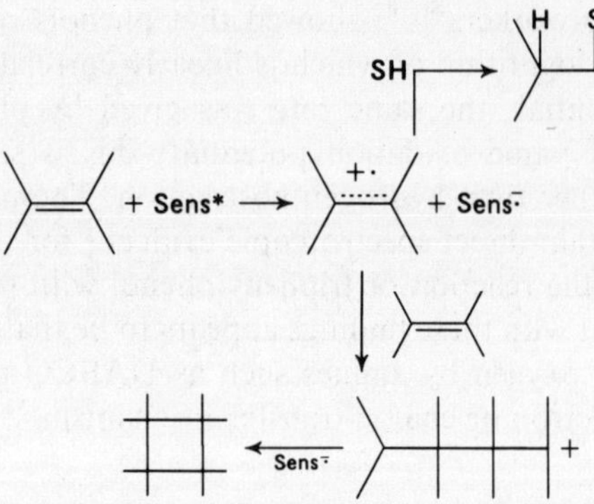

$\phi_2S \xrightarrow{DCA,\ ,\ O_2} \phi_2SO$

$Et_2S \longrightarrow Et_2SO + Et_2SO_2$

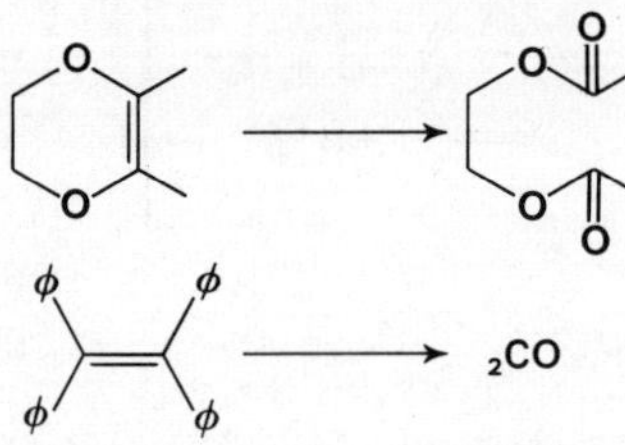

$\longrightarrow$ $_2CO$

Using dicyanoanthracene, photoexcited by long-wave UV light, it has been possible to sensitize the photo-oxidation of several electron-rich olefins and sulphides in acetonitrile.[28] So far, the following reactions have been carried out. Substantial yields of the products are obtained.

The following observations have been made in preliminary studies:

(1) The reaction is sometimes, but not always, faster under air than under oxygen; oxygen inhibition suggests a Type I pathway.

(2) The reaction is faster in acetonitrile than in benzene, suggesting polar intermediates.

(3) β-Carotene, at concentrations up to 10^{-4} M, either has no effect or actually promotes the reaction (evidence for a non-singlet oxygen pathway).

(4) At least some of the reactions do not occur in the absence of oxygen and the presence of water, so that hydrolysis of radical cation alone does not explain the results.

(5) The reactions appear to go by way of the *singlet* state of dicyanoanthracene, since fluorescence is quenched, and with the same Stern–Volmer constant found for the formation of reaction products. These preliminary studies are consistent with a Type I reaction proceeding via singlet DCA, with the reduced DCA being re-oxidized by oxygen. However, the system is very complex, and

much more detailed studies will be necessary in order to establish this mechanism. Nevertheless, it is already clear that it is a photosensitized oxygenation which does not proceed via singlet oxygen.

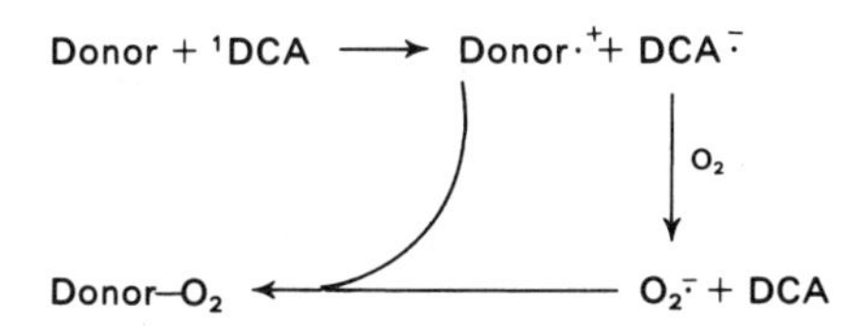

REFERENCES

1. (a) K. Gollnick, *Adv. Photochem.*, **6** (1968) 1; (b) R. Livingston, in *Autoxidation and Antioxidants* (Ed. W. O. Lundberg), Vol. 1, Wiley, New York, 1961, p. 249; (c) K. Gollnick and G. O. Schenck, in *1,4-Cycloaddition Reactions* (Ed. J. Hamer), Academic Press, New York, 1967, p. 255.
2. (a) J. D. Spikes, *Ann. Rev. Phys. Chem.*, **18**, 409 (1967); (b) J. D. Spikes and R. Livingston, *Adv. Radiat. Biol.*, **3**, 29 (1969); (c) J. D. Spikes, *Photophysiology*, **3**, 33 (1968); (d) J. D. Spikes and M. L. MacKnight, *Ann. N.Y. Acad. Sci.*, **171**, 149 (1970).
3. (a) C. S. Foote, *Science. N.Y.*, **162**, 963 (1968); (b) C. S. Foote, in *Free Radicals in Biology* (Ed. W. Pryor), Academic Press, New York, 1976, p. 85.
4. G. O. Schenck and E. Koch, *Z. Elektrochem.*, **64**, 170 (1960).
5. (a) C. S. Foote, *Accounts Chem. Res.*, **1**, 104 (1968); (b) D. R. Kearns, *Chem. Rev.*, **71**, 395 (1971).
6. (a) A. Kepka and L. I. Grossweiner, *Photochem. Photobiol.*, **14**, 621 (1972); (b) V. Kasche and L. Lindquist, *Photochem. Photobiol.*, **4**, 923 (1965).
7. W. F. Smith, Jr., *J. Amer. Chem. Soc.*, **94**, 186 (1972).
8. (a) I. Saito, S. Kato, and T. Matsuura, *Tetrahedron Lett.*, **1970**, 239; (b) I. Matsuura, N. Yoshimura, A. Nishinaga, and I. Saito, *Tetrahedron*, **1972**, 4933; (c) M. Thomas and C. S. Foote, in preparation; (d) C. S. Foote, M. Thomas, and T.-Y. Ching, *J. Photochem.*, **5**, 172 (1976).
9. L. K. Patterson, G. Porter, and M. R. Topp, *Chem. Phys. Lett.*, **7**, 612 (1970).
10. (a) J. S. Bellin, *Photochem. Photobiol.*, **4**, 33 (1965); (b) K. J. Youtsey and L. I. Grossweiner, *Photochem. Photobiol.*, **6**, 721 (1967).
11. G. O. Schenck, H.-D. Becker, K.-H. Schulte-Elte, and C.-H. Krauch, *Chem. Ber.*, **96**, 509 (1963).
12. R. F. Bartholemew and R. S. Davidson, *J. Chem. Soc., D*, **1970**, 1174.
13. K. Gollnick, this book, Chapter 10.
14. C. S. Foote, R. W. Denney, L. Weaver, Y. Chang, and J. Peters, *Ann. N.Y. Acad. Sci.*, **171**, 139 (1970).
15. (a) R. Nilsson, P. B. Merkel, and D. R. Kearns, *Photochem. Photobiol.*, **16**, 117 (1972); (b) P. B. Merkel and D. R. Kearns, *J. Amer. Chem. Soc.*, **94**, 1029 (1972); (c) P. B. Merkel, R. Nilsson, and D. R. Kearns, *J. Amer. Chem. Soc.*, **94**, 1030 (1972).
16. C. S. Foote and M. P. Easton, unpublished work.
17. C. S. Foote and T.-Y. Ching, *J. Amer. Chem. Soc.*, **97**, 2609 (1975).
18. R. H. Young and R. L. Martin, *J. Amer. Chem. Soc.*, **94**, 5183 (1972).
19. (a) C. S. Foote, T. T. Fujimoto, and Y. C. Chang, *Tetrahedron Lett.*, **1972**, 45; (b) N. Hasty, P. B. Merkel, P. Radlick, and D. R. Kearns, *Tetrahedron Lett.*, **1972**, 49.
20. (a) A. Knowles and G. N. Mautner, *Photochem. Photobiol.*, **15**, 199 (1972); (b) R. Nilsson, P. B. Merkel, and D. R. Kearns, *Photochem. Photobiol.*, **16**, 117 (1972); (c) I. Saito, K. Inoue, and T. Matsuura, *Photochem. Photobiol.*, **21**, 27 (1975).

21. (a) C. S. Foote, A. A. Dzakpasu, and J. W.-P. Lin, *Tetrahedron Lett.*, **1975**, 1247; (b) C. S. Foote and J. W.-P. Lin, *Tetrahedron Lett.*, **1968**, 3267.
22. P. D. Bartlett and A. P. Schaap, *J. Amer. Chem. Soc.*, **92**, 3223 (1970).
23. S. Mazur and C. S. Foote, *J. Amer. Chem. Soc.*, **92**, 3225 (1970).
24. P. D. Bartlett, G. D. Mendenhall, and A. P. Schapp, *Ann. N.Y. Acad. Sci.*, **171**, 79 (1970).
25. A. P. Schaap and P. D. Bartlett, *J. Amer. Chem. Soc.*, **92**, 6055 (1970).
26. (a) K. Gollnick and J. H. E. Lindner, *Tetrahedron Lett.*, **1973**, 1903; (b) W. F. Smith, Jr., *J. Amer. Chem. Soc.*, **94**, 186 (1972).
27. (a) C. S. Foote and J. W. Peters, *J. Amer. Chem. Soc.*, **93**, 3795 (1971); (b) C. S. Foote and J. W. Peters, *XXIIIrd International Congress of Pure and Applied Chemistry, Special Lectures*, **4**, 129 (1971).
28. C. S. Foote and T. Parker, unpublished work.
29. S. Mazur, *Ph.D Thesis*, University of California, Los Angeles, 1971.
30. C. S. Foote, S. Mazur, P. A. Burns, and D. Lerdal, *J. Amer. Chem. Soc.*, **95**, 586 (1973).
31. (a) R. H. Young and R. L. Martin, *J. Amer. Chem. Soc.*, **94**, 5183 (1972); (b) E. A. Ogryzlo and C. W. Tang, *J. Amer. Chem. Soc.*, **92**, 5034 (1970); (c) C. S. Foote, R. W. Denny, L. Weaver, Y. Chang, and J. Peters, *Ann. N.Y. Acad. Sci.*, **171**, 139 (1970).
32. (a) D. Rehm and A. Weller, *Ber. Bunsenges. Phys. Chem.*, **73**, 834 (1969); (b) H. Knibbe, D. Rehm, and A. Weller, *Ber. Bunsenges. Phys. Chem.*, **73**, 839 (1969); (c) H. Knibbe and A. Weller, *Z. Phys. Chem., N.F.*, **56**, 95, 99 (1967).
33. J. B. Guttenplan and S. G. Cohen, *J. Amer. Chem. Soc.*, **94**, 4040 (1972); *Tetrahedron Lett.*, **1972**, 2163.
34. Y. Shigemitsu and D. R. Arnold, *Chem. Commun.*, **1975**, 407.
35. H. D. Roth and M. L. Manion, *VII International Conference on Photochemistry, Edmonton, August 1975, Abstracts*, U2.

12

Photo-stimulated Oxidation and Peroxidation in Crystalline Anthracene and some Related Derivatives

S. E. MORSI and J. O. WILLIAMS

Chemistry Department, Faculty of Science, University of Tanta, Egypt
Edward Davies Chemical Laboratories, University College of Wales, Aberystwyth, Wales

INTRODUCTION

Problems of the photo-oxidation of polymers have aroused considerable interest in the past few years. In the absence of air, photolysis involves chain scission and de-crosslinking in polymers, but in the presence of air the solar UV radiation should suffice to excite oxygen from the ground triplet state to the excited singlet state. The excited singlet oxygen atoms can readily be added to ethylenic π-bonds producing peroxides. On the other hand, when polynuclear aromatic compounds are present in the polymer, which is frequently the case when UV stabilizers are added, then singlet oxygen addition may take place, leading to *trans*-annular peroxides.

The peroxide linkages are usually weak, with bond energies of about 120 kJ mol^{-1}, and thus they can easily be decomposed both thermally and photochemically, and initiate a series of oxidative degradation reactions.

Unlike dialkyl peroxides in which O—O bond homolysis takes place,[1] some *trans*-annular peroxides involve a loss of oxygen upon thermolysis or photolysis.[2] It has been argued[3] that a *trans*-annular peroxide will decompose to the parent hydrocarbon if the resulting gain in resonance energy by the hydrocarbon is sufficiently great to make this course the preferred decomposition pathway; otherwise, O—O bond homolysis occurs. It is expected that the *trans*-annular peroxides will lose oxygen in the singlet multiplicity and that oxygen readily reacts with singlet oxygen-accepting substrates.[2]

The interest in the solid-state photochemistry of anthracene (**1**) and its derivatives is due to the fact that the topochemical pre-formation theory of Cohen[4] and Schmidt[5] does not apply to these compounds. The 9-substituted and 9,10-disubstituted anthracenes[6] yield centrosymmetric dimers instead of the mirror symmetrical units predicted by the theory. In several instances, notably in anthracene itself and its 9-cyano derivative, defects have been shown to exert a controlling influence on the reaction. However, the photoreactivity of anthracene is very sensitive to the experimental conditions applied and the photodimer is only one of many products that can be formed by varying the atmosphere surrounding the crystals.[7] Numerous photo-oxidation products can be formed and in several cases the full crystal structure data for these compounds have been reported. One of these products, anthracene-9,10-epiperoxide (**2**), was found on exposure to X-ray irradiation[8] to undergo a single crystal → single crystal transformation to a mixture of anthraquinone (**3**) and anthrone (**4**). The space group remains invariant but the products are always twinned in a manner which is related to the unit cell of the parent.[9] Since the structures of anthracene, anthracene-9,10-epiperoxide, anthraquinone, and anthrone are closely interrelated (see Table 1), it is not unreasonable to expect the interconversion of these compounds through a photochemical cycle. In this paper, following fluorescence microscopical investigations, we establish evidence for such photoinduced solid-state interconversion that also involves di-*p*-anthracene (**5**). The central pivotal role of an excimeric state in this cycle is also indicated.

Table 1 Crystal data

Type of crystal	a(Å)	b(Å)	c(Å)	(Å)	Z	Space group
Anthracene	8·561	6·036	11·163	124	2	$P2_{2/a}$
Anthracene-9,10-epiperoxide	15·94	5·86	11·43	108·2	4	$P2_{1/a}$
Anthraquinona/anthrone	15·8	4·0	7·9	102	2	$P2_{1/a}$
Di-*p*-anthracene	8·127	12·08	18·85		4	P*bca*

EXPERIMENTAL

Anthracene was purified by zone refining and thin crystals (ca. 1–100 μm) were grown from a benzene solution on microscope slides. Anthraquinone and anthrone (BDH) were recrystallized from ethanol and thin crystals were grown from ethanolic solutions. Anthracene-9,10-epiperoxide was prepared by the photo-oxidation of anthracene solution in carbon disulphide[10] and purified by crystallization from benzene. The product was characterized by differential scanning calorimetry and IR spectroscopy. Di-*p*-anthracene was prepared by irradiation of anthracene solution in benzene contained in a Pyrex vessel using light from a high-pressure mercury arc. The photodimer crystals were washed with benzene and checked by melting-point determinations and electron diffraction measurements. Diels–Alder adducts of anthracene with 2-methylbutene and cyclopentadiene were prepared by UV irradiation of benzene solutions of anthracene containing the olefinic compound. No analysis is provided for these adducts.

All investigations were carried out with a Reichert Zetopan fluorescence microscope and irradiations were carried out in air using the filtered light (<400 nm) from a high-pressure mercury lamp. Fluorescence was observed from the irradiated surface. To minimize interference in the fluorescence spectra and to achieve maximum resolution, the spectra were recorded in four regions (blue, green, yellow, and orange), employing cut-off filters on the emission side to record spectra in the ranges at 400–460, 460–490, 490–560, and 560–600 nm, respectively. All spectra were recorded in the following sequence: (a) non-polarized light, (b) polarized light; and (c) crossed nicols.

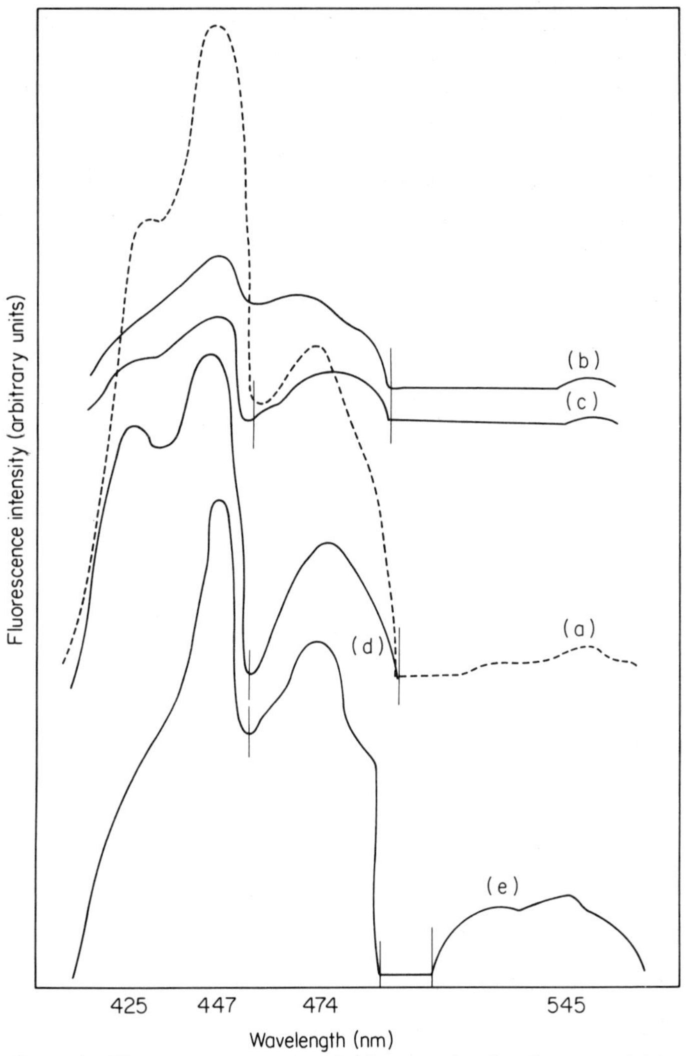

Figure 1 Fluorescence spectra of thin crystals of anthracene: (a) in non-polarized UV radiation; (b) in polarized UV radiation; (c) in crossed nicols; (d) spectra of blue-emitting region in an aged crystal, non-polarized radiation; (e) spectra of green-emitting region in an aged crystal, exposure to non-polarized radiation. Note the difference in blue- and green-emitting regions with respect to the existence of the yellow band.

Anthracene

The Fluorescence spectra of anthracene are shown in Figure 1. There are three main peaks at 425 and 474 nm together with a minor peak at 545 nm. The same peaks are observed in polarized light [Figure 1(b)] and in crossed nicols [Figure 1(c)] with the exception that in crossed nicols a red shift occurs in the 474 nm peak to 482 nm. The first two peaks in the spectra of anthracene correspond to its characteristic blue fluorescence while the last two peaks correspond to green and yellow emissions, respectively.

Blue and green areas in anthracene are observed upon fluorescence ageing for ca. 10^4 s (compare the report of Donati *et al.*[11]). The green fluorescence areas are normally associated with edges and cracks. With the aid of a diaphragm built into the microscope, it is possible to record spectra from the blue and green areas separately [Figures 1(d) and 1(e)]. It can be seen that the positions of the three main peaks remain unchanged whereas the fourth minor peak is submerged in the case of the blue areas, but is resolved and split into peaks at 520 and 545 nm in the green areas.

Anthracene-9,10-epiperoxide

The spectra of anthracene-9,10-epiperoxide (Figure 2) are very similar to that of anthracene [Figure 1(a)] and its behaviour on fluorescence ageing also follows the anthracene pattern.

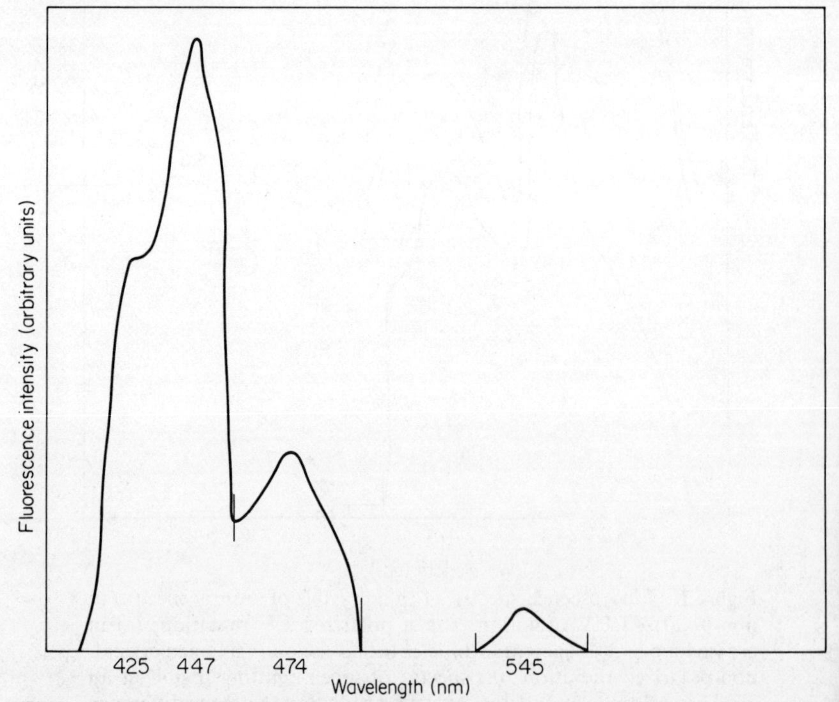

Figure 2 Fluorescence spectra of anthracene-9,10-epiperoxide thin crystals when irradiated with non-polarized UV radiation. Note the correspondence between this spectrum and that of anthracene [Figure 1(a)].

Anthraquinone

Anthraquinone fluoresces in the yellow region when irradiated with non-polarized UV radiation. The main peak is at 545 nm, with a shoulder at 575 nm

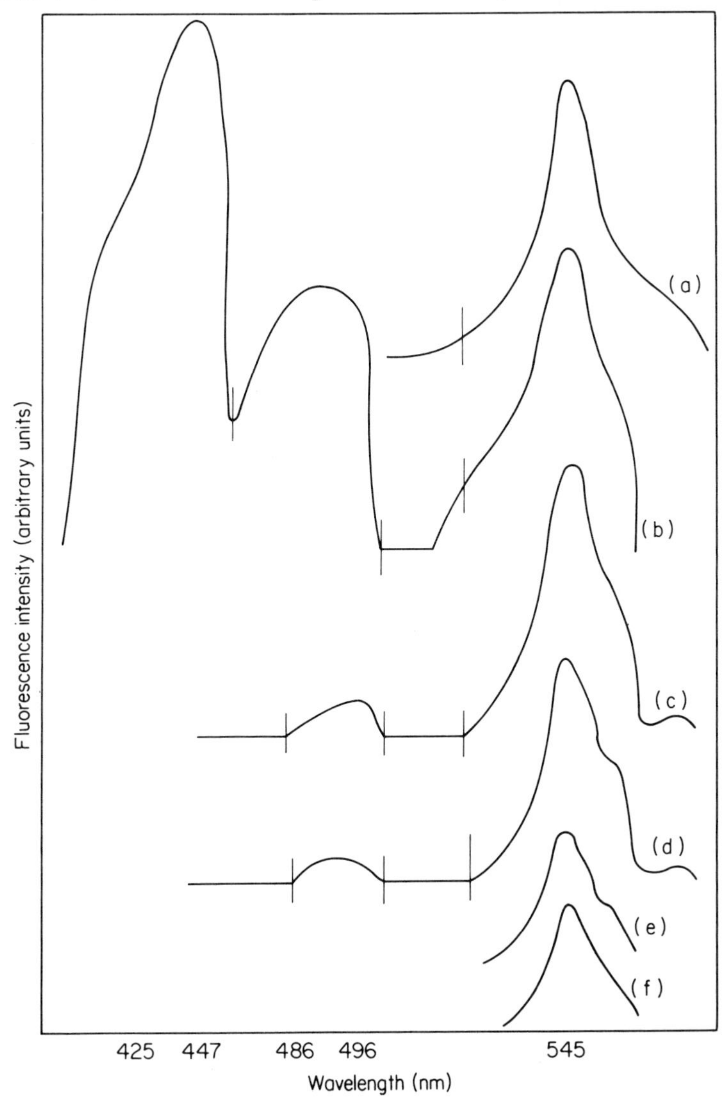

Figure 3 Fluorescence spectra of anthraquinone: (a) in non-polarized UV radiation (in polarized UV radiation this band disappears); (b) spectra of a blue-emitting region in an aged crystal, exposure to non-polarized radiation; (c) spectra of a yellow-emitting region in an aged crystal, exposure to non-polarized radiation; (d) spectra of a green-fluorescing region in an aged crystal, exposure to non-polarized radiation; (e) spectra after ageing in polarized radiation; (f) spectra after ageing, crystal between crossed nicols. Note that the spectra (e) and (f) do not exist in fresh crystals.

[Figure 3(a)]. Anthraquinone does not fluoresce when irradiated with polarized UV radiation. On prolonged irradiation of anthraquinone (ca. 5×10^3 s), green and blue fluorescence appear [Figure 3(b)]. Fluorescence-aged crystals of

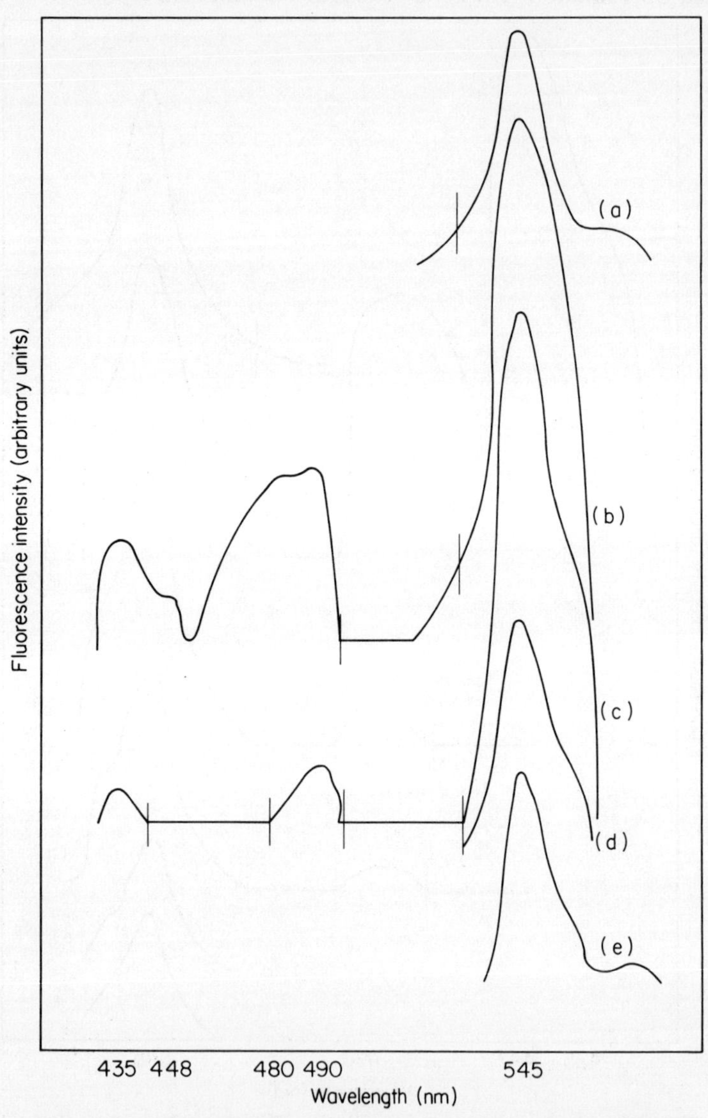

Figure 4 Fluorescence spectra of anthrone: (a) in non-polarized UV radiation (this band disappears in polarized UV radiation); (b) spectra of a yellow-fluorescing region in an aged crystal, exposure to non-polarized radiation; (c) spectra of blue-fluorescing region in an aged crystal, exposure to non-polarized radiation; (d) and (e) spectra of an aged crystal in polarized radiation and crossed nicols, respectively. Note that as in (a) a fresh crystal does not fluoresce in polarised radiation.

anthraquinone have a polarized yellow band (see Figures 3(c), 3(e), and 3(f), while the blue and green bands are not polarized [Figures 3(e) and 3(f)].

Anthrone

The spectrum and general behaviour of anthrone (Figure 4) are very similar to those of anthraquinone, showing a yellow fluorescence at 545 nm with a shoulder at 573 nm in non-polarized UV radiation. After prolonged irradiation, unpolarized blue and green areas are formed.

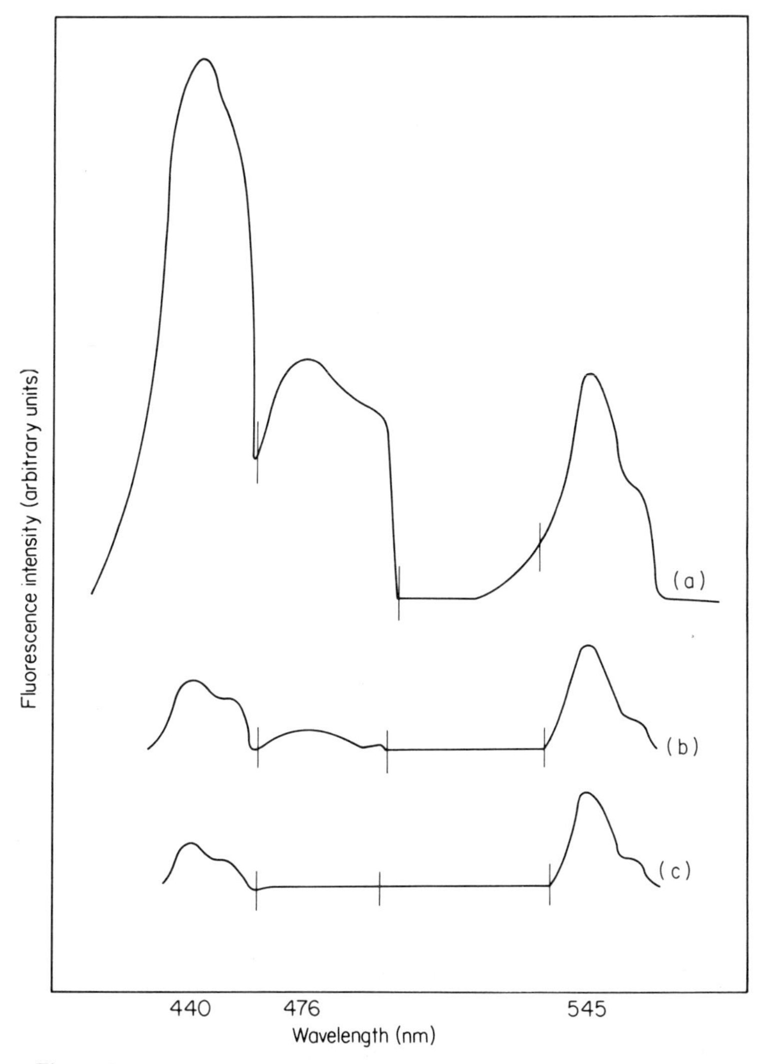

Figure 5 Fluorescent spectra of an aged di-*p*-anthracene crystal: (a) in non-polarized UV radiation; (b) and (c) in polarized radiation and crossed nicols, respectively. Note that the polarized spectra in (b) and (c) do not appear except at later stages of ageing.

Di-*p*-anthracene

The anthracene photodimer does not fluoresce but on irradiation for ca. 5×10^3 s fluorescence appears with peaks at 440, 470, and 545 nm, as shown in Figure 5(a). The blue and green peaks are unpolarized initially but upon further ageing they become polarized.

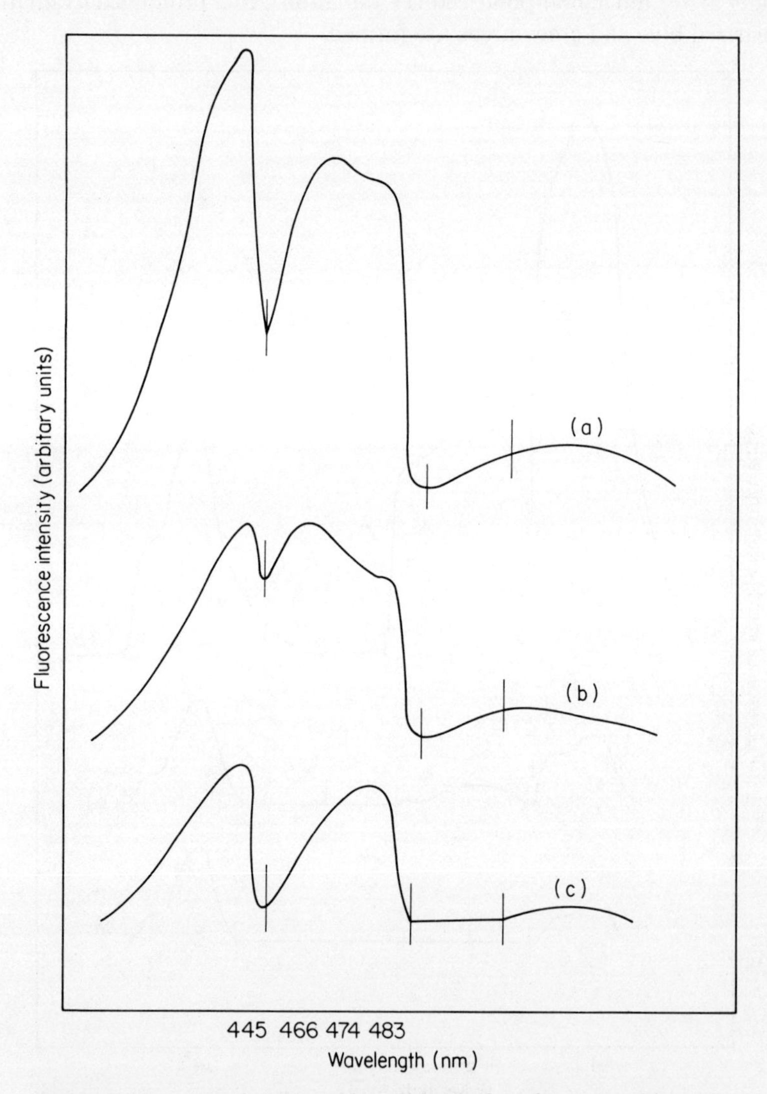

Figure 6 Spectra of Diels–Alder adduct of anthracene with cyclopentadiene: (a) in non-polarized radiation; note the major green band which leads to the green fluorescence of the adduct; (b) and (c) spectra in polarized radiation and in crossed nicols, respectively. Note that the polarized bands do not appear except after fluorescence ageing

Diels–Alder adducts with anthracene

Diels–Alder adducts of cyclopentadiene with anthracene fluoresce in the green region, with some blue and yellow fluorescence [Figure 6(a)]. Polarized fluorescence is not observed except after fluorescence ageing.

DISCUSSION

The spectra of the compounds studied can be classified into four groups:

(1) anthracene and anthracene- 9,10-epiperoxide;
(2) anthraquinone and anthrone;
(3) di-*p*-anthracene;
(4) Diels–Alder adducts.

The first three bands in group (1) spectra are inherent in the anthracene structure and compare with the bands observed in solution fluorescence spectra. The yellow peak can be attributed to trace amounts of anthraquinone and/or anthrone and the green peak to an excimeric emission.[12] Group (2) spectra show green and blue fluorescence upon ageing, which is attributed to photoreduction of anthraquinone and anthrone to anthracene-9,10-epiperoxide, di-*p*-anthracene, and anthracene. During such a photoreaction the anthraquinone/anthrone acquires the orientation of a mesomorphic phase which probably has an orientation similar to that of the products. This follows from the observation that the polarized yellow fluorescence of anthraquinone/anthrone does not occur except after ageing. The nascently generated blue and green bands are not polarized, suggesting the random arrangement of the products (probably close to structural imperfections).

The anthracene dimer [group (3)] spectra show fluorescence in the blue, green, and yellow regions after prolonged irradiation. This reveals the possibility of the photodecomposition of the dimer to the monomer, and subsequently to its photo-oxidation products. The fact that nascently generated blue and green emissions are submerged in polarized emissions may be rationalized in terms of the association of such fluorescence with localized regions on the surface.

The spectra of Diels–Alder adducts [group (4)] are related to the above spectra in that they possess a mainly green (excimer) emission but prolonged irradiation shows the presence of blue and yellow emissions attributable to photo-decomposition and subsequent photo-oxidation.

Scheme 1 suggests a mechanism for the photostimulated chemical interconversions of the above-mentioned four groups.

The photodimerization of anthracene occurs preferentially along several crystallographic dierctions permitted in the anthracene crystal[13] and the decomposition of anthracene-9,10-epiperoxide to a mixture of anthraquinone and anthrone[8] was found to occur along main crystallographic directions which remains an invariant direction during the transformation. The preferential

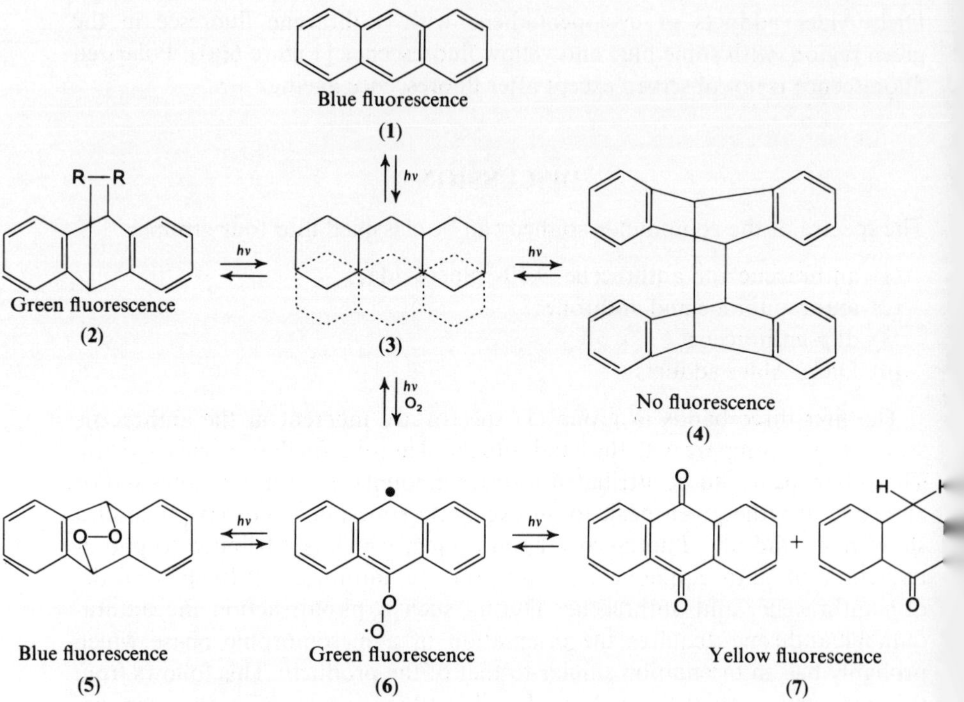

Blue fluorescence
(1)
hν
R—R
Green fluorescence
(2)
hν
(3)
hν
No fluorescence
(4)
hν
O2
hν
O—O
hν
O
O
O
·O
H
+
O
Blue fluorescence
(5)
Green fluorescence
(6)
Yellow fluorescence
(7)

Scheme 1 Mechanism of photo-stimulated interconversions

orientation of anthracene-9,10-epiperoxide formed by irradiation of the underlying anthracene matrix in an air/carbon disulphide atmosphere[14] is shown in Figure 7. The preferred orientation of anthraquinone/anthrone on an anthra-

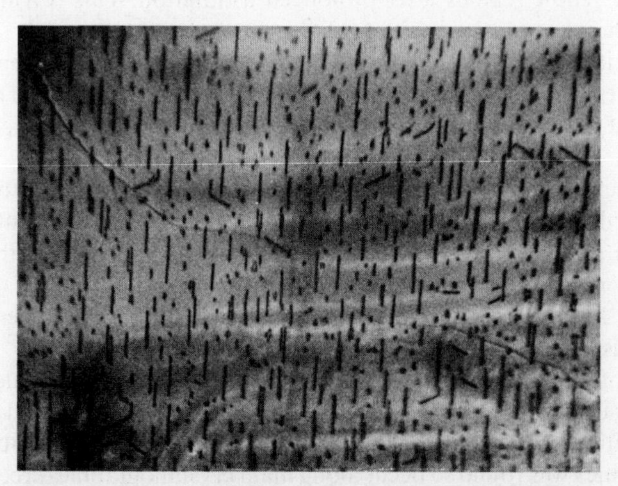

Figure 7 Cleavage surface of anthracene showing preferential growth of the 9,10-epiperoxide upon UV irradiation in air/CS_2 atmosphere for 2×10^3 s.

cene matrix is shown in Figure 8, which was obtained after the irradiation in an air/water atmosphere.[14] Consequently, it appears that in all of the photo-induced reactions involving the anthracene compounds considered the orientation of the products is controlled by the crystallographic structure of the reactant, i.e. the reactions are topotactic.

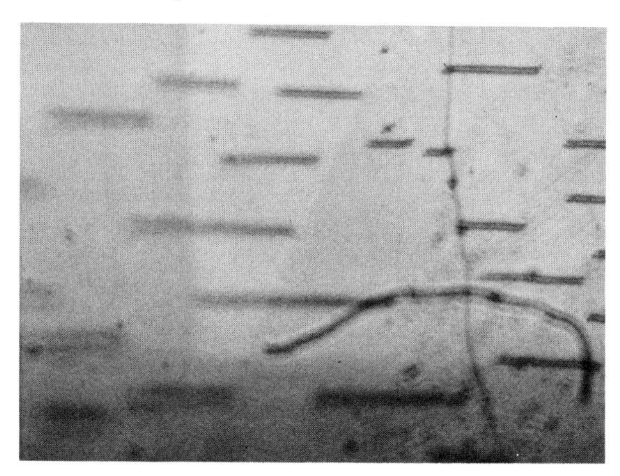

Figure 8 Cleavage surface of anthracene showing preferential growth of anthraquinone anthrone on the peroxide matrix. UV irradiation applied in air/H_2O atmosphere for 2×10^3 s.

Finally, we should emphasize here that the observations of the growth of product (daughter) phases on parent molecular phases in preferred crystallographic directions resemble a similar behaviour during phase transformations in such materials[15,16] and dismiss the generality of isotropic growth at an interface.[17]

REFERENCES

1. A. G. Davies, *Organic Peroxides*, Butterworths, London, 1961.
2. H. H. Wasserman, J. R. Scheffer, and J. L. Cooper, *J. Amer. Chem. Soc.*, **94**, 4991 (1972).
3. D. R. Kearns and A. U. Khan, *Photochem. Photobiol.*, **10**, 193 (1969).
4. M. D. Cohen, *Pure Appl. Chem.*, **9**, 567 (1964).
5. G. M. J. Schmidt, *13th Solvay Conference, New York*, 1967, p. 227.
6. M. D. Cohen, Z. Ludmer, J. M. Thomas, and J. O. Williams, *Proc. Roy. Soc. (Lond.)*, **A324**, 459 (1971).
7. D. Donati, G. G. T. Guarini, and P. Sarti-Fantoni, *Mol. Cryst. Liq. Cryst.*, **21**, 289 (1973).
8. K. Lonsdale, E. Nave, and J. F. Stephens, *Phil. Trans. Roy. Soc. (Lond.)*, **A261**, 1 (1966).
9. K. Lonsdale, *Acta Crystallogr.*, **21**, 5 (1966).
10. C. Duffraisse and M. Gerand, *C.R. Hebd. Séanc. Acad. Sci. Paris*, **201**, 428 (1935).
11. D. Donati, G. G. T. Guarini, and P. Sarti-Fontoni, *Mol. Cryst. Liq. Cryst.*, in press.

12. J. B. Birks in *Organic Molecular Photophysics* (Ed. J. B. Birks), Wiley, London, Vol. 2, p. 409.
13. M. M. Julian, *J. Chem. Soc. Dalton Trans.*, **1972**, 558.
14. W. Jones and S. E. Morsi, unpublished results.
15. J. O. Williams, *J. Mater. Sci.*, **8**, 1361 (1973).
16. W. Jones, J. M. Thomas, and J. O. Williams, *Phil. Mag.*, **32**, 1 (1975).
17. Yn. V. Mnyukh and N. A. Panfilova, *J. Phys. Chem. Solids*, **34**, 159 (1973).

13

Oxidation Reactions of Substrates Containing Vinyl Halide Structures

K. GRIESBAUM, H. KEUL, M. P. HAYES and M. HAJI JAVAD

Lehrstuhl für Petrochemie und Organische Technologie, Universität Karlsruhe, D-7500 Karlsruhe, West Germany

INTRODUCTION

In contrast to the extensive information on the course and on the result of the oxidation of hydrocarbon substrates, surprisingly little is known about the oxidation of halogenated hydrocarbons. In recent years, however, interest in this field has been growing for such varied reasons as the protection of useful products against oxidative degradation and deliberate oxidative degradation of chlorinated waste materials.

Our research in this field originated from an interest in the synthesis, stability, and chemistry of epoxides containing chlorine substituents attached to the carbon atoms of the epoxide ring.[1] Since then, our activities have expanded into a general programme on the oxidation behaviour of saturated, olefinic, and aromatic chlorinated hydrocarbons. In this paper we report on the liquid-phase oxidation of substrates containing vinyl chloride moieties under conditions involving singlet oxygen as well as triplet oxygen and ozone.

OXIDATIONS UNDER SINGLET OXYGEN CONDITIONS

These oxidations were carried out with three differently structured substrates, *viz*. *trans*-1,2-dichloroethylene (**1**), which contains only vinylic hydrogen, 2-chloro-3,3-dimethyl-1-butene (**8**), which contains vinylic and aliphatic hydrogen, and *trans*-2,3-dichloro-2-butene (**11**), which contains only allylic hydrogen. The reactions were carried out at ambient temperature for 72 h in the presence of tetraphenylporphine (TPP) as a sensitizer and under irradiation with a 450-W Hanovia lamp. The reaction mixtures were analysed by gas–liquid chromatography and subsequently distilled in order to assess the amount of non-volatile reaction products; these products comprised between 5 and 10% of the total reaction products in all three cases. The natures of the volatile reaction products are discussed separately below for each substrate.

The major organic products from the reaction of *trans*-1,2-dichloroethylene

(**1**) were the *trans*- (**3**) and, due to isomerization of (**1**) to (**2**), the *cis*-epoxide (**4**), which were formed in yields of approximately 50% and 20%, respectively. Additionally, about 25% of the ester (**5**), small amounts of 1,1,2,2-tetrachloroethane (**6**) and of dichloroacetaldehyde (**7**) as well as the cleavage products HCl and CO were also formed:

$$\text{ClHC=CHCl (trans) } (\mathbf{1}) \xrightarrow{O_2,\ h\nu,\ \text{TPP}} \text{ClHC=CHCl (cis) } (\mathbf{2}) + \text{trans-ClHC(O)CHCl } (\mathbf{3}) + \text{cis-ClHC(O)CHCl } (\mathbf{4})$$

$$+\ Cl_2CHCHClOCH(=O)\ (\mathbf{5}) + Cl_2CHCHCl_2\ (\mathbf{6}) + Cl_2CHCHO\ (\mathbf{7})$$

Oxidation of 2-chloro-3,3-dimethyl-1-butene (**8**) gave epoxide (**9**) and its rearrangement product (**10**) in a total yield of about 75%, together with some other unidentified products:

$$CH_2{=}CCl{-}C(CH_3)_3\ (\mathbf{8}) \xrightarrow[h\nu,\ \text{TPP}]{O_2} CH_2{-}CCl(O){-}C(CH_3)_3\ (\mathbf{9}) + ClCH_2COC(CH_3)_3\ (\mathbf{10})$$

Oxidation of *trans*-2,3-dichloro-2-butene (**11**) also resulted in predominant epoxidation of the chlorinated double bond: the *trans*-epoxide (**13**), *cis*-epoxide (**14**) [derived from simultaneously formed *cis*-2,3-dichloro-2-butene (**12**)] and ketone (**15**) [derived from isomerization of (**13**) and (**14**)] were obtained in yields of 70, 10 and 5%, respectively. Additionally, about 10% of acetyl chloride (**16**) was detected in the reaction mixture. There was, however, no evidence for the formation of products derived from oxidation at the allylic positions, i.e. at the methyl groups:

$$\text{Cl(CH}_3\text{)C=C(CH}_3\text{)Cl (trans) } (\mathbf{11}) \xrightarrow{O_2} \text{Cl(CH}_3\text{)C=C(CH}_3\text{)Cl (cis) } (\mathbf{12}) + \text{trans-epoxide } (\mathbf{13}) + \text{cis-epoxide } (\mathbf{14})$$

$$+\ CH_3CCl_2COCH_3\ (\mathbf{15}) + CH_3COCl\ (\mathbf{16})$$

OXIDATIONS WITH TRIPLET OXYGEN

hese oxidations were carried out under the same conditions as the singlet xygen reactions, except that no sensitizer was used. Major differences between he sensitized and the non-sensitized reactions were observed only in the eaction rates: the former reactions proceeded at approximately double the ate of the latter. The products were, however, the same in both types of eaction and the product distributions were also very similar.

In contrast, the thermal oxidation of *trans*-2,3-dichloro-2-butene (**11**) in the bsence of a source of radiation resulted in different products and different roduct distributions: oxidation at 90 °C and in the absence of a solvent gave e dichloroketone (**15**) as the major reaction product, together with the ep- xides (**13**) and (**14**), acetyl chloride (**16**) and compounds (**18**)–(**20**). The preferred ormation of the dichloroketone (**15**) is due to the accelerated rearrangement of poxides (**13**) and (**14**) at elevated temperatures, whereas the formation of ompounds (**18**)–(**20**) is due to partial allylic oxidation of the substrate (**11**) to orm the alcohol (**17**), which in turn reacts further with acetyl chloride and with xygen to form esters (**18**) and oxidation products (**19**) and (**20**), respectively:

$CH_3CCl{=}CClCH_2OH$

(**17**) (*cis, trans*)

$CH_3CCl{=}CClCH_2OC({=}O)CH_3$

(**18**) (*cis, trans*)

$CH_3CCl{=}CClCH{=}O$

(**19**) (*cis, trans*)

$CH_3CCl{=}CClCOOH$

(**20**) (*cis, trans*)

The most striking result of the above oxidation experiments is that, in contrast o the behaviour of non-chlorinated olefins, epoxide formation was the pre- ominant reaction in each case. The excess of the epoxide products over other roducts such as allylic oxidation products or chain-cleavage products is ndicative of a direct attack of oxygen on the double bond, whereas less direct outes to epoxide formation, such as the intermediacy of hydroperoxides[2] or of ligomeric peroxides,[3] would produce balanced amounts of by-products. nother unexpected result is that under both singlet and triplet oxygen con- itions the same products were formed in comparable amounts. Furthermore, here was no direct evidence for the occurrence of typical singlet oxygen eactions such as the formation of dioxetanes or ene-type reactions. Although he occurrence of double-bond cleavage products could in principle be ex- lained by such singlet oxygen reactions, they can also be rationally explained y free-radical reaction paths. It is, therefore, questionable whether singlet xygen plays an important role in the above sensitized reactions of oxygen vith the chlorinated substrates (**1**), (**8**), and (**11**). It is conceivable and it has not een ruled out that, owing to the low reaction rate, a considerable degree of lecay of singlet oxygen may have occurred prior to its attack on the halo- enated substrates.

OXIDATIONS WITH OZONE[4]

These oxidations were carried out with *trans*-2,3-dichloro-2-butene (**11**) in inert and in reactive solvents, and with 2-chloro-3,3-dimethyl-1-butene (**8**) in reactive solvents. The results of these two types of ozonolysis reactions were very different and will, therefore, be discussed separately.

Ozonolysis of (**11**) in inert solvents[5] (pentane, chloroform, and methyl formate) gave essentially three types of products: (a) cleavage products (**16**) and (**21**); (b) peroxides (**22**), (**23**) and (**24**), in which the carbon atoms of the original double bond are still linked together via labile peroxide bonds; and (c) compounds (**13**) and (**15**), in which the original double bond has been converted into single bonds which are resistant towards further ozone attack. The formation of these ozonolysis products can readily be explained by a Criegee-type cleavage.[6]

$$CH_3-C(Cl)=C(Cl)-CH_3 \xrightarrow{O_3} \begin{cases} CH_3COCl\ (\mathbf{16}) + CH_3COOH\ (\mathbf{21}) \\ CH_3C(=O)OOC(=O)CH_3\ (\mathbf{22}) + CH_3C(=O)OOCCl_2CH_3\ (\mathbf{23}) + CH_3(Cl)C\langle{}^{O-O}_{O-O}\rangle C(Cl)CH_3\ (\mathbf{24}) \\ (CH_3)(Cl)C\overset{O}{-}C(Cl)(CH_3)\ (\mathbf{13}) + CH_3CCl_2COCH_3\ (\mathbf{15}) \end{cases}$$

Ozonolysis of either (**11**) or (**8**) in reactive solvents (alcohols, water) resulted in the complete cleavage of the double bonds and in the formation of the corresponding acyl derivatives (**25**) and (**26**)/(**27**), respectively.

$$\underset{(\mathbf{11})}{CH_3CCl=CClCH_3} \xrightarrow[ROH]{O_3} \underset{(\mathbf{25})}{2\,CH_3COOR}$$

$$\underset{(\mathbf{8})}{CH_2=C(Cl)-C(CH_3)_3} \xrightarrow[ROH]{O_3} \underset{(\mathbf{26})}{(CH_3)_3CCOOR} + \underset{(\mathbf{27})}{HCOOR}$$

Similar results were obtained with a number of additional olefins containing one or two chloro substituents in vinylic positions. The results can again be explained by a Criegee-type cleavage of the substrates. In the case of the dichloro-substituted olefins [e.g. (**11**)], the products are formed by straightforward solvolysis reactions of the primary fragments (**16**) and (**28**):

$$CH_3CCl{=}CClCH_3 \ (\mathbf{11}) \xrightarrow{O_3} \begin{cases} CH_3COCl \ (\mathbf{16}) \xrightarrow{ROH} CH_3COOR \ (\mathbf{25}) + HCl \\ CH_3\overset{+}{C}ClO\bar{O} \ (\mathbf{28}) \xrightarrow{ROH} CH_3{-}C(Cl)(OR){-}O{-}OH \longrightarrow CH_3COOR \ (\mathbf{25}) + HOCl \end{cases}$$

In the case of monochloro-substituted olefins [e.g. **(8)**] the quantitative formation of acyl fragments [e.g. **(26)** and **(27)**] is due to (a) the selective cleavage of the double bond into an acyl chloride fragment **(29)** and into a non-chlorinated zwitterion fragment **(30)**, and (b) an acid-catalysed internal redox reaction at the methoxy hydroperoxide **(31)**:[7]

$$CH_2{=}C(Cl)C(CH_3)_3 \ (\mathbf{8}) \xrightarrow{O_3} \begin{cases} (CH_3)_3CCOCl \ (\mathbf{29}) \xrightarrow{ROH} (CH_3)_3CCOOR \ (\mathbf{26}) \\ H{-}\overset{+}{C}H{-}O\bar{O} \ (\mathbf{30}) \xrightarrow{ROH} H{-}CH(OR){-}OOH \ (\mathbf{31}) \xrightarrow{H^+} H{-}CH(OR){-}O_2H_2^+ \longrightarrow H{-}C({=}O){-}OR \ (\mathbf{27}) \end{cases}$$

REFERENCES

1. K. Griesbaum, R. Kibar, and B. Pfeffer, *Justus Liebigs Ann. Chem.*, 214 (1975).
2. W. F. Brill, *Adv. Chem. Ser.*, No. 51, 70 (1965).
3. D. E. van Sickle, F. R. Mayo, R. M. Arluck, and M. G. Syz, *J. Amer. Chem. Soc.*, **89**, 967 (1967).
4. R. W. Denny and A. Nickon, *Organic Reactions*, Vol. 20 (Ed. W. G. Dauben *et al.*), Wiley, New York, 1973, p. 141.
5. K. Griesbaum and P. Hofmann, *J. Amer. Chem. Soc.*, **98**, 2877 (1976).
6. R. Criegee, *Angew. Chem.*, **87**, 765 (1975); *Angew. Chem. Int. Ed. Engl.*, **14**, 745 (1975).
7. K. Griesbaum and H. Keul, *Angew. Chem.*, **87**, 748 (1975); *Angew. Chem. Int. Ed. Engl.*, **14**, 716 (1975).

14

Singlet Oxygen Reactions with Polyisoprene Model Molecules: 4-Methyl-4-Octene and 4,8-Dimethyl-4, 8-Dodecadiene

J. CHAINEAUX and C. TANIELIAN

Laboratoire de Chimie Organique Appliquée,
Laboratoire Associé au C.N.R.S. No. 81,
Université Louis Pasteur, Institut de Chimie,
1 rue Blaise Pascal, 67008 Strasbourg, France

INTRODUCTION

It was recently pointed out that singlet oxygen may have great importance in the photodegradation of polymers,[1] for example by reacting with unsaturated bonds to give allylic hydroperoxides:

$$R{-}CH_2{-}CH{=}CH{-}R' + {}^1O_2 \longrightarrow R{-}CH{=}CH{-}\underset{\displaystyle OOH}{\underset{|}{CH}}{-}R' \qquad (1a)$$

These unsaturated bonds may already exist, or may be formed in the course of a previous degradation.

Our aim was to investigate the behaviour of a polymer containing unsaturated bonds (polyisoprene) when it is exposed to the action of singlet oxygen produced by photosensitization.

Some work has been carried out on polydienes[1,2] but the results were very general and some assumptions still remain to be tested. This is the reason why we decided to study the sensitized photo-oxygenation of polyisoprene model molecules in order to distinguish the elementary processes and to characterize possible reaction products and the reactivity of the olefinic compounds.

As the polyisoprene model molecules with a 1,4-unit, we used particularly 4-methyl-4-octene (M4O4), and 4,8-dimethyl-4,8-dodecadiene (DMDD), corresponding to polymerization numbers of one and two, respectively:

$$Et{-}\!\!\left[CH_2{-}\overset{\displaystyle CH_3}{\overset{|}{C}}{=}CH{-}CH_2\right]_n\!\!Et \qquad n=1: M4O4 \quad n=2: DMDD \qquad (1b)$$

These molecules represent most closely the environment of the isoprene unit as it occurs in the macromolecule since the ethyl groups introduce methylene groups on each side of the unit and furthermore bring conditions of quasi-neutrality owing to their low inductive and electromotive forces.

We describe here some results on the sensitized photo-oxygenation of these model molecules and we especially attempt to answer some questions, for instance:

(a) Can several singlet oxygen molecules add to the same unit and, if so, what is the result of the first addition on the reactivity towards a second addition?

(b) What is the effect of the addition of singlet oxygen to a unit on the reactivity of a neighbouring unit towards $^1O_2$?

(c) Is it possible to characterize monoendodihydroperoxide corresponding to the triaddition of oxygen to two neighbouring groups (Kaplan and Kelleher's hypothesis[3]?

EXPERIMENTAL

Synthesis of Model Molecules

A method described by Pinazzi and Reyx[4] was used on the basis of Carroll and Wittig reactions.

Irradiations

Reactions were performed in the presence of Rose Bengal (3 g l^{-1}) in solutions containing from 5×10^{-3} to 5×10^{-2} mol l^{-1} of olefinic compound in methanol, at -20 °C so that degradation of the resultant hydroperoxides did not occur.

The UV lamps used were mercury vapour lamp (middle pressure Philips HPK, 125 W). The irradiation apparatus was of the immersion type, a special device allowing oxygen consumption to be followed.

We confirmed that processes which occur at the time of a UV irradiation in an inert atmosphere are very slow in comparison with photo-oxygenation and therefore can be neglected. We also confirmed that, under our experimental conditions, no photolysis of the hydroperoxides occurred.

Analytical Techniques

The usual analytical techniques were found to be ineffective and we applied a method based on analysis by gas chromatography, mass spectrometry, and combined gas chromatography–mass spectrometry using the trimethysilyl ethers of the resultant hydroperoxides.

β-Determination

The simplest course of the oxidation of a substrate A with singlet oxygen can be written as

$$^{1}O_2 + A \xrightarrow{k_R} A\,O_2$$

$$^{1}O_2 \xrightarrow{k_D} {}^{3}O_2$$

$\beta = k_D/k_R$ values characterize the reactivity of a substrate towards singlet oxygen. In order to obtain the values of β we used the method of Schenck *et al.*[5] based on a comparison of substrate A with a very reactive reference olefin, the quantum yield of which is well known: α-terpinene was chosen for this purpose.

RESULTS AND DISCUSSION

4-Methyl-4-octene

Monohydroperoxide formation

As 4-methyl-4-octene is a trisubstituted olefin, and considering the number of allylic hydrogen atoms, it might be expected that its reaction with singlet oxygen would lead to three different monohydroperoxides. In fact, the photo-oxygenation of a mixture of *cis*- and *trans*-isomers leads to a mixture of allylic hydroperoxides, *trans*-5-methyl-4-hydroperoxy-5-octene (**1**), 5-methylene-4-hydroperoxyoctane (**2**) and *trans*-4-methyl-4-hydroperoxy-5-octene (**3**):

(**1**) (**2**) (**3**)

The results are given in the Table 1. It can be stated that the distribution of the products depends on the initial isomeric composition of M4O4, and that product (**1**), which contains a trisubstituted double bond, is present in small amounts in

Table 1 Distribution of products formed in the photo-oxygenation of mixtures of *cis*- and *trans*-4-methyl-4-octene

M4O4 Isomers	β (mol l^{-1})	Distribution of products formed (%)		
		(**1**)	(**2**)	(**3**)
100% *trans*	0·20	11	48	41
46% *trans*, 54% *cis*	0·27	9	38	53
25% *trans*, 75% *cis*	0·29	7	33	60

every case. It can also be seen that the *trans*-isomer is the most reactive. Moreover, a complete kinetic study indicated that the reactivity of the *trans*-isomer towards singlet oxygen is 1·7 times greater than that of the *cis*-isomer.[6]

Dihydroperoxide formation. Influence of Oxygenated Functions on the Reactivity of Unsaturated Bonds

It is known that only tri- or tetrasubstituted double bonds show good reactivity towards singlet oxygen and that disubstituted double bonds are almost unreactive. As shown in reaction (**1**) the photo-oxygenation of an olefin results in the formation of a hydroperoxide and in a 'displacement' of the double bond which is at first trisubstituted and then may become mono- or disubstituted or again trisubstituted in the hydroperoxide molecule; in the third case one may wonder whether this double bond may react once more with singlet oxygen.

Thus, in the case of M4O4, 5-methyl-4-hydroperoxy-5-octene (**1**) has a trisubstituted double bond and differs from 4-methyl-3-octene only by the presence of the hydroperoxy group. Therefore, the problem becomes that of the influence of a hydroperoxy group (or more generally of an oxygenated group) in the α-position to the double bond on the rate of photo-oxygenation of an olefin and on the nature of the resulting products.

Formation of dihydroperoxides. We carried out a prolonged photo-oxygenation of M404 and monitored its progress by measuring the volume of oxygen consumed. After a relatively long period we observed the following results: a considerable decrease in the concentration of 5-methyl-4-hydroperoxy-5-octene (**1**) whereas that of both other monohydroperoxides remained constant; and the appearance of two new products, namely 4-methyl-4,5-dihydroperoxy-2-octene (**4**) and 4-methylene-3,5-dihydroperoxy octane (**5**) in the proportions 1:4.

OOH $\xrightarrow{^1O_2}$ HOO OOH + OOH OOH (3)

(**1**) (**4**) (**5**)

Thus, a second addition of singlet oxygen to an olefin is possible by the intermediacy of a monohydroperoxide with a trisubstituted double bond. However, it is much slower than the first addition. In fact, a kinetic study showed that the reactivity of (**1**) towards $^1O_2$ is about 20 times weaker than that of *cis*-M4O4.[6]

Influence of oxygenated groups on the reactivity of a double bond. Most studies described in the literature are related to the influence of ester or alcohol functional groups in the α-position to a double bond. Acetate and benzoate esters are completely inert towards singlet oxygen (for instance, Dunphy[7] found only two hydroperoxides as products of the photo-oxygenation of geranyl acetate, the double bond in the α-position to the ester remaining untouched); and in photo-oxygenation, the esterification of an allylic alcohol function may be considered as a good protection.

The double bond in an allylic alcohol is indeed not completely unreactive even, according to Nickon and Mendelson[8,9] when it is considerably deactivated, as in cholest-4-en-3β-ol. They believe that this result is not due to a steric effect but rather to an electroattractive effect of the oxygenated function pointing out the statement of Schenck *et al.* that the active species ($^1O_2$) shows an unquestionable electrophilic character.

As to the nature of the products, Nickon and Mendelson reported the formation of α,β-epoxyketones (**7**) and of α,β-ethylenic ketones (**8**):

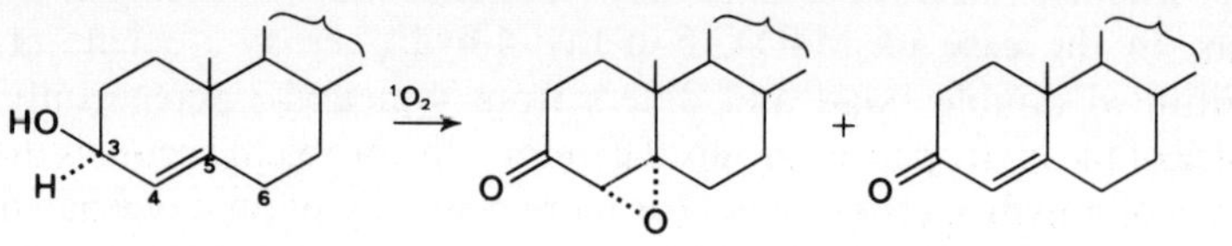

(4)

Cholest-4-en-3β-ol
(**6**) (**7**) (**8**)

It should be noted that these products involve the reaction of only the allylic hydrogen on C_3 and not that of the two hydrogen atoms on C_6 which should yield an α-hydroperoxy-β-ol-ene (whatever the conformation may be, one of these hydrogen atoms has a quasi-axial position).

Therefore, with regard to the nature of the products, a difference appears between Nickon and Mendelson's results and ours concerning the hydroperoxide (**1**). This may be due to the nature of the substrates (cyclic sterol and aliphatic hydroperoxide) and to the difference between the two oxygenated functions (hydroxy and hydroperoxy).

It seems, however, that this second explanation can be discounted on account of the results that we obtained for the photo-oxygenation of the allylic alcohol 2-methyl-2-penten-4-ol (**9**). This substrate yields two products relatively slowly: 4-methyl-3-hydroperoxy-4-penten-2-ol (**10**) in a large amount (98%) and 4-methyl-4-hydroperoxy-2-penten-2-ol (**11**) in equilibrium with the tautomeric 4-methyl-4-hydroperoxy-2-pentanone (**12**):

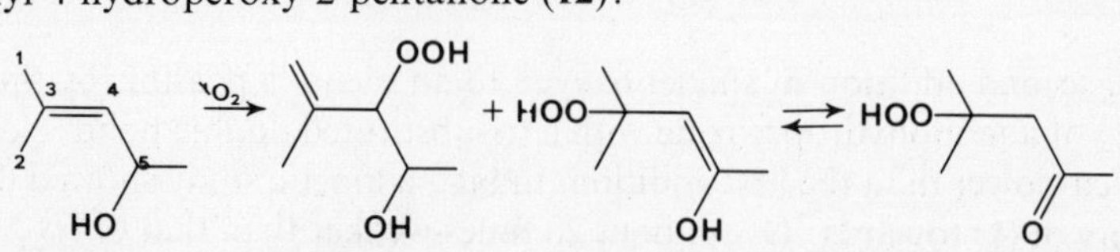

(5)

$\beta = 8{\cdot}4$ 98% 2%
(**9**) (**10**) (**11**) (**12**)

When considering only the main product we observe that, as in the case of M404 and contrary to Nickon and Mendelson's results, the photo-oxygenation products of 2-methyl-2-penten-4-ol are formed by a reaction of the allylic hydrogens of C_1 and C_2, and not by a reaction of the C_5 hydrogen, C_5 bearing the alcohol group. Concerning the distribution of the products, we found that 2-methyl-2-pentene (**13**), in comparison with 2-methyl-2-penten-4-ol (**9**), yields relatively rapidly secondary (**14**) and tertiary hydroperoxides (**15**) in equal amounts.

$$\text{(13)}\ \beta = 0.16 \xrightarrow{^1O_2} \text{(14)}\ 49\% + \text{(15)}\ 51\% \qquad (6)$$

The deactivation induced by the presence of the OH group in the allylic position can be characterized quantitatively. The measurement of β for 2-methyl-2-penten-4-ol gives a value of 8·4 mol l^{-1}, which shows a deactivation factor of 50 with respect to 2-methyl-2-pentene ($\beta = 0{\cdot}16$ mol l^{-1}).

This result can be compared with that obtained with hydroperoxide (**1**) for which the deactivation factor with respect to *cis*-M4O4 is 20. This difference may result from a lower electroattractive effect for the hydroperoxy group than for the hydroxyl group.

4,8-Dimethyl-4,8-dodecadiene

Mono- and dihydroperoxide formation

The study of 4,8-dimethyl-4,8-dodecadiene (DMDD) as a model compound with two isoprene units in a 1,4–1,4 addition mode is complex. This product has four geometrical isomers and the synthetic route used leads to a mixture of these isomers in approximately equal amounts. Moreover, the molecule has no symmetry and there are six possible products for the first oxygen addition (each of them giving *cis-trans* isomerism), and nine possible products for the second addition, with a unique *trans* structure because of conformational factors.

The products of the photo-oxygenation of DMDD are hydroperoxides; the consumption of oxygen is slightly more than 2 mol per mole of DMDD.

An incomplete photo-oxygenation of the substrate allows one to conclude that each of the following expected monohydroperoxides is actually formed: 4,8-dimethyl-4-hydroperoxy-5*tr*,8-dodecadiene (**16**), 4-methylene-8-methyl-5-hydroperoxy-8-dodecene (**17**), 4-methyl-8-methylene-9-hydroperoxy-4-dodecene (**18**), 4,8-dimethyl-5-hydroperoxy-3*tr*,5-dodecadiene (**19**), 4,8-dimethyl-8-hydroperoxy-4,9*tr*-dodecadiene (**20**), and 4,8-dimethyl-9-hydroperoxy-4,7*tr*-dodecadiene (**21**), where in each instance *tr* following a position number indicates a *trans* double bond at that position. Hydroperoxides with a trisubstituted double bond are less abundant.

(7)

NOT IDENTIFIED

(22) (23) (24) (25) (26) (27) (28) (29)

$^1O_2$

(16) (17) (18) (19) (20) (21)

$^1O_2$

Among the diaddition products we have identified the following: 4-methylene-8-methyl-5,8-dihydroperoxy-9*tr*-dodecene (**22**), 4,8-dimethyl-4,8-dihydroperoxy-5*tr*,9*tr*-dodecadiene (**23**), 4,8-dimethylene-5,9-dihydroperoxydodecane (**24**), and 4-methyl-8-methylene-4,9-dihydroperoxy-5*tr*-dodecene (**25**).

We did not identify five other products that are probably formed, which contain at least one trisubstituted double bond: 4-methyl-8-methylene-5,9-dihydroperoxy-3*tr*-dodecene (**26**), 4-methylene-8-methyl-5,9-dihydroperoxy-7*tr*-dodecene (**27**), 4,8-dimethyl-5,8-dihydroperoxy-3*tr*,9*tr*-dodecadiene (**28**), 4,8-dimethyl-4,9-dihydroperoxy-5*tr*,7*tr*-dodecadiene (**29**), and 4,8-dimethyl-5,9-dihydroperoxy-3*tr*,7*tr*-dodecadiene (**30**).
We assume, however, that these compounds are actually formed in small amounts, for reasons that we shall discuss below.

Thus, in the photo-oxygenation of DMDD, 1 mol of oxygen reacts with 1 mol of DMDD to give a monohydroperoxide and consecutively a second mole of oxygen adds to the intact double bond to give a dihydroperoxide. Moreover, in a kinetic study[6] we assumed the following course:

$$\text{DMDD} \xrightarrow[k_1]{^1O_2} \text{DMDD(OOH)} \xrightarrow[k_2]{^1O_2} \text{DMDD(OOH)}_2 \tag{8}$$

The results show that $k_2/k_1 \approx 0{\cdot}4$; this value indicates that an addition of oxygen to a unit deactivates further additions to other non-oxygenated units.

Further Addition of Oxygen (more than one molecule on a unit)

Bi-addition to the same unit. We carried out a photo-oxygenation of DMDD until oxygen was no longer consumed, and we observed the formation of the following products: 8-methyl-4-methylene-5,8,9-trihydroperoxy-6*tr*-dodecene (**16**), 4,8-dimethyl-4,5,8-trihydroperoxy-2*tr*,9*tr*-dodecadiene (**17**), and 4-methyl-8-methylene-4,5,9-trihydroperoxy-2*tr*-dodecene (**18**).

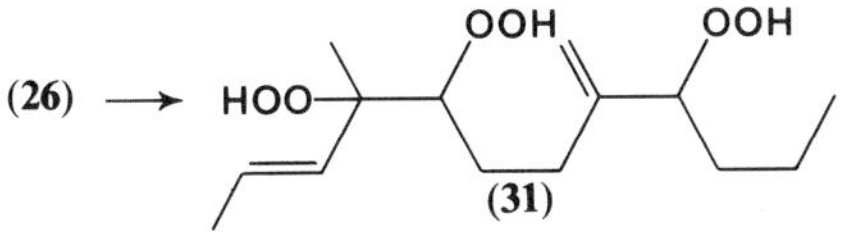

(**26**) → (**31**) (9)

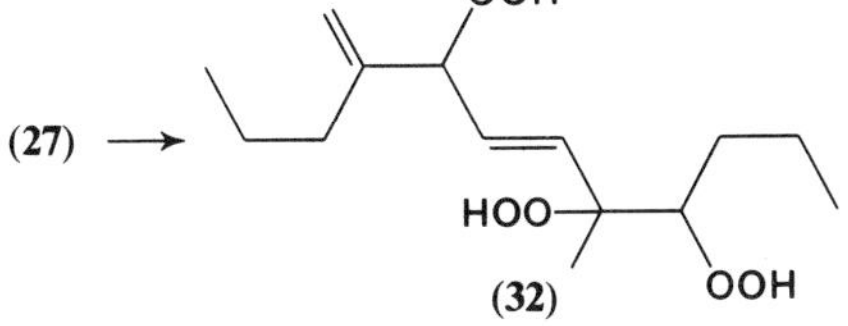

OOH OOH HOO

(**27**) → (**32**) (10)

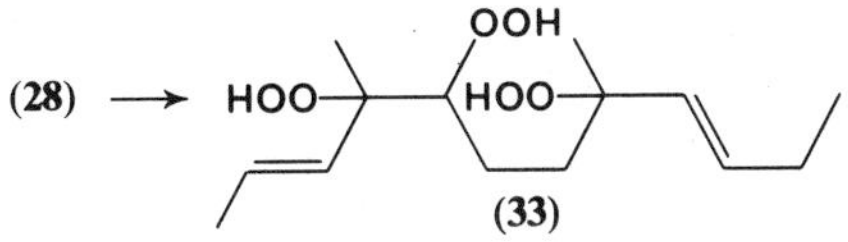

OOH HOO OOH

(**28**) → (**33**) (11)

OOH HOO HOO

This corroborates the formation of dihydroperoxides (**26**), (**27**), and (**28**) that we have observed previously and indicates that a diaddition to the same unit is possible, but to a hydroperoxide that has a trisubstituted double bond, which is formed in relatively small amounts.

We point out that we did not observe the formation of dihydroperoxides corresponding to a diaddition of oxygen to the same unit, the other remaining untouched. This result seems normal, considering that, as we stated for M404, an allylic hydroperoxide deactivates an unsaturation.

Tri-addition to two adjoining groups. It is well known that singlet oxygen reacts with a conjugated diene by 1,4-cycloaddition to yield an endoperoxide and this is the reason why Kaplan and Kelleher[3] proposed the following mechanism for singlet oxygen oxidation of *cis*-polybutadiene:

```
—CH2      H2C—CH2      H2C—CH2      H2C—
    \    /       \    /       \    /
    HC=CH        HC=CH        HC=CH
               | 1O2
               ↓
—CH2      HC—CH        H2C—CH2      H2C—
    \    //            /      \    /
    HC—CH        HC—CH        HC=CH
    |                |
    OOH              OOH
               | 1O2
               ↓
—CH2      HC=CH        H2C—CH2      H2C—
    \    /     \      /       \    /
    HC—CH        HC—CH        HC=CH          (12)
    |    \      /    |
    OOH   O—O        OOH
```

Now, among the preceding photo-oxygenation products, the dihydroperoxide (**29**) has the structure of a conjugated diene and the question arises of whether such a reaction is possible with this compound. In fact, we did not observe the formation of an endoperoxide, contrary to Kaplan and Kelleher's hypothesis, according to which a triaddition on two adjoining groups might take place. This observation, confirmed with other model molecules,[6] may be explained by the weak reactivity of aliphatic conjugated systems and by the deactivating effect of —OOH allylic groups.

CONCLUSIONS

Singlet oxygen may have great importance in the photodegradation of polymers and especially of polyisoprene. For this reason, we studied the photosensitized oxygenation of 4-methyl-4-octene (M4O4) and of 4,8-dimethyl-4,8-dodecadiene (DMDD), model molecules of polyisoprene with a 1,4-addition mode corresponding to polymerization numbers of one and two, respectively.

The study of M4O4 indicated that the isoprene unit may yield three types of hydroperoxides, their distribution depending on the stereochemistry of the double bond, but a hydroperoxide having a trisubstituted double bond is always formed in small amounts. Studies on the model with two units led to the same conclusions; each 1,4-unit may fix singlet oxygen to give three types of allylic hydroperoxides, one of them with a trisubstituted double bond always in small amount (neighbouring groups may have reacted or not). A second addition of singlet oxygen to the same unit is again possible, with the hydroperoxide with a trisubstituted double bond.

We have shown that the fixation of an oxygen molecule to a unit deactivates the double bond shifted to the α-position with respect to the hydroperoxy group formed, and also deactivates the double bond of unattacked neighbouring units.

We were however, unable to characterize monoendodihydroperoxide corresponding to a triaddition of oxygen on a group of two adjoining units.

REFERENCES

1. B. Rånby and J. F. Rabek, *Photodegradation, Photo-oxidation and Photostabilization of Polymers*, Wiley, London, 1975, and references therein.
2. D. J. Carlsson and D. M. Wiles, *Rubb. Chem. Technol.*, **47**, 991 (1974).
3. M. L. Kaplan and P. G. Kelleher, *J. Polym. Sci. A1*, **8**, 3163 (1970).
4. C. Pinazzi and D. Reyx, *Bull. Soc. Chim. Fr.*, **10**, 3930 (1972).
5. G. O. Schenck, K. Gollnick, and O. A. Neumuller, *Justus Liebigs Ann. Chem.*, **603**, 46 (1957).
6. J. Chaineaux and C. Tanielian, unpublished results.
7. P. J. Dunphy, *Chem. Ind., Lond.*, **1971**, 731.
8. A. Nickon and W. L. Mendelson, *J. Amer. Chem. Soc.*, **85**, 1894 (1963).
9. A. Nickon and W. L. Mendelson, *J. Amer. Chem. Soc.*, **87**, 3921 (1965).

15

Reactions of Ketenes with Molecular Oxygen. Formation of Peroxylactones and a Polymeric Peroxide. Catalysed Thermal Generation of Singlet Oxygen

N. J. TURRO, MING-FEA CHOW, and Y. ITO

Chemistry Department, Columbia University, New York, N.Y. 10027, USA

INTRODUCTION

Scattered reports of reactions of molecular oxygen with ketenes appear in the literature (e.g. refs. 1–3). The only systematic study of this class of reactions appears to be that of Staudinger *et al.*[4] who reported that aryl and alkyl ketenes are oxidized to ketones and CO_2 by air. They also proposed that in the case of dimethyl ketene, a peroxide of structure (**2**) is formed from the liquid ketene and air. This material was found to decompose violently and explosively when solidified at room temperature. In solution, however, it undergoes smooth and quantitative decomposition to acetone and CO_2.

$$(CH_3)_2C{=}C{=}O + \text{air} \longrightarrow \left[-C(=O)-C(CH_3)_2-O-O- \right]_n \xrightarrow{\text{Solid}} \text{explosion} \quad (1)$$

$$\mathbf{(1)} \qquad\qquad \mathbf{(2)} \xrightarrow{\text{Solution}} (CH_3)_2C{=}O + CO_2 \quad (2)$$

If structure (**2**) were correct, this polymeric peroxide would be a potential precursor to α-hydroxy peracids, compounds which are difficult to obtain by alternative routes, but which are the only known and available precursors to isolatable peroxylactones. Several peroxy lactones have been prepared in an elegant fashion and characterized by Adam and co-workers;[5,6] they all show a strong C=O absorption near 1875 cm^{-1}. We have re-investigated the reaction of dimethyl ketene with molecular oxygen in order to establish the structure of the product of Staudinger *et al.* convincingly by modern spectroscopic techniques. In addition, we wanted to extend their work to reactions of ketenes with another form of molecular oxygen that was not readily available to Staudinger *et al.*, namely singlet molecular oxygen. We report here the results of our studies on the reactions of triplet and singlet molecular oxygen with dimethyl and diphenyl ketenes.

RESULTS

Treatment of solutions of either dimethyl ketene (**1**) [or diphenyl ketene (**3**)] with air results in the quantitative formation of acetone (or benzophenone) and CO_2.

$$R_2C{=}C{=}O + \text{air} \longrightarrow R_2C{=}O + CO_2 \tag{3}$$

(**1**) $R=CH_3$
(**3**) $R=Ph$

The same products are obtained when singlet oxygen generators (phosphite ozonides or low-temperature photosensitization) are decomposed in solutions containing (**1**) or (**3**):

$$R_2C{=}C{=}O + (R\,O)_3PO_3 \longrightarrow R_2C = O + CO_2 \tag{4}$$

Detectable intermediates are involved in all of these reactions except for the reaction of air or oxygen with (**3**). The structures assigned to these intermediates are (**2**), (**4**), and (**5**). The assignments were based on (1) the structure of products produced upon warming to room temperature, (2) chemiluminescence measurements, (3) chemical properties, and (4) spectroscopic (low-temperature infrared with NMR) analysis.

$$(CH_3)_2C = C = O + {}^3O_2 \longrightarrow (\mathbf{2}) \tag{5}$$

$$(CH_3)_2C = C = O + {}^1O_2 \longrightarrow CH_3{-}\underset{CH_3}{C}{-}C{=}O \text{ (with O–O bridge: 1,2-dioxetanone)} \tag{6}$$

(**4**)

$$Ph_2C = C = O + {}^1O_2 \longrightarrow Ph{-}\underset{Ph}{C}{-}C{=}O \text{ (with O–O bridge: 1,2-dioxetanone)} \tag{7}$$

(**5**)

Reaction of neat (**1**) with oxygen produced a white solid which is stable to filtration below -20 °C. Upon warming to room temperature, this material is prone to undergo sudden, violent, and unpredicatable explosions. We consider this behaviour to be sufficient evidence to identify the material to which Staudinger *et al.*[4] assigned a polymeric peroxide structure. The properties of this material in solution (Freon, $CDCl_3$) are completely consistent with the Staudinger *et al.* assignment:

(1) Warming of $CDCl_3$ solutions of the white solid to room temperature produced acetone quantitatively (NMR). CO_2 was detected as the co-product of reaction by IR analysis (2350 cm^{-1}).

(2) The infrared spectrum of solutions (below -30 °C) of the white solid show a strong absorption at 1780 cm^{-1} and the NMR (-30 °C) shows only a single unsplit part at $\delta 1{\cdot}60$. There is no evidence for the formation of a $C{=}CH_2$ (NMR) or OH group (IR).

(3) Warming of a solution of the white solid produced a chemiluminines-cence spectrum identical with that of acetone (fluorescence).

Reaction of dilute (10^{-1} M) solutions at -20 °C of (**1**) with phosphite ozonides leads to completely different results. Firstly (**2**) is not formed (NMR, IR at below -30 °C). Secondly, in addition to the appearance of acetone (IR at below -30 °C) an IR band at 1870 cm^{-1} appears immediately.[5, 6] The samples are chemiluminescent (acetone fluorescence).

These data are uniquely consistent with the formation of (**4**). Addition of $^1O_2$ to an equimolar mixture of (**1**) and tetramethylethylene (**6**) with phosphite ozonide leads mainly to oxidation of the ketene. This means that $^1O_2$ is much more reactive toward (**1**) than (**6**).

Treatment of solutions of diphenyl ketene (**3**) with air leads to formation of benzophenone. The reaction mixtures are chemiluminescent, analogous to the situation observed for (**1**). Treatment of solutions of (**3**) with phosphite ozonides (at below -30 °C) results in the formation of a transient species possessing an IR absorption at 1878 cm^{-1}. This species disappears in a few minutes at room temperature. By analogy with the results with (**1**), we presumed that this transient species was the peroxy lactone (**5**). In agreement with this premise, even after total destruction of the phosphite ozonide (>2 h at -23 °C for triphenyl phosphite ozonide), a chemiluminescence is observed (perylene acceptor) which decays with a half-life at room temperature similar to the decay of the 1878 cm^{-1} IR band.

If (**5**) is formed as indicated in reaction (7), then it may also be formed at -78 °C using photosensitized generation of singlet oxygen. Photosensitized oxidation (polymer Rose Bengal) of (**3**) at -78 °C (filter cutting off below 520 nm), results in nearly exclusive formation of (**5**) (IR, no benzophenone). In the reaction with singlet oxygen, (**3**) reacts three times faster than 9,10-dimethylanthracene (competing photo-oxidations in CH_2Cl_2 at -50 °C or treating with triphenyl phospite ozonides at 0 °C).

Addition of 2-methyl-2-butene to a solution of (**5**), prepared by treatment of (**3**) with triphenyl phosphite ozonide, followed by warming to room temperature in degassed solution, results in formation of oxetanes (**7**) and (**8**). Quantitative comparison with the formation of (**7**) and (**8**) from photochemical excitation of benzophenone[7] indicates that excited benzophenone is formed in 35% yield from (**5**).

(**5**) + 2-methyl-2-butene $\xrightarrow{\Delta}$ (**7**) + (**8**) + CO_2 (8)

DISCUSSION

The salient features of the results obtained in these studies are:

(1) the dichotomy of reaction of $^3O_2$ and $^1O_2$ with dimethyl ketene;
(2) the formation of a peroxy ester polymer upon reaction of $^3O_2$ with dimethyl ketene;
(3) the formation of peroxylactones when dimethyl or diphenyl ketene is treated with $^1O_2$;
(4) the production of a high yield of excited states when (**5**) is decomposed.

We shall discuss two features of the mechanism for reaction of molecular oxygen with ketenes: (1) the basis for attack of an electrophilic species (molecular oxygen) on a class of compounds whose chemistry usually defines them as electrophiles (ketenes); (2) the mechanism of the spin inversion which must occur at some stage in the reactions of $^3O_2$ with ketenes.

The reaction of molecular oxygen and ketenes probably occurs via a stereoelectronic course similar to the reaction of ozone with ketenes.[8, 9] The latter class of compounds normally behave as electrophiles (rationalized as the result of developing allyl anion character when nucleophiles add to the carbonyl carbon by attack in the plane of the molecule) (Figure 1).[10]

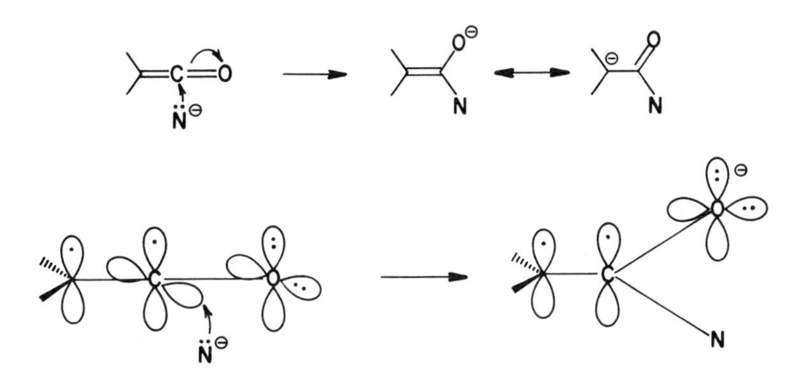

Figure 1 Model for the stereochemistry of addition of nucleophiles to ketenes

On the other hand, electrophiles should attack above (or below) the molecular plane (Figure 2).

A mechanism of the latter type is probably also operating in the addition of ozone to ketenes.[8, 9]

The reaction of ground-state oxygen to produce benzophenone and a chemiluminescent species, possibly (**5**) [equation (9)] may be explained in terms of a

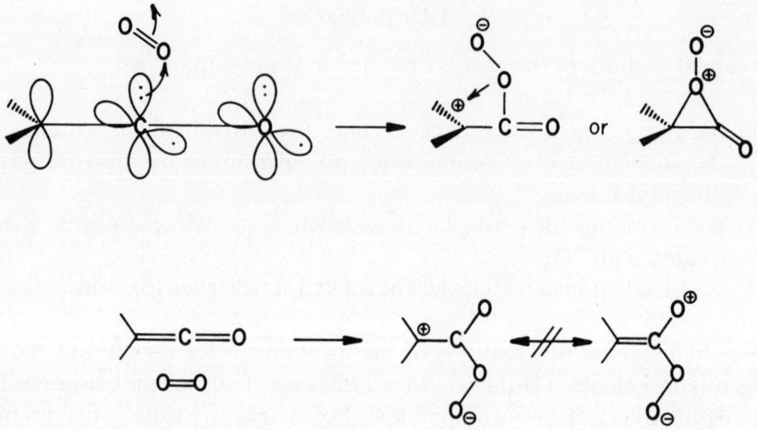

Figure 2 Model for the stereochemistry of addition of molecular oxygen to ketenes

catalytic thermal generation of $^1O_2$ or its equivalent. Such a mechanism has been proposed to explain the reaction of $^3O_2$ with strained acetylenes.[11] The principle of this mechanism is as follows:

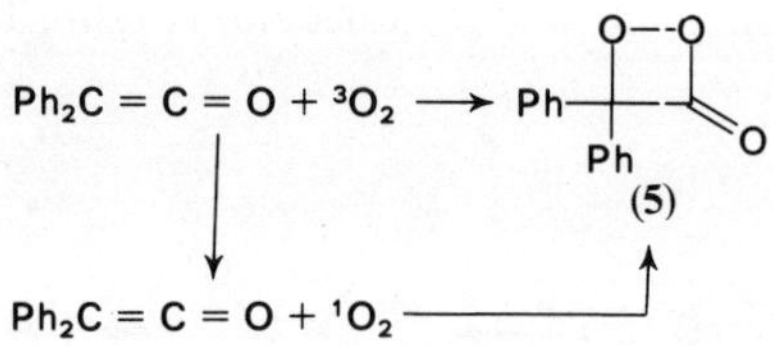

$$Ph_2C=C=O + {}^3O_2 \longrightarrow Ph_2C(OO)C{=}O \; \textbf{(5)}$$
$$\downarrow$$
$$Ph_2C=C=O + {}^1O_2 \longrightarrow \textbf{(5)} \qquad (9)$$

According to the theoretical analyses of the addition of singlet oxygen to ethylenes (due to Kearns,[12–14] Dewar and co-workers,[15,16] and Fukui and co-workers[17,18]) the most probable geometry of approach is as shown in Figure 3.

According to Kearns' correlation diagrams, the electronic behaviour of the state arising from the initial interaction of $^3\Sigma$ oxygen with a ground-state olefin is exactly parallel to that of the interaction of the corresponding state arising from the interaction of $^1\Delta$ oxygen with ground-state olefin, except that the latter state lies 22 kcal higher in energy and possesses a different spin multiplicity. Thus, the conversion of the triplet collision complex (**9**) to a weakly bound singlet complex is expected to require an activation of ca. 22 kcal mol^{-1}. The activation energy for the chemiluminescent pathway from the reaction of diphenyl ketene with $^3O_2$ is found to be ca. 20 kcal mol^{-1}, in good agreement with the energy required to produce free $^1O_2$ with little activation required beyond that required to achieve the reaction endo thermicity (Figure 4). The energetic requirement to produce $^1\Delta$ oxygen or its chemical equivalent is met However, in addition to the energetic requirement, a mechanism for spi inversion must be available and must operate effectively during the lifetime o the complex.

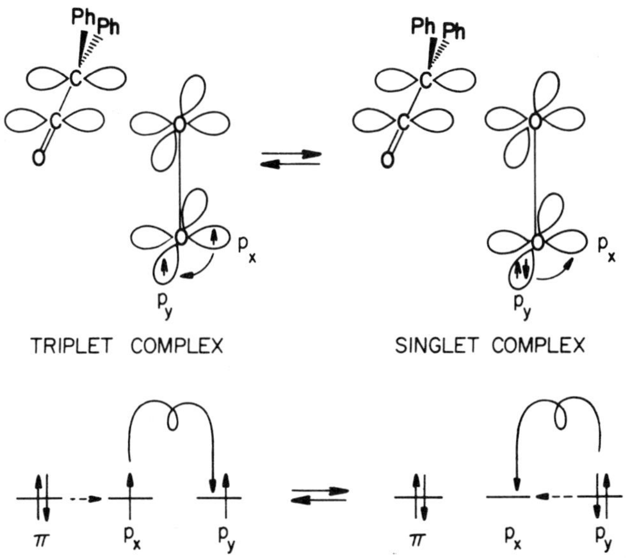

Ph Ph

p_x

p_y

TRIPLET COMPLEX

SINGLET COMPLEX

π p_x p_y

Figure 3 Schematic representation of the initial interaction of (3) and $^3\Sigma$ oxygen. Top: orbital interactions show how a $p_x \rightarrow p_y$ jump occurs in a charge-transfer complex of (3) + $^3\Sigma$ oxygen. Bottom: simplified orbital level scheme indicating the coupled nature of the $p_x \rightarrow p_y$ jump induced by charge-transfer interaction of (3) and $^3\Sigma$. According to this mechanism, $^1\Delta$ oxygen should be produced selectively

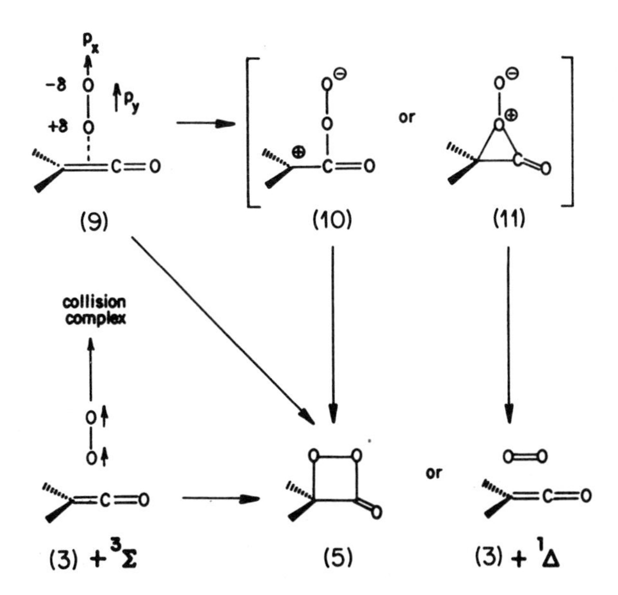

Figure 4 Proposed mechanism for catalytic thermal generation of singlet oxygen

We propose that the addition of $^3\Sigma$ oxygen to (**3**) probably occurs via a similar geometry to that for addition of $^3\Sigma$ to acetylene[11] (Figure 3). The latter allows the most effective charge transfer from the π_y orbital of the acetylene to a π^* orbital of oxygen. A complex of the ketene and ground-state oxygen would result from this charge-transfer interaction. The electronic structure of the complex may be qualitatively represented as one in which an electron of the π_y orbital of the ketene is transferred to an oxygen π^* orbital. This geometry is similar to that nominated as the most probable one for the addition of $^1\Delta$ to ethylenes. With closer proximity of the molecular components, a tendency toward full bond formation from carbon to oxygen occurs. This bond formation has the effect of producing two orbitals of comparable energy on the second oxygen atom. Such a situation is precisely what is needed to promote a favourable situation for spin–orbit coupling, i.e. the one-centre $p_x \rightarrow p_y$ jump on an oxygen atom (Figure 3). Furthermore, the state produced by such a $p_x \rightarrow p_y$ jump may possess substantial ionic character, a situation which further favours a rapid rate of spin inversion.

Once the spin inversion has occurred, either an open zwitterion (**10**) or peropoxide (**11**) may result as the initial product. Whichever species is formed, generation of peroxylactone, (**5**) or fragmentation to singlet oxygen and (**3**) are plausible reaction pathways (Figure 4). Thus, the geometry of approach that we suggest appears to provide (1) a particularly favourable electronic pathway for spin inversion, (2) for formation of ionic species that are theoretically implicated in the addition of $^1\Delta$ oxygen to ethylenes, and (3) a natural rationalization of all of the salient chemical observations related to the reactions of ketenes with molecular oxygen. With regard to point (3), we note that our mechanism provides a resolution to the problem of spin inversion and suggests that the reactive oxidizing intermediate which is produced during reaction (9) is probably an ionic adduct or is free singlet oxygen generated by fragmentation of the ionic adduct.

The latter process, if it occurs, would regenerate the starting material and therefore would represent a catalytic thermal generation of $^1\Delta$ oxygen from $^3\Sigma$ oxygen. Such a catalysis has far-reaching implications in many oxidation reactions which involve $^3\Sigma$ as a reagent.

An interesting analogy can be made between our proposed mechanism for a catalyzed $^3\Sigma \rightarrow {}^1\Delta$ reaction and the radical pair theory of chemically induced nuclear polarization (CIDNP). For CIDNP to be observed, the time scale available for hyperfine nuclear–electron spin interaction relative to that for cage recombination and cage escape must have the proper relationship. (The factors involved in CIDNP were lucidly reviewed by Kaptein.[19]) If recombination is too fast or if escape is too fast relative to the time scale of hyperfine mixing of singlet and triplet states, no net CIDNP is observed. Similarly, in the observation of singlet oxygen, the time scale would also be the crucial point, if the collision complex of ketene and $^3\Sigma$ is too short-lived or if the ketene – $^1\Delta$ collision complex always reacts, no net of $^1\Delta$ is observed, analogous to the observation of net polarization in CIDNP. Thus, to observe 'free' $^1\Delta$

oxygen, conditions must be carefully selected so as to avoid either complete trapping or non-productive spin mixing. Finally, we might speculate that the reaction of (**1**) with phosphorus ozonides might involve direct attack of the ozonide on (**1**) (or a competition between direct attack and attack by $^1O_2$). Such a possibility has analogy in the reaction of enol ethers with phosphorus ozonides.[20]

ACKNOWLEDGEMENTS

The authors thank the Air Force Office of Scientific Research (Grant AFOSR-74-2589) and the National Science Foundation (Grants NSF-CHE70-02165 and NSF-MPS73-04672) for their generous support of this research. They also thank Dr. Charles Angell of the Union Carbide Company for his generous loan of low-temperature infrared equipment.

REFERENCES

1. W. Adam, *J. Chem. Educ.*, **52**, 138 (1975).
2. D. J. Bogan, R. S. Sheinson, R. G. Gann, and F. W. Williams, *J. Amer. Chem. Soc.*, **97**, 2560 (1975).
3. L. J. Bollyky, *J. Amer. Chem. Soc.*, **92**, 3230 (1970).
4. H. Staudinger, K. Dycherhoff, H. W. Klever, and L. Ruzicka, *Chem. Ber.*, **58**, 1079 (1975), and references therein.
5. W. Adam and J. C. Liu, *J. Amer. Chem. Soc.*, **94**, 2894 (1972).
6. W. Adam and H.-C. Steinmetzer, *Angew. Chem., Int. Ed. Engl.*, **11**, 540 (1972).
7. William H. Richard, F. C. Montgomery, M. B. Yelvington, and G. Ranney, *J. Amer. Chem. Soc.*, **96**, 4045 (1974).
8. R. Wheland and P. D. Bartlett, *J. Amer. Chem. Soc.*, **92**, 6057 (1970).
9. J. K. Crandall, S. A. Sojka, and J. B. Komin, *J. Org. Chem.*, **39**, 2172 (1974).
10. N. J. Turro and R. B. Gagosian, *J. Amer. Chem. Soc.*, **92**, 2036 (1970).
11. N. J. Turro, V. Ramamurthy, K.-C. Liu, A. Krebs, and R. Kemper, *J. Amer. Chem. Soc.*, **98**, 6758 (1976).
12. D. R. Kearns, *Chem. Rev.*, **71**, 395 (1971).
13. D. R. Kearns, *Ann. N.Y. Acad. Sci.*, **171**, 34 (1970).
14. D. R. Kearns, *J. Amer. Chem. Soc.*, **91**, 6554 (1969).
15. M. J. S. Dewar and W. Thiel, *J. Amer. Chem. Soc.*, **97**, 3978 (1975).
16. M. J. S. Dewar, A. C. Griffin, W. Thiel, and I. J. Turchi, *J. Amer. Chem. Soc.*, **97**, 4439 (1975).
17. S. Inagaki and K. Fukui, *J. Amer. Chem. Soc.*, **97**, 7480 (1975).
18. S. Inagaki, S. Yamabe, H. Fujimoto, and K. Fukui, *Bull. Chem. Soc. Japan*, **45**, 3510 (1972).
19. R. Kaptein, *Adv. Free-Radical Chem.*, **5**, 319 (1975).
20. A. P. Schaap and P. D. Bartlett, *J. Amer. Chem. Soc.*, **92**, 6055 (1970).

16

Formation of Dioxetanes and Related Species

C. W. JEFFORD, A. F. BOSCHUNG, and C. G. RIMBAULT

Département de Chimie Organique, Université de Genève, 30 quai Ernest Ansermet, 1211 Genève 4, Switzerland

Singlet oxygen possesses the unique characteristic of being able to undergo both 4+2 and 2+2 cycloaddition with suitable olefins [reactions (1) and (2)]. In the second case, the resulting dioxetane usually reacts further to give the cleavage

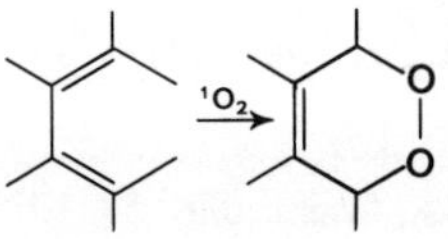

4+2 cycloaddition (1)

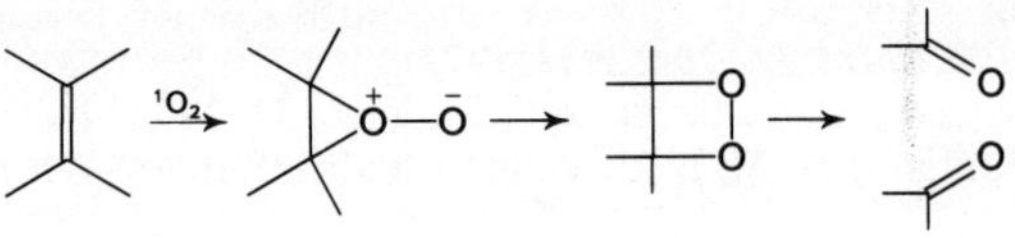

2+2 cycloaddition of singlet oxygen (2)

products, namely two carbonyl fragments. It has been suggested that the first event is the formation of a fugitive perepoxide.[1,2] Although *S*-oxides and *N*-oxides are commonplace species, attempts to characterize perepoxides have been unsuccessful. It has been reported that under special circumstances, where the double bond is hindered, for example, the lifetime of the perepoxide can be lengthened so that it can be chemically intercepted.[3] In principle, the olefin itself, singlet oxygen and apparently pinacolone should be capable of capturing the additional oxygen of the perepoxide to yield the corresponding epoxide [reactions (3) and (4)].

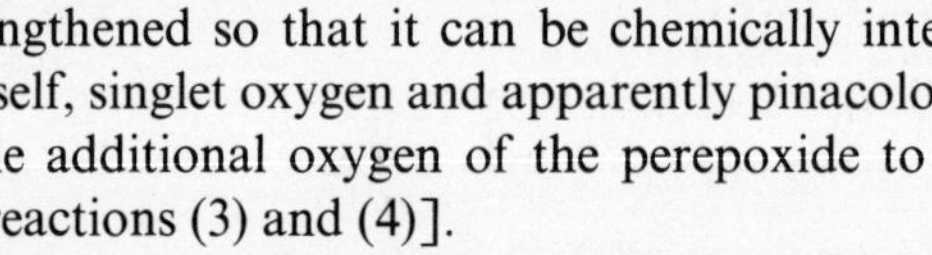

(3)

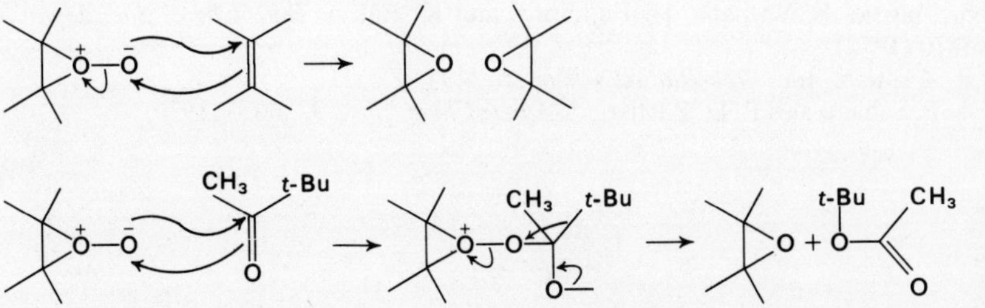

(4)

Interception of perepoxide by (3) olefin and (4) pinacolone to give epoxide

Indeed, hindered olefins such as adamantylideneadamantane, bisnorbornylidene and norbornene give high yields of the corresponding epoxides on reaction with singlet oxygen, even in solvents which are completely inert.[4, 5] It transpires that the reported Baeyer–Villiger type reaction for the perepoxide with pinacolone is incorrect[3] and that the origin of the epoxide must be sought elsewhere.

In order to answer this question and to check further on the intermediacy of the dipolar perepoxide intermediate, we have carried out a series of oxidations of 2-phenylnorbornene (**1**) with singlet and triplet oxygen. When the reaction is carried out in aprotic solvents, 3-formylcyclopentylphenyl ketone (**2**) is formed in 10% yield, accompanied by 90% of uncharacterized polymer. However, when methanol is used as the solvent the main product formed is *endo*-2-phenyl-*exo*-2-methoxy-*exo*-3-hydroperoxynorbornane (**3**) in 63% yield, together with 7% of the aforementioned ketoaldehyde (**2**) and about 6% of the product of addition of two molecules of oxygen, namely the double epoxy peroxide (**4**) [reactions (5) and (6)].

(**1**) $\xrightarrow[CCl_4]{^1O_2}$ (**2**) + polymer (5)

(**1**) $\xrightarrow[MeOH]{^1O_2}$ (**2**) + (**3**) + (**4**) (6)

The reaction of triplet oxygen with 2-phenylnorbornene was found to be about 47 times slower than the singlet oxygen reaction. Moreover, the composition of the product was entirely different. The products obtained in benzene as solvent were 20% of the ketoaldehyde (**2**), 5% of the *exo*-epoxide (**5**) and 75% of the macrocyclic trimer (**6**) [reaction (7)]. The proportion of products was solvent-dependent, and it was found that although the amount of ketoaldehyde remains constant, the ratio of trimer to epoxide decreases with increasing polarity of the solvent.

These results have been rationalized in the light of three primary 1:1 adducts, the dioxetane (**7**), the perepoxide (**8**), and the Diels–Alder adduct (**9**). Although these adducts are formally related, complete reversibility between them is unlikely. Moreover, they should have entirely different chemical properties,

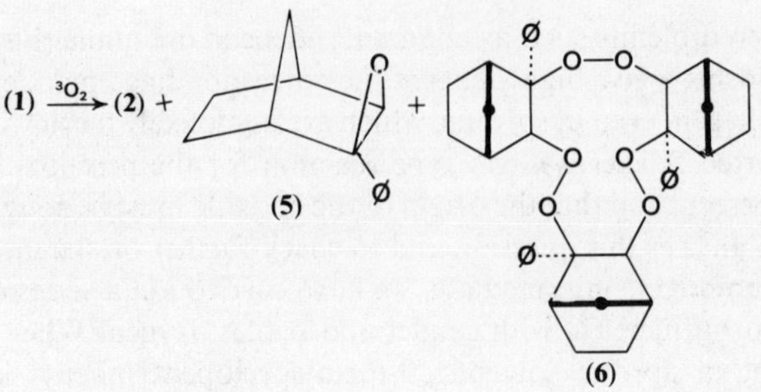

(1) $\xrightarrow{^3O_2}$ (2) + (5) + (6) (7)

especially with respect to nucleophiles. In the absence of reactant, the perepoxide will be expected to expand to the dioxetane or the Diels–Alder adduct. The latter process should be reversible, but once the dioxetane has been formed it should simply cleave. In the presence of nucleophiles, for example methanol, the perepoxide (**8**) and the Diels–Alder adduct (**9**) should be attacked to give the same methoxyhydroperoxide (**3**). It has already been suggested that singlet oxygen would be able to add to the perepoxide to give the tetraoxy zwitterion (**10**), which should then cleave to ozone and the epoxide (**3**) [reactions (8) and (9)]. The absence of epoxide from the singlet oxygen experiment is striking. On

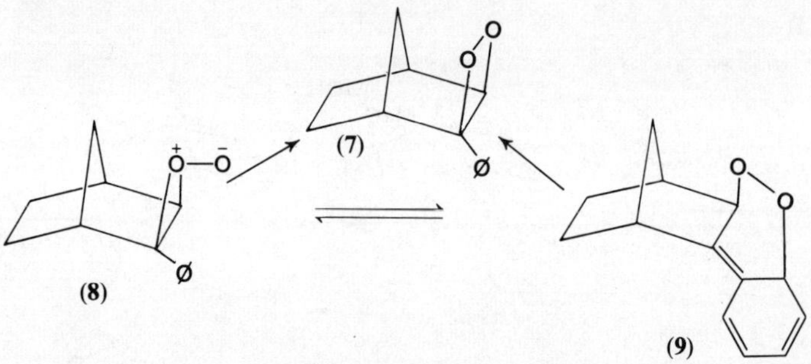

(8) 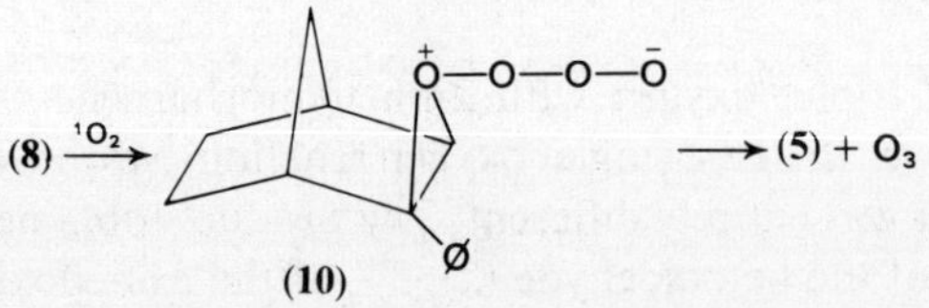 ⇌ (9); (8) → (7) ← (9) (8)

(8) $\xrightarrow{^1O_2}$ (10) ⟶ (5) + O_3 (9)

the other hand, the evidence for the intermediacy of the Diels–Alder adduct (**9**) is convincing. We now consider that the origin of epoxides is due to the reaction of triplet oxygen with olefin to give the corresponding peroxy radical (**11**) which, on abstraction of a hydrogen atom from some source, yields the hydroperoxy radical (**12**), which then gives the epoxide by cleavage of the oxygen–oxygen bond [reaction (10)].

$$(1) \xrightarrow{^3O_2} \underset{(11)}{\text{[O—O•, •, Ø]}} \xrightarrow{RH} \underset{(12)}{\text{[O—O—H, •, Ø]}} \longrightarrow (5) + \bullet OH \qquad (10)$$

Another conclusion from this work is that the behaviour of the trimer (**6**) is very similar to that of the dioxetane in that it explodes to give the ketoaldehyde (**2**) quantitatively.

REFERENCES

1. D. R. Kearns, *Chem. Rev.*, **71**, 395 (1971).
2. M. J. S. Dewar and W. Thiel, *J. Amer. Chem. Soc.*, **97**, 3978 (1975).
3. A. P. Schaap and G. R. Faler, *J. Amer. Chem. Soc.*, **95**, 3381 (1973).
4. P. D. Bartlett, *Chem. Soc. Rev.*, **5**, 149 (1976).
5. C. W. Jefford and A. F. Boschung, *Helv. Chim. Acta*, **57**, 2257 (1974).

17

Reaction of Singlet Oxygen in Solids with Electron-rich Aromatics. Peroxidic Intermediates in Indole Oxidation†

I. SAITO and T. MATSUURA

Department of Synthetic Chemistry, Faculty of Engineering, Kyoto University, Kyoto 606, Japan

INTRODUCTION

Recently we have investigated the photosensitized oxygenation of electron-rich aromatic systems bearing —C═C—XH (X = O, NR) moieties such as phenols indoles, imidazoles, and purines, in connection with the chemical mechanism of the photodynamic degradation of amino acids and nucleic acids.[1] In th photo-oxygenation of such readily oxidizable substrates, a Type I mechanism may play an important role as well as the singlet oxygen process.[2,3] We presen here the results on solid-state reactions of electron-rich aromatic systems particularly of indoles, with singlet oxygen generated by the microwave discharge method. The chemical behaviour of peroxidic intermediates formed in solution-phase photo-oxygenations of indoles, including tryptophan derivatives, is also discussed.

REACTION OF INDOLES IN SOLIDS

Many reports have appeared on the photoproducts as well as on the mechanism of the primary step of the dye-sensitized photo-oxygenations of tryptophan.[1,4,] The complex photoproducts so far obtained are considered to be derived from the secondary reactions of an initially formed 3-hydroperoxyindolenine derivative, which is suggested to be formed by a Type I process[6] or by a singlet oxygen process.[7] Photo-oxygenation of substituted indoles usually gives normal C_2—C_3 ring cleavage products and/or 3-hydroperoxyindolenines.[8–11] For example, Rose Bengal-sensitized photo-oxygenation of (**1a**) and (**1b**) in aceton at room temperature gives (**3a**) (90%) and (**26**) (80%), respectively. *N*-Acetyl-L tryptophan methyl ester (**1c**) in methanol under similar conditions gives formylkynurenine-type product (**2c**) (90%).

In order to ascertain the participation of singlet oxygen in these photo oxygenations, we carried out the oxidation using microwave discharge-generated

† Photoinducted Reactions. Part 96.

singlet oxygen by adsorbing the substrates (3% by weight) on silica gel. The experimental arrangement is essentially the same as that of Scheffer and Ouchi.[12] After the electric discharge had been continued for ca. 40 h, the products were eluted with acetone. The same products (**2a**), (**2b**), and (**2c**) were obtained from (**1a**), (**1b**), and (**1c**), respectively [equation (1)]. These results indicate that a pure singlet oxygen reaction with indole (**1**) in the absence of solvent can produce the ring cleavage products.

(**1**) $\xrightarrow[\text{in solid}]{^1O_2}$ (**2**) + (**3**)

(**a**) $R_1=H$, $R_2=Ph$, $R_3=Me$
(**b**) $R_1=Me$, $R_2=Ph$, $R_3=Me$
(**c**) $R_1=R_2=H$, $R_3=CH_2CH(NAc)CO_2Me$

(**2**): (**a**) (39%) (**b**) (30%) (**c**) (9%)

(**3**): (**a**) X=OOH (**b**) X=OH (49%)

PEROXIDIC INTERMEDIATES OF *N*-METHYLINDOLES

We previously reported that the dye-sensitized photo-oxygenation of *N*-substituted indoles gives the C_2—C_3 ring cleavage products.[10,11] For example, photo-oxygenation of 1,3-dimethylindole (**4a**) in methanol or CH_2Cl_2 gives (**5a**) in good yield. However, when (**4a**) was photo-oxygenated in methanol at −70 °C by irradiation with a tungsten–bromine lamp through a $CuCl_2$–$CaCl_2$ filter solution, unstable 3-hydroperoxyindoline (**6a**) was obtained in almost quantitative yield. A similar type of intramolecular trapping reaction of the intermediate peroxide has been observed with (**7**). Photo-oxygenation of *N*-methyltryptophol (**7a**) in methanol at −70 °C under the same conditions gave (**8a**) (95%), whereas at room temperature (**7a**) gave (**9a**) exclusively.[10] The

Table 1 Solvent and temperature effects on the product ratio (**8**):(**9**) in the photo-oxygenation of (**7a**)

Solvent	Sensitizer	Temperature (°C)	Yield (%)[a] (**8**)	Yield (%)[a] (**9**)
Methanol	Rose Bengal	r.t.	0	95[b]
Methanol	Rose Bengal	−35	78	17
Methanol	Rose Bengal	−70	95	3
Acetonitrile	Rose Bengal	r.t.	0	85
Acetone	Rose Bengal	r.t.	18	50
Acetone	Rose Bengal	−70	62	26
Dichloromethane	Methylene blue	r.t.	37	45

[a] Determined by NMR analysis.
[b] Isolated yield.

product ratio [(**8**):(**9**)] is highly dependent not only on the reaction temperature but also on the polarities of the solvents (Table 1). Photo-oxygenation in non-polar solvents at low temperature preferentially gives (**8**). Low-temperature photo-oxygenation of *N*-methoxycarbonyl-*N*-methyltryptamine (**7b**) has also been shown to give exclusively (**8b**).[13] The 3-hydroperoxyindolines [(**6a**), (**8**)] thus formed undergo, in the presence of acid, a rearrangement to give the corresponding 2,3-dihydro-1,4-benzoxazine [e.g. (**10**)].[10]

On the other hand, low temperature (– 70 °C) photo-oxygenation of 1,2,3-trisubstituted indole (**4b**) in methanol gave mainly the cleavage product (**5b**). The presence of dioxetanes in the reaction mixture obtained in the low-temperature photo-oxygenation of (**4**) or (**7**) was not detected by NMR spectrometry (– 70 °C). Since the dioxetanes, including enamine dioxetanes, so far obtained are not known to undergo such a nucleophillic displacement reaction with alcohols or secondary amines,[14–16] it seems most likely that (**6**) and (**8**) are formed from zwitterionic peroxides (**11**).

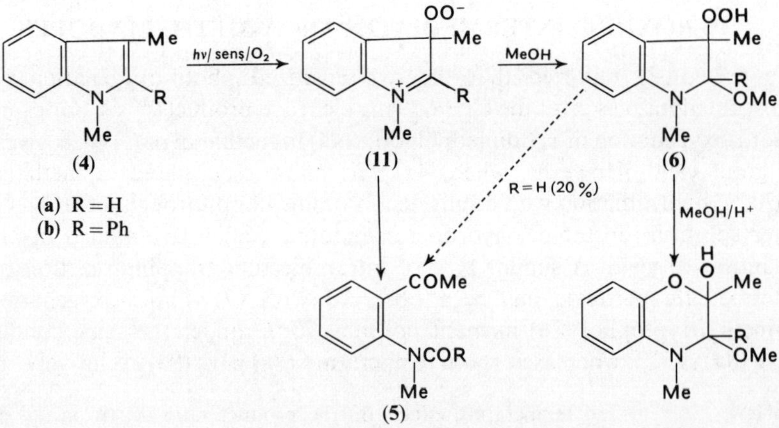

(**a**) R = H
(**b**) R = Ph

(2)

(**a**) X = O
(**b**) X = NCO_2Me

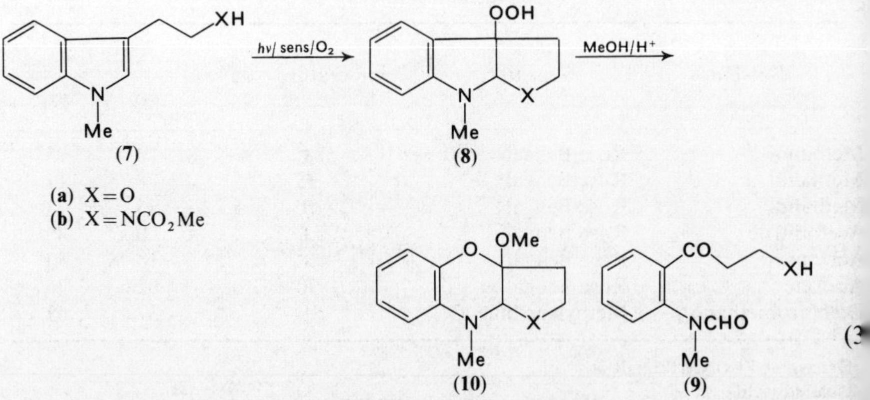

(3

PHOTOSENSITIZED OXYGENATIONS OF TRYPTOPHAN DERIVATIVES

As mentioned earlier, singlet oxygen reaction of *N*-acetyl-L-tryptophan methyl ester (**1c**) gives a formylkynurenine type product (**2c**). However, photo-oxygenation under acidic conditions produces a 2,3-dihydro-1,4-benzoxazine derivative.[17] Methylene blue-sensitized photo-oxygenation of L-tryptophan methyl ester hydrochloride (**12**) in methanol gave *cis*- and *trans*-isomers of (**13**) (60%). On the other hand, Nakagawa *et al.*[18] have shown that Rose Bengal-sensitized photo-oxygenation of (**14a**) and (**14b**) at 0 °C in pyridine–methanol gave 3*a*-hydroperoxypyrroloindole derivatives (**15a**) and (**15b**), respectively. We also observed that photo-oxygenation of tryptophol (**14c**) at −70 °C with filtered light yielded (**15c**) in quantitative yield. The hydroperoxide (**15c**) thus obtained is very sensitive to light. Short irradiation of (**15c**) with a halogen lamp or a high-pressure mercury lamp without a filter gave (**16c**) and (**17c**). Treatment of (**15c**) with alkali (0·1 N NaOH) yielded (**17c**), which was identical with the compound obtained from dimethyl sulphide reduction of (**15c**). Under acidic conditions, (**15a**)–(**15c**) undergo a rearrangement to benzoxazines (**18c**).

(**12**) $\xrightarrow{\text{MB}/h\nu/O_2,\ \text{MeOH}}$ (**13**) (4)

(**14**) $\xrightarrow{h\nu/\text{sens}/O_2}$ (**15**) $\xrightarrow{h\nu}$ (**16**)

(**15**) $\xrightarrow{\text{MeOH}/H^+}$ (**18**)

(**15**) $\xrightarrow{Me_2S \text{ or } 0\cdot1\ \text{NaOH}}$ (**17**) (5)

(**a**) X = NCO_2Me, Y = H
(**b**) X = NCO_2Me, Y = CO_2Me
(**c**) X = O, Y = H

Rose Bengal-sensitized photo-oxygenation in methanol of the cyclic tryptophan derivative pyrrolo[2,3-*b*]indole (**19**), which is readily available from the reaction of (**1c**) with *N*-bromosuccinimide,[19] gave the eight-membered ring product (**20**) (46%). Likewise, Rose Bengal-sensitized photo-oxygenation of (**21**)[20] in methanol examined by silica-gel chromatography produced a ten-membered ring lactum (**22**) (50%) together with 1,4-benzoxazine (**23**) (25%). Thus the photo-oxygenation reactions provide a convenient and simple synthetic route to medium-sized ring compounds from the corresponding indole derivatives.[17] Photo-oxygenation of *N*-methylindole-3-propionic acid (**24**) in methanol gave a spirolactone (**25**) (95%), as shown in equation (8).[10]

N H N Ac CO_2Me (**19**) $\xrightarrow[MeOH]{RB/h\nu/O_2}$ O CO_2Me N—Ac N H O (**20**) (6)

Me N O N Me (**21**) $\xrightarrow[MeOH]{RB/h\nu/O_2}$ O Me O N N O Me (**22**)

\+ O O O N Me OMe NH Me (**23**)

$CH_2CH_2CO_2H$ N Me (**24**) $\xrightarrow[MeOH/O_2]{RB/h\nu}$ H H O O C=O O CH_2 CH_2 N Me OMe → O O O N Me OM (**25**)

ACKNOWLEDGEMENT

This work was supported by a Grant-in-Aid for Scientific Research from the Ministry of Education and the Japan Society for the Promotion of Science.

REFERENCES

1. I. Saito and T. Matsuura, in *Photochemistry of Heterocyclic Compounds* (Ed. by O. Buchardt), Wiley, New York, 1976, p. 456.

2. C. S. Foote, *Science, N.Y.*, **162**, 963 (1968).
3. T. Matsuura, N. Yoshimura, A. Nishinaga, and I. Saito, *Tetrahedron*, **28**, 4933 (1972).
4. J. D. Spikes and M. L. MacKnight, *Ann. N.Y. Acad. Sci.*, **171**, 149 (1970).
5. C. A. Benassi, E. Scoffone, and G. Galiazza, *Photochem. Photobiol.*, **6**, 857 (1967).
6. L. I. Grossweiner and A. G. Kepka, *Photochem. Photobiol.*, **16**, 305 (1972).
7. R. Nilsson, P. B. Merkel, and D. R. Kearns, *Photochem. Photobiol.*, **16**, 117 (1972).
8. I. Saito, M. Imuta, and T. Matsuura, *Chem. Lett.*, **1972**, 1173.
9. I. Saito, M. Imuta, and T. Matsuura, *Chem. Lett.*, **1972**, 1199.
10. I. Saito, M. Imuta, S. Matsugo, and T. Matsuura, *J. Amer. Chem. Soc.*, **97**, 7191 (1975).
11. I. Saito, M. Imuta, S. Matsugo, H. Yamamoto, and T. Matsuura, *Synthesis*, **1976**, 255.
12. J. R. Scheffer and M. D. Ouchi, *Tetrahedron Lett.*, **1970**, 223, and references therein.
13. J. Saito, M. Imuta, Y. Takahashi, S. Matsugo, and T. Matsuura, *J. Amer. Chem. Soc.*, **99**, 2005 (1977).
14. K. R. Kopecky, J. E. Filby, C. Mumford, P. A. Lockwood, and J.-Y. Ding, *Can. J. Chem.*, **53**, 1104 (1975), and references therein.
15. C. S. Foote, A. A. Dzakpasu, and J. P. Lin, *Tetrahedron Lett.*, **1975**, 1247.
16. H. H. Wasserman and S. Terao, *Tetrahedron Lett.*, **1975**, 1735.
17. I. Saito, Y. Takahashi, M. Imuta, S. Matsugo, H. Kaguchi, and T. Matsuura, *Heterocycles*, **5**, 53 (1976).
18. M. Nakagawa, H. Okajima, and T. Hino, *J. Amer. Chem. Soc.*, **98**, 635 (1976).
19. M. Ohno, T. F. Spande, and B. Witkop, *J. Amer. Chem. Soc.*, **92**, 343 (1970).
20. J. B. Hester, Jr., *J. Org. Chem.*, **32**, 3904 (1967).

18

Reaction of Singlet Oxygen with α-Ketocarboxylic Acids of Biological Interest

C. W. JEFFORD, A. F. BOSCHUNG, and P. A. CADBY

Département de Chimie Organique, Université de Genève, 30 quai Ernest Ansermet, 1211 Genève 4, Switzerland

There is an important class of monooxygenases which bring about hydroxylation of a substrate (SH) with molecular oxygen which requires iron(II) ions and ascorbate, but which also needs α-ketoglutaric acid as reductant.[1] The stoichiometry of the process is that one molecule of oxygen oxidizes the substrate and α-ketoglutarate in equimolar amounts so that the latter undergoes decarboxylation to succinate [equation (1)].

$$SH + O_2 + HO_2CCH_2CH_2COCO_2H \longrightarrow SOH + HO_2CCH_2CH_2CO_2H + CO_2 \quad (1)$$

How this happens is not known. In principle, molecular oxygen, through the agency of the metalloenzyme, could attach itself to the substrate to give the hydroperoxy derivative, SOOH, which could subsequently be reduced to the alcohol by α-ketoglutaric acid [equation (2)].

$$SH + O_2 \longrightarrow SOOH$$
$$SOOH + HO_2CCH_2CH_2COCO_2H \longrightarrow SOH + HO_2CCH_2CH_2CO_2H + CO_2 \quad (2)$$

Another possibility is that α-ketoglutaric acid itself is oxidized to persuccinic acid which then selectively hydroxylates the substrate in some way [equation (3)].

$$O_2 + HO_2CCH_2CH_2COCO_2H \longrightarrow HO_2CCH_2CH_2COO_2H + CO_2$$
$$HO_2CCH_2CH_2COO_2H \xrightarrow{+SH} HO_2CCH_2CH_2CO_2H + SOH \quad (3)$$

The latter suggestion is plausible, more so than the former, but suffers from the disadvantage that molecular oxygen itself does not react with α-keto acids.[2] This is not unexpected in view of the spin-forbiddeness of the reaction. However, the intervention of a transition metal cation may overcome this restriction.

We recently showed that the plausibility of this idea depends upon the reaction of singlet oxygen with α-keto acids. In fact, singlet oxygen reacts readily with a variety of α-keto acids and the parent acid and carbon dioxide are the sole products.[3] We have proposed that the first step is the formation of the corresponding peracid in an oxidative decarboxylative step and that this, in the presence of unreacted α-teto acid, gives the Baeyer–Villiger intermediate which cleaves to two molecules of carboxylic acid and carbon dioxide [equation (4)].

$$
\begin{array}{l}
RCOCO_2H + {}^1O_2 \longrightarrow R{-}COO_2H + CO_2 \\
RCOO_2H + RCOCO_2H \longrightarrow RCO_2{-}O{-}\underset{\displaystyle CO_2H}{\overset{\displaystyle OH}{\overset{|}{\underset{|}{C}}}}{-}R \longrightarrow 2RCO_2H + CO_2
\end{array}
\qquad (4)
$$

This finding opens up a new area of singlet oxygen chemistry which may be relevant to the mechanism of biological oxidation. Indeed, it becomes of interest to know if the Udenfriend conditions will oxidize α-keto acids. Moreover, the suggestion has been made that a superoxide radical anion, which is formed from Fe(II)-containing enzymes and aerial oxygen, could be the reagent responsible for creating peracid from α-keto acid.[4] In this work we made a comparison of the chemistry of singlet oxygen, superoxide and the Udenfriend conditions on a series of α-keto acids (Table 1). As can be seen, the chemistry of singlet oxygen and the superoxide ion are entirely different, a case in point being the inertness of phenylglyoxalic acid with singlet oxygen while it reacts completely with superoxide ion. Another surprise is the general ineffectiveness of the Udenfriend conditions. Accordingly, whatever the mechanisms may be for these oxidations, it is certain that they are mechanistically distinct for each class.

REFERENCES

1. M. T. Abbott and S. Udenfriend, in *Molecular Mechanisms of Oxygen Activation* (Ed. O. Hayaishi), Academic Press, New York, 1974, p. 167.
2. G. A. Hamilton, in *Progress in Bio-Organic Chemistry* (Ed. E. T. Kaiser and F. J. Kézdy), Wiley-Interscience, New York, 1971, Vol. 1, pp. 83–157.
3. C. W. Jefford, A. F. Boschung, T. A. B. M. Bolsman, R. M. Moriarty, and B. Melnick, *J. Amer. Chem. Soc.*, **98**, 1017 (1976).
4. J. San Filippo, Jr., C.-I. Chern, and J. S. Valentine, *J. Org. Chem.*, **41**, 1077 (1976).

Table 1 Relative reactivities of activated oxygen and some α-keto-carboxylic acids

Substrate	Results			
	Superoxide	Udenfriend	$^1O_2$	Control[e]
$HO_2C(CH_2)_2COCO_2H$	$HO_2C(CH_2)_2CO_2H$, 100%	N.R.	$HO_2C(CH_2)_2CO_2H$, 100%	N.R.
t$BuCOCO_2H$	*t*$BuCO_2H$, 100%	N.R.	*t*$BuCO_2H$, 80%	N.R.
n$PrCOCO_2H$	*n*$PrCO_2H$, 100%	idem 36%	idem 95%	N.R.
$EtMeCHCOCO_2H$	$EtMeCHCO_2H$, 100%	idem 56%	idem 90%	N.R.
$ØCOCO_2H$	$ØCO_2H$, 100%	N.R.	N.R.	N.R.
$MeOC_6H_4COCO_2H$	$MeOC_6H_4CO_2H$, 100%	N.R.	$MeOC_6H_4CO_2H$, 52%	N.R.
$ØCH_2CO_2H$	100%[a]	75%[b]	100%[c]	N.R.
$ØCH(OH)CO_2H$	$ØCO_2H$, 100%	N.R.	N.R.	N.R.
ØCOCOØ	100%[d]	N.R.	N.R.	N.R.
ØCOCH(OH)Ø	$ØCO_2H$, 100%	N.R.	N.R.	N.R.

[a] $ØCH_2CO_2H$, 11%, $ØCO_2H$, 89%.
[b] $ØCO_2H$, 75%
[c] In MeOH, $ØCO_2H$, 100% in pH7 buffered H_2O, $ØCO_2H$, 70%, $ØCH_2CO_2H$, 30%.
[d] $ØCO_2H$, 84%, $Ø_2C(OH)CO_2H$, 16%.
[e] Without $FeSO_4$, without MB, N.R. = no reaction.

19

Mechanism of the Dye-sensitized Photodynamic Inactivation of Lysozyme

P. ROSENKRANZ, AHMED AL-IBRAHIM, and H. SCHMIDT

Institut für physikalische Biochemie der Universität Frankfurt am Main, West Germany

Hen egg-white lysozyme is a small globular protein with a molecular weight of 14 500.[1] Its bacteriolytic activity is identified by its attack on the 1,4-glycosidic bonds of polyaminosaccharides in bacterial cell walls. The cleavage of the saccharide occurs in the enzymatic centre, which is located in a cleft on the surface of the protein. This cleft is capable of binding a hexasaccharide residue of the cell wall polyamino sugars. Therefore, the centre of the enzymatic activity is divided into six subsites designated as A, B, C, D, E, and F. The hydrolysis of the polysugars (the enzymatic reaction) takes place at sub-site D and is catalysed by a proton originating from the carboxyl group of the glutamic acid side-chain 35. The three tryptophan residues 62, 63, and 108 are also essential for the enzymatic activity because the saccharide residues are hydrogen-bonded to these side-chains.[2] They play an important role in photodynamic reactions.[3,4]

The enzymatic activity of lysozyme decreases by an overall pseudo-first-order rate process during the irradiation with visible light of a solution of the protein containing acridine orange.[5,6] When the solvent water is replaced with D_2O at low protein concentrations, the experimental pseudo-first-order rate constant of the photo-oxidation increases by a factor of about 10. On the other hand, the reaction is completely quenched by sufficiently high concentrations of azide ions. Both of these results indicate that singlet oxygen is involved in the photodynamic reaction. (These experiments are not restricted to acridine orange as sensitizer. We have observed similar effects with other sensitizers such as eosine, acridine yellow, acriflavine, and methylene blue.)

The deactivation of lysozyme in the presence of acridine orange can be described kinetically by the following scheme:[5]

$$
\begin{aligned}
&D_0 \longrightarrow {}^1D && I_{abs} \\
&{}^1D \longrightarrow D_0 && \tau = 2{\cdot}4 \text{ ns (ref. 7)} \\
&{}^1D \longrightarrow {}^3D && \phi_T = 0{\cdot}10 \text{ (ref. 8)}
\end{aligned}
\qquad (1)
$$

$$^3D \longrightarrow D_0 \qquad k_{2a}=4\times10^3\ \mathrm{s}^{-1}\ \text{(ref. 9) or}$$
$$k_{2b}=300\ \mathrm{s}^{-1}\ \text{(ref. 10)} \qquad (2)$$

$$^3D+O_2 \longrightarrow D_0+O_2 \qquad k_{3a}=1{\cdot}4\times10^8\ \mathrm{M}^{-1}\ \mathrm{s}^{-1}$$
$$k_{3b}=1{\cdot}0\times10^7\ \mathrm{M}^{-1}\ \mathrm{s}^{-1} \qquad (3)$$

$$^3D+O_2 \longrightarrow D_0+{}^1\Delta \qquad k_{4a}=2{\cdot}2\times10^8\ \mathrm{M}^{-1}\ \mathrm{s}^{-1}$$
$$k_{4b}=1{\cdot}7\times10^7\ \mathrm{M}^{-1}\ \mathrm{s}^{-1} \qquad (4)$$

$$^1\Delta \longrightarrow O_2 \qquad k_5(H_2O)=5\times10^5\ \mathrm{s}^{-1}\ \text{(ref. 11)}$$
$$k_5(D_2O)=5\times10^4\ \mathrm{s}^{-1}\ \text{(ref. 11)} \qquad (5)$$

$$P+{}^1\Delta \longrightarrow PO_2 \qquad k_6(H_2O)=2{\cdot}9\times10^7\ \mathrm{M}^{-1}\ \mathrm{s}^{-1}$$
$$k_6(D_2O)=4{\cdot}7\times10^7\ \mathrm{M}^{-1}\ \mathrm{s}^{-1} \qquad (6)$$

$$P+{}^1\Delta \longrightarrow P+O_2 \qquad k_7(H_2O)=4{\cdot}1\times10^8\ \mathrm{M}^{-1}\ \mathrm{s}^{-1}$$
$$k_7(D_2O)=5{\cdot}9\times10^8\ \mathrm{M}^{-1}\ \mathrm{s}^{-1} \qquad (7)$$

$$^3D+P_0 \longrightarrow D_0+P_0 \qquad k_{8a}=4{\cdot}7\times10^7\ \mathrm{M}^{-1}\ \mathrm{s}^{-1}$$
$$k_{8b}=3{\cdot}5\times10^6\ \mathrm{M}^{-1}\ \mathrm{s}^{-1} \qquad (8)$$

where D_0, ^1D, ^3D denote the ground, excited singlet and triplet states of the dye, respectively, $^1\Delta$ excited oxygen of the electronic configuration $^1\Delta_g$, P native lysozyme (P_0 = initial concentration), PO_2 the photo-oxidation product, and ϕ_T the quantum yield of step (1). Unfortunately, the kinetic constants for process (2) found in the literature[9,10] differ considerably [k_2 is important for the calculation of the rate constants of steps (3), (4), and (8)]. In this paper we restrict ourselves to considering only reactions (6) and (7) in more detail.

By definition, reaction (6) is the chemical reaction between singlet oxygen and the enzyme responsible for the observed decrease in the enzymatic activity. However, it is not clear at first whether reaction (7) is also a chemical reaction but without influencing the enzymatic activity. Such a reaction could take place outside the enzymatic centre with an amino acid side-chain not essential for the enzymatic activity, such as histidine. Alternatively, reaction (7) could be a pure physical process of singlet oxygen quenching. Such processes are well known with molecules that contain amino functions.[12] Proteins do contain such groups.

A decision between both possibilities can be made by quantitative analysis of our kinetic data. Applying quasi-stationary conditions for the singlet state of oxygen and the triplet state of the dye, the overall pseudo-first-order rate constant for the photodeactivation of the enzyme is given by the following expression:

$$k_{exp} = I_{abs} \cdot \phi_T \cdot \frac{k_4[O_2]}{k_2 + (k_3 + k_4)[O_2] + k_8[P_0]} \cdot \frac{k_6}{k_5 + k_6[P] + k_7[P]}$$

$$-\frac{d[P]}{dt} = k_{exp}[P]$$

It is seen that k_{exp} is a 'constant' only under special conditions, when the sum $k_5 + k_6[P] + k_7[P]$ is approximately independent of time. This is always the case at sufficiently low protein concentrations when $k_5 \gg k_6[P] + k_7[P]$. However, this condition is not necessarily fulfilled at higher protein concentrations, especially not in D_2O where k_5 is smaller by a factor of 10 compared with the value in water. Because reaction (7) is considerably faster than reaction (6), the sum at high protein concentrations depends primarily on the nature of step (7).

Reaction (7) does not depend on whether the lysozyme is still native or photodeactivated, if it is a pure physical quenching process. In this case, reaction (7) is independent of the duration of illumination. Therefore, k_{exp} is always a pseudo-first-order rate constant.

If, on the contrary, step (7) describes a chemical reaction, the product $k_7[P]$ is dependent on the part of lysozyme that has not yet reacted in process (7). Therefore, the kinetics at high protein concentrations, especially in D_2O, do not follow a pseudo-first-order rate equation.

For both cases we have solved the corresponding differential equations by analogue computation using the numerical values for the rate constants taken from our kinetic scheme. The results are shown in Figures 1 and 2. If step (7) is a pure physical process the kinetics can always be described by a first-order rate equation (Figures 1a and 1b). If reaction (7) is a chemical process that does not influence the enzymatic activity, the deviations from a first-order process at higher protein concentrations are clearly seen (Figures 2a and 2b). By comparison with our experimental data, it is obvious that step (7) is a physical process.

Additional chemical reactions which do not affect the enzymatic activity cannot, of course, be excluded generally by our calculations. They would not noticeably influence the kinetics of the deactivation when their second-order rate constant is smaller than 1×10^8 M^{-1} s^{-1}, which is lower by a factor of about 5 than that in the physical process, but faster than the actual photodynamic reaction (6).

In order to identify the place where singlet oxygen reacts in our system, we investigated the kinetics of the deactivation in the presence of *N*-acetylglucosamine (GlcNAc), a substance which is part of the oligosaccharides in the bacterial cell walls. GlcNAc is bound at sub-site C at the enzymatic centre, the two anomers being slightly different.[2] It therefore inhibits the enzymatic reaction competitively. The three tryptophan residues 62, 63, and 108, which might be sensitive to singlet oxygen, are located in the environment of sub-site C.

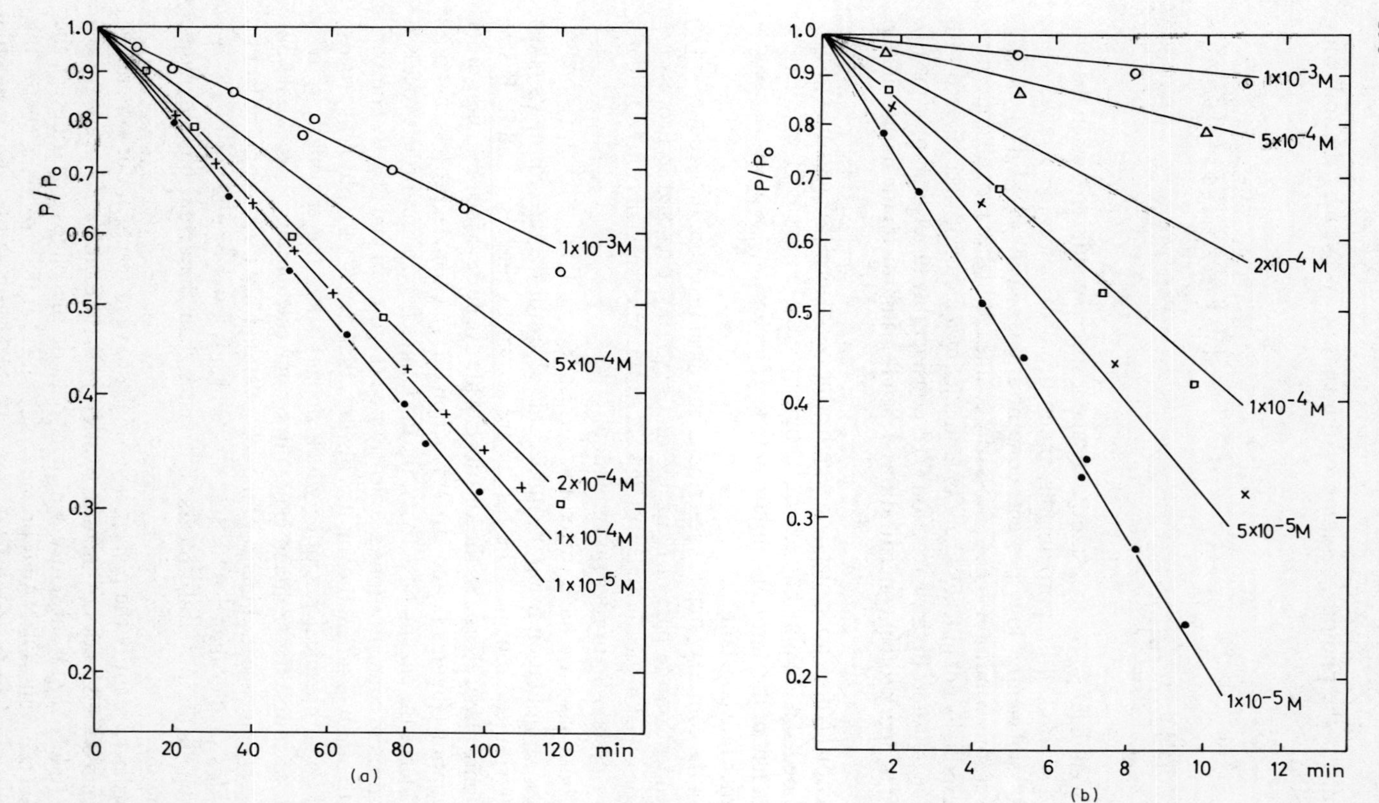

Figure 1 Decrease of the enzymatic activity with time of irradiation (492 nm). Dependence of the kinetics on the protein concentration. The solid lines were calculated under the assumption that step (7) is a physical quenching process. Solvent: (a) phosphate buffer/H_2O; (b) (*see opposite*) phosphate buffer/D_2O. Dye concentration, 1×10^{-4} M

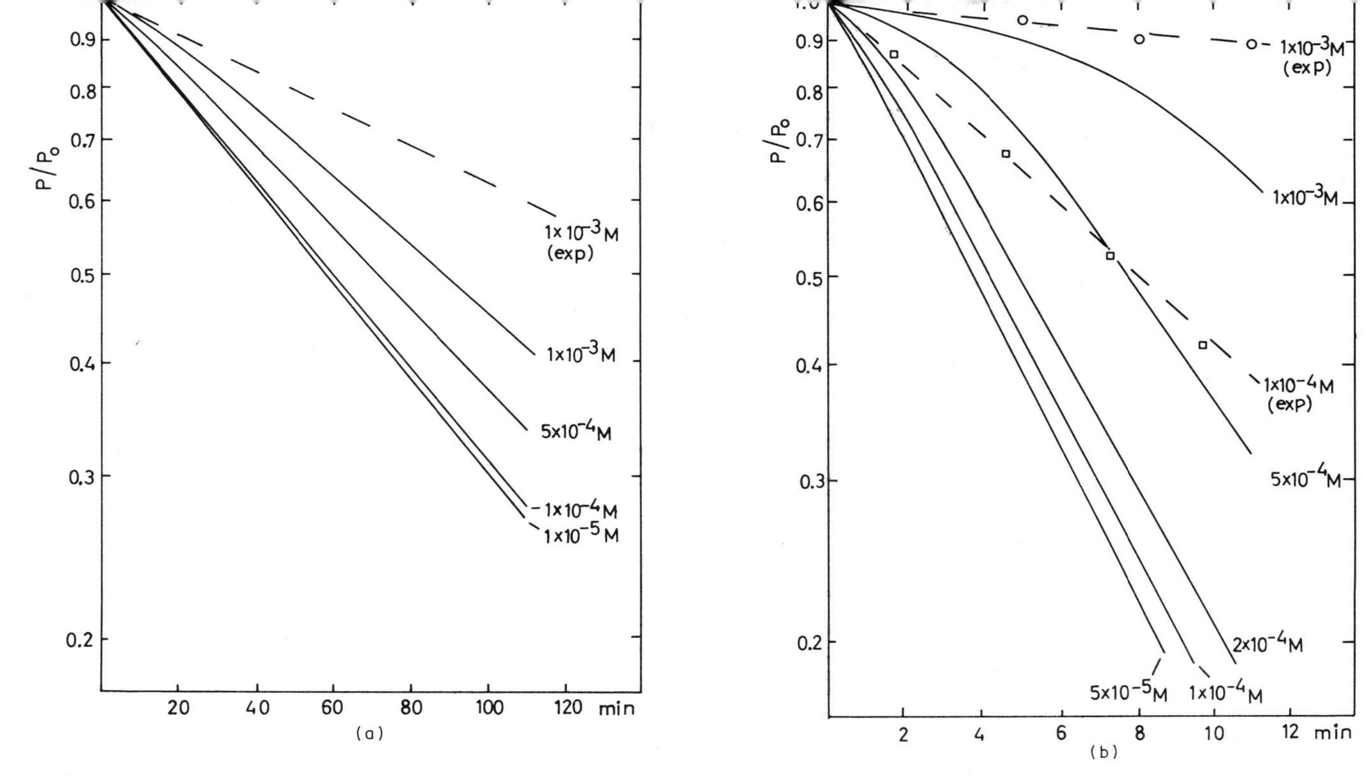

Figure 2 Decrease of the enzymatic activity with time of irradiation calculated under the assumption that step (7) is a chemical reaction that does not influence the enzymatic activity. (a) For phosphate buffer/H_2O; (b) (*see opposite*) for phosphate buffer/D_2O

If the photodynamic reaction (6) occurs in the absence of the inhibitor in it binding region it is expected that in the presence of the saccharide the reactio would become slower because the sugar protects sterically the surrounding of it binding position against singlet oxygen.

In Figure 3, it is seen that the overall photodynamic reaction becomes slowe with increasing saccharide concentrations. However, in order to draw un equivocal conclusions from this type of experiment concerning the location of th reaction of singlet oxygen, first we must rule out that the cause of this effect i simply a physical quenching process of singlet oxygen or of the dye triple state by the saccharide.

The triplet quenching mechanism can be excluded by measuring the kinetic of the photodynamic deactivation as a function of the oxygen concentration If the saccharide quenches physically the dye triplet state, the influence of th oxygen concentration on the kinetics of the photosensitized reaction would b described by the following expression:

$$\frac{k^{s}_{exp}}{k^{o}_{exp}-k^{s}_{exp}}=\frac{k_2+k_8[P_0]}{k_q[S]}+\frac{k_3+k_4}{k_q[S]}[O_2]$$

where k^{s}_{exp} and k^{o}_{exp} are the overall rate constants in the presence and in th absence of the saccharide, respectively, and k_q refers to the reaction $^3D+S \rightarrow D_0+S$. This means that a graph of $k^{s}_{exp}/(k^{o}_{exp}-k^{s}_{exp})$ *versus* oxygen con centration should be a straight line. However, experiments showed that th ratio is independent of the oxygen content of the solution. Hence, a quenchin process of the dye triplet state cannot be responsible for the observed effect.[1]

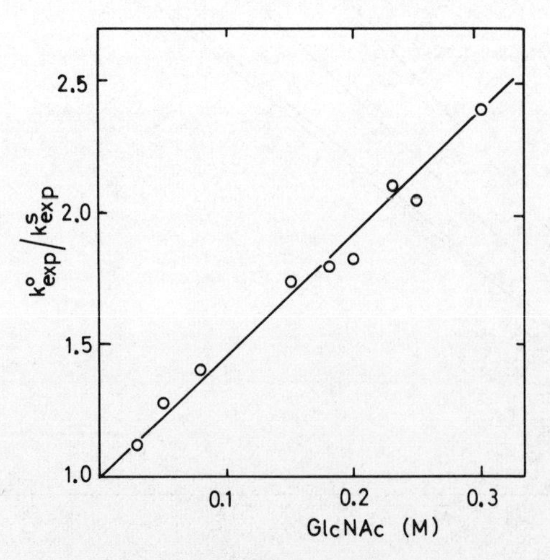

Figure 3 k^{0}_{exp}/k^{s}_{exp} as a function of the saccharide concentration. Protein concentration, 1×10^{-4} M; dye concentration, 1×10^{-4} M; phosphate buffer/H_2O

A decision between the other two possibilities—quenching of singlet oxygen or steric protection of the enzymatic centre—needs further experiments because in both cases the ratio k^{o}_{exp}/k^{s}_{exp} *versus* saccharide concentration should give a straight line (cf. figure 3):

$$\frac{k^{o}_{exp}}{k^{s}_{exp}} = 1 + \frac{k_s}{k_5 + k_6[P] + k_7[P_0]}[S] \quad \text{(quenching mechanism)}$$

$$\frac{k^{o}_{exp}}{k^{s}_{exp}} = 1 + \frac{(k_5 + k_7[P_0])K}{k_5 + k_6[P] + k_7[P_0]}[S] \quad \text{(protection mechanism)}$$

(k_s: ${}^{1}\Delta + S \rightarrow O_2 + S$; $K = [PS]/[P][S]$). However, the ratio of the experimental rate constants in water and in D_2O depends on the actual mechanism:

$$\frac{k^{s,D_2O}_{exp}}{k^{s,H_2O}_{exp}} = \frac{k_6^{D_2O}(k_5 + k_6[P] + k_7[P_0])^{H_2O} + k_s[S]}{k_6^{H_2O}(k_5 + k_6[P] + k_7[P_0])^{D_2O} + k_s[S]} \quad \text{(quenching)}$$

$$\frac{k^{s,H_2O}_{exp}}{k^{s,H_2O}_{exp}} = \frac{k_6^{D_2O}(k_5 + k_7[P_0])^{H_2O}(1 + K[S]) + k_6^{H_2O}[P]}{k_6^{H_2O}(k_5 + k_7[P_0])^{D_2O}(1 + K[S]) + k_6^{D_2O}[P]} \quad \text{(protection)}$$

Our results agree satisfactorily with the protection mechanism of the inhibitor binding place (Table 1).

Table 1 Comparison of the experimental ratio $k^{s,D_2O}_{exp}/k^{s,H_2O}_{exp}$ with the calculated value, assuming that GlcNAc acts as a steric protector of the enzymatic centre and as a singlet oxygen quencher respectively. Lysozyme concentration, 1×10^{-4} M

GlcNAc (M)	$k^{s,D_2O}_{exp}/k^{s,H_2O}_{exp}$		
	Observed	Mechanism (calc.)	
		Protection	Quenching
0	8·0	7·8	7·8
0·1	7·6	7·8	3·5
0·2	8·4	7·9	2·8
0·3	8·6	7·9	2·4

Although we still do not understand the reactions in detail, it can be said that here are two processes responsible for the quenching of singlet oxygen in our ystem: a fast physical quenching reaction and a chemical reaction with one of he three tryptophan side-chains at the enzymatic centre, which results in a ecrease in the lysozyme activity.

REFERENCES

1. P. Jollès, *Angew. Chem.*, **76**, 20 (1964).
2. T. Imoto, L. N. Johnson, A. C. T. North, D. C. Phillips, and J. A. Rupley, in *The Enzymes*, Vol. VII, p. 123, Third Edition (Ed. P. D. Boyer), Academic Press, New York, London, 1972.
3. T. R. Hopkins and J. D. Spikes, *Photochem. Photobiol.*, **12**, 175 (1970).
4. A. G. Kepka and L. I. Grossweiner, *Photochem. Photobiol.*, **18**, 49 (1973).
5. H. Schmidt and P. Rosenkranz, *Z. Naturforsch.*, **31c**, 29 (1976).
6. H. Schmidt and P. Rosenkranz, *Z. Naturforsch.*, **27b**, 1436 (1972).
7. A. Schmillen, *Z. Naturforsch.*, **9a**, 1036 (1954).
8. B. Soep, A. Kellmann, M. Martin, and L. Lindqvist, *Chem. Phys. Lett.*, **13**, 241 (1972).
9. G. Blauer and H. Linschitz, *J. Chem. Phys.*, **33**, 937 (1960).
10. A. Kellmann, *Photochem. Photobiol.*, **20**, 103 (1974).
11. P. B. Merkel, R. Nilsson, and D. R. Kearns, *J. Amer. Chem. Soc.*, **94**, 1030 (1972).
12. R. H. Young, R. L. Martin, D. Feriozi, D. Brewer, and R. Kayser, *Photochem. Photobiol.*, **17**, 233 (1973).
13. P. Rosenkranz and H. Schmidt, *Z. Naturforsch.*, **31c**, 679 (1976).

20

Chemical Behaviour of Singlet Oxygen Produced by Energy Transfer from Dye Adsorbed on a Micelle

Y. USUI

Department of Chemistry, Faculty of Science, Ibaraki University, Mito, Japan

The investigation of dye-sensitized photo-oxidation by singlet oxygen where the acceptor is dissolved in a micelle provides fundamental information for the study of photodynamic inactivation of enzymes and cells.[1–5] Surfactant micelles serve as both structural and functional models for complex bioaggregates of proteins and biomembranes. This paper describes a kinetic investigation of photo-oxygenation reactions of singlet oxygen ($^1O_2$) with an acceptor (A) dissolved in a micelle. In this case, singlet oxygen is produced by energy transfer from a dye adsorbed on micelles or dissolved in the liquid phase. Comparing the concentration effect of A and the mixed solvent effect upon the quantum yield of the disappearance of $^1O_2$ acceptor in the micellar phase ($\Phi_{-A/\text{micelle}}$) with the effect in a homogeneous solution (Φ_{-A}), it is found that singlet oxygen can readily penetrate into the micelles of compounds such as dodecyl trimethylammonium chloride (DTA) and sodium dodecyl sulphate (SDS), and reacts with the acceptor in solution to a similar extent.

DYE-SENSITIZED PHOTO-OXYGENATION IN SOLUTION

Dependence of Acceptor Concentration on Φ_{-A}

The essential reaction course is represented and the quantum yield of disappearance of A is expressed as follows:[6,7]

$$^1\text{Sens} \xrightarrow{h\nu} \longrightarrow {}^3\text{Sens} \qquad \Phi_{st} \qquad (1)$$

$$^3\text{Sens} \longrightarrow {}^1\text{Sens} \qquad k_d \qquad (2)$$

$$^3\text{Sens} + {}^3O_2 \longrightarrow {}^1\text{Sens} + {}^1O_2 \qquad (k^r_{O_2}) \qquad (3)$$

$$^3\text{Sens} + {}^3O_2 \longrightarrow {}^1\text{Sens} + {}^3O_2,\ \text{Sens}^+ \cdots O_2^- \qquad \Big\} k^s_{O_2} \qquad (4)$$

$$^3\text{Sens} + \text{A} \longrightarrow {}^1\text{Sens} + \text{A},\ \text{Sens}^- + \text{A}^+ \qquad k_{TA} \qquad (5)$$

$$^1O_2 \longrightarrow {}^3O_2 \qquad k'_d \tag{6}$$

$$^1O_2 + A \longrightarrow AO_2 \qquad (k'^{r}_{A}) \tag{7}$$

$$\longrightarrow {}^3O_2 + A \qquad k'^{s}_{A} \tag{8}$$

$$\Phi_{-A} = \Phi_{^1O_2}\gamma_A = \Phi_{^1O_2}\frac{[A]}{\beta+[A]}\cdot\alpha r \tag{9}$$

where $\Phi_{^1O_2} = \Phi_{st}\gamma_{^1O_2}$, $r = k'^{r}_{A}/k'^{s}_{A}$, and $\beta = k'_d/k'^{s}_{A}$. $\gamma_{^1O_2}$ and α express the probability of $^1O_2$ formation from triplet sensitizer and the reaction probability of A in equation (10), respectively. When methylene blue (MB) triplet sensitizer reacts predominantly with dissolved oxygen in air-saturated aqueous and ethanolic solutions, i.e., $k^{s}_{O_2}[O_2] \gg (k_d + k_D[MB] + k_{TA}[A])$, the value of $\Phi_{^1O_2}$ may be the same as that of $\Phi_{st}(=0{\cdot}52)$, since $\gamma_{^1O_2}(k^{r}_{O_2}/k^{s}_{O_2})$ is approximately unity.[8,9] Thus, with the efficient 9,10-dimethylanthracene (DMA), 1,3-diphenylisobenzofuran (DPBF), and 2,5-dimethylfuran (DMF) acceptors, the maximum quantum yield of the reaction (Φ^{max}_{-A}) at higher concentration of A is equal to the $\Phi_{^1O_2}$ value, i.e. Φ_{st} (0·52). However, the observed value is about twice the Φ_{st} value shown in Figure 1. In the case of MB-sensitized photo-oxidation of DMF in

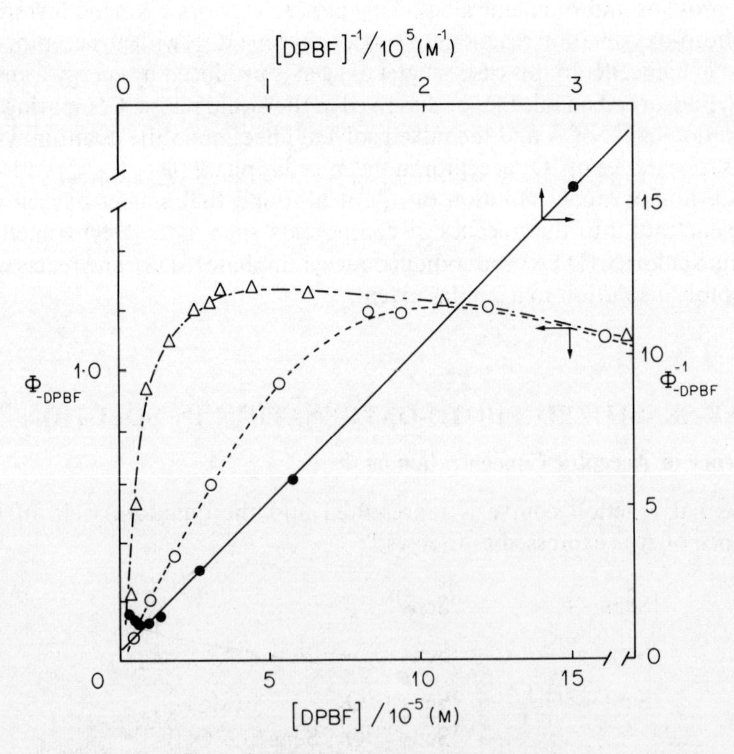

Figure 1 Effect of [DPBF] on Φ_{-DPBF} in methanol (○ and ●) and in tetrachloroethane (△). [MB] = 10 μM

aqueous solutions, an isosbestic point appears on the absorption spectra and the primary product is identified as *cis*-3-hexene-2,5-dione. These results support a successive reaction at higher concentrations of some efficient acceptors[9,10], i.e.

$$AO_2 \text{ (endoperoxide)} + A \longrightarrow 2AO \text{ (diketone)} \qquad (10)$$

Since reaction (5) at even higher concentration of DPBF becomes competitive, the concentration dependence of Φ_{-DPBF} is affected. The Φ_{-DPBF} value decreases from a given concentration of DPBF as shown in Figure 1, in which the concentration coincides with the value obtained by the kinetic examination of the quantum yield.

Effect of Mixed Solvent on Φ_{-A}

The lifetime of singlet oxygen ($^1\Delta$) depends on the solvent period Kearns and co-workers related it to the optical density (O.D.) of the solvent at 7880 cm^{-1} and 6280 cm^{-1} corresponding to 0–0 and 0–1 components of the $^1O_2(^1\Delta)$–$^3O_2(^3\Sigma)$ transition.[8,11,12] When $\beta \gg [A]$ in equation (9), the solvent effect on the value of Φ_{-A} reflects this sensitivity of the lifetime of singlet oxygen ($1/k'_d$).

In the case of a mixed system of solvents 1 and 2 at molar fractions χ_1 and χ_2, respectively, it may be assumed that the Kearns type of equation for the spontaneous decay constant of $^1O_2$ (k'_d) is expressed as

$$k'_d(\mu s^{-1}) = D_{mix} = \chi_1 D_1 + \chi_2 D_2 \qquad (11)$$

where D_1 and D_2 represent the sum of 0·5 $(O.D.)_{7880} + 0{\cdot}05\ (O.D.)_{6280}$ for solvent 1 or 2, i.e., Kearns and co-workers, estimation. When the condition of $\beta \gg [A]$ is still maintained in the mixed solvent, Φ_{-A} is expressed as

$$\Phi_{-A} = \Phi_{^1O_2}\, \beta^{-1}[A]\alpha r = \Phi_{^1O_2}(k'_d/k'^s_A)^{-1}[A]\alpha r \qquad (12)$$

Substituting equation (11) into equation (12) and neglecting the solvent effect on the values of αr, k'^s_A and $\Phi_{^1O_2}$, the ratio of $\Phi^{(2)}_{-A}$ in pure solvent 2 to Φ_{-A} in the mixture is given as a linear function of χ_1:

$$\Phi^{(2)}_{-A}/\Phi_{-A} = 1 - \left(1 - \frac{D_1}{D_2}\right)\chi_1 \qquad (13)$$

In several examples shown in Figure 2 the experimental $\Phi^{(2)}_{-A}/\Phi_{-A}$ points coincide with the predicted curves of equation (13).

The relation $D_{mix} = \chi_1 D_1 + \chi_2 D_2$ is not maintained with the mixture of water and ethanol. In such a case the k'_d value must be calculated from Kearns and co-workers' equation, measuring the values of D_{mix} for each ratio of mixing. Then Φ_{-A} below the region of $\beta \gg [A]$ can be expressed by the equation

$$\Phi_{-A}/\Phi^{(1)}_{-A} = k'^{(1)}_d/k'_d \qquad (14)$$

In the systems MB–O_2–DMF and MB–O_2–DPBF in water–ethanol mixtures, the quantum yields obtained differ completely from the values calculated from

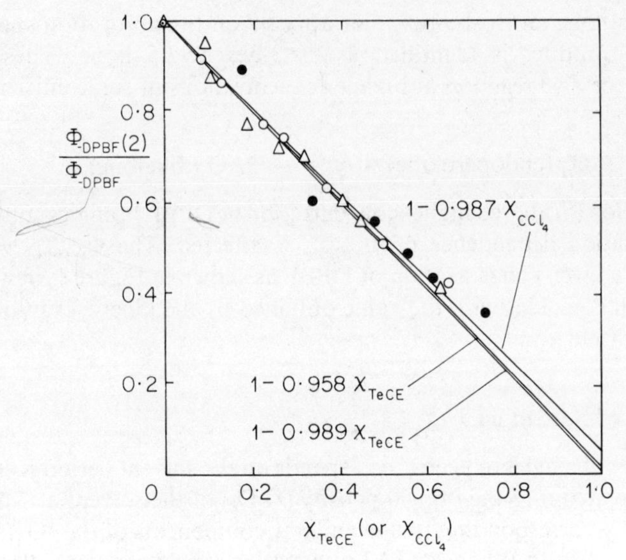

Figure 2 Effect of mixed solvent on $\Phi_{-\text{DPBF}}$. [MB] = 10 μM; [DPBF] = 35 μM. ○, Methanol (solvent 2)–tetrachloroethane (solvent 1); △, *n*-hexane (solvent 2)–tetrachloroethane (solvent 1); ●, *n*-hexane (solvent 2)–carbon tetrachloride (solvent 1). Curves and equation (13) calculated from D_1 and D_2 values of solvents 1 and 2 are shown.

equation (14). A maximum around $\chi_{\text{EtOH}} = 0{\cdot}3$ is found. Since the effect of χ_{EtOH} (or χ_{MeOH}) on $\Phi_{-\text{A}}^{\text{max}}$ is also observed in the mixture of water and ethanol (or methanol), the value of $\Phi_{^1\text{O}_2}\ \alpha r$ must be correlated. The corrected k_{d}' value after the above treatment still deviates from the values from equation (14) obtained with Kearns and co-workers' equation. A maximum at $\chi_{\text{EtOH}} \approx 0{\cdot}2$–$0{\cdot}3$ is found but not in water and methanol mixtures.

DYE-SENSITIZED PHOTO-OXYGENATION OF $^1O_2$ ACCEPTOR IN A MICELLE

Dependence of Acceptor Concentration on $\Phi_{-\text{A/micelle}}$

In the system MB–O_2–DMA/DTA (or DPO/DTA) micelle, most of the MB is not bound to the micelle but remains in the bulk phase. This effect is due to an electrostatic repulsion between both the cationic DTA micelles and the dye DMA or DPO, which is introduced into Hartley's cubic micelle of DTA by an ordinary method, reacts with singlet oxygen produced by energy transfer from the dye in the bulk phase. The concentration effect of DMA/DTA on $\Phi_{-\text{DMA/DTA}}$ is shown in Figure 3 and the reciprocal plot of $1/\Phi_{-\text{DMA/DTA}}$ against 1/[DMA/DTA] exhibits a linear relationship similar to the case of homogeneous solution. The β value obtained from equation (9) in water ($6{\cdot}6 \times 10^{-4}$ M) is about ten times higher than that in deuterium oxide ($6{\cdot}8 \times 10^{-5}$ M). Since the same

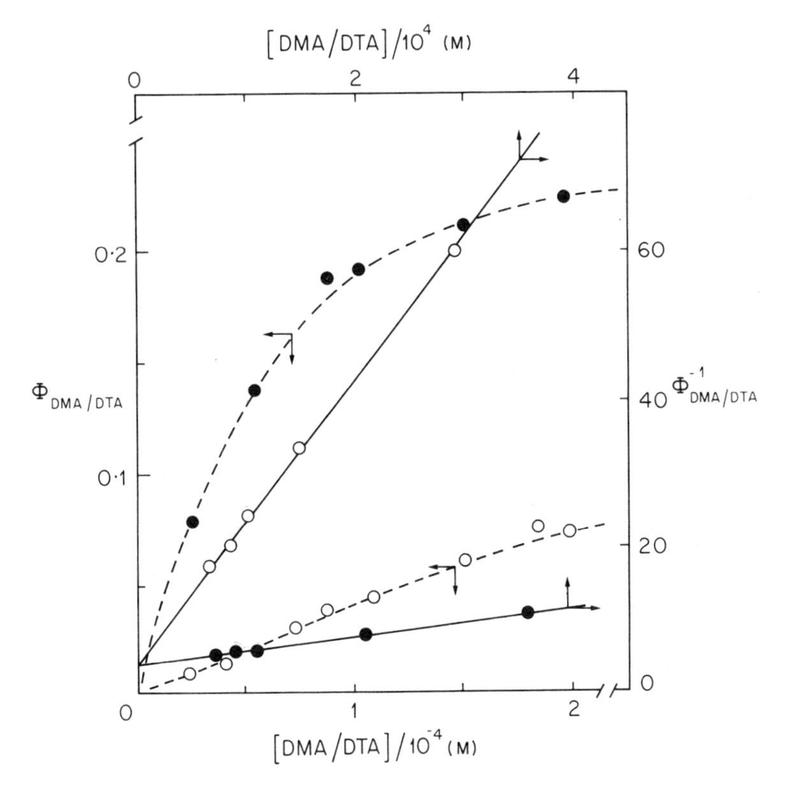

Figure 3 Effect of [DMA/DTA] on $\Phi_{-DMA/DTA}$ in H_2O (○) and in D_2O (●). [MB] = 10 μM; [DTA] = 50 mM

$\Phi^{max}_{-DMA/DTA}$ values are obtained in water and deuterium oxide, the k'_d value in water is ten times that in deuterium oxide. The $\Phi^{max}_{-DmA(DTA}$ value of 0·3–0·5, which is about half that obtained from the concentration effect in ethanol solution, may be considered as accurate since the photo-oxygenated DMA in the micelle is not reactive towards another DMA in the course of reaction, and thus process (10) cannot occur. Similar results for β and $\Phi^{max}_{-DPO/DTA}$ values were obtained in the case of DPO/DTA oxygenation sensitized by MB in bulk.

The dependence of the DMA/DTA concentration on $\Phi_{-DMA/DTA}$ was investigated to test whether DMA/DTA is photo-oxygenated by singlet oxygen produced with the sensitization by anionic eosine (EOS) dye adsorbed on the micelle. The result exhibits a linear relationship between $1/\Phi_{-DMA/DTA}$ and 1/[DMA/DTA]. A similar β value of $5{\cdot}5 \times 10^{-4}$ M and $\Phi^{max}_{-DMA/DTA}$ of 0·3 ($=\Phi_{^1O_2}$) as in the case of MB sensitization is obtained in water. For comparison of $\Phi_{-DMA/DTA}$ with Φ_{-DMA} values sensitized by EOS and MB in ethanol, β and k'^{s}_{A} values are both shown in Table 1. The efficient oxygenation of DMA in a DTA micelle can occur whether or not the dye is electrostatically bound on the ionic surfactant micelle. Thus, assuming the scheme of equations (1)–(8) for these results of quantum yield, it is concluded that singlet oxygen readily penetrates into micellar phase and reacts efficiently with DMA.

Table 1 Kinetic constants for the reaction of singlet oxygen with DMA and DMA/DTA

System	Solvent	β (M)	k'^{s}_{A} (M^{-1} s^{-1})
MB–O_2–DMA	EtOH	$2{\cdot}1 \times 10^{-3}$	$4{\cdot}1 \times 10^{7}$
EOS–O_2–DMA	EtOH	$1{\cdot}9 \times 10^{-3}$	$4{\cdot}5 \times 10^{7}$
MB–O_2–DMA/DTA	H_2O	$6{\cdot}6 \times 10^{-4}$	$7{\cdot}5 \times 10^{8}$
MB–O_2–DMA/DTA	D_2O	$6{\cdot}8 \times 10^{-5}$	$7{\cdot}4 \times 10^{8}$
MB–O_2–DPO[a]/DTA	H_2O	$2{\cdot}2 \times 10^{-3}$	$1{\cdot}6 \times 10^{8}$
EOS/DTA–O_2–DMA/DTA	H_2O	$5{\cdot}5 \times 10^{-4}$	$9{\cdot}1 \times 10^{8}$

[a] DPO = 2,5-diphenyloxazole.

Effect of Mixed Solvent on $\Phi_{-A/\text{micelle}}$

Since β and $\Phi_{-DMA/DTA}$ values were interpreted as accurate and based on the lifetime of singlet oxygen in pure water and deuterium oxide solvents, the mixed solvent effect on the quantum yield for a micellar system can also be represented by equation (13), i.e. it will be determined by the lifetime of singlet oxygen in the bulk phase in a similar manner to that in a homogeneous solution. As expected, the results of the effect of water and deuterium oxide on $\Phi_{-DMA/DTA}$ (or $\Phi_{-DPO/DTA}$) sensitized by MB or EOS/DTA are explained by equation (13), as shown in Figure 4.

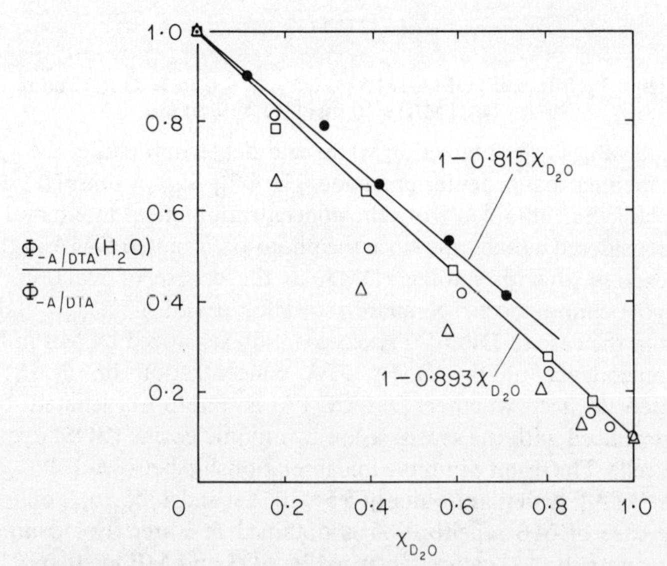

Figure 4 Effect of mixed H_2O–D_2O solvent on Φ_{-DMA} in water (EtOH) and $\Phi_{-DMA/DTA}$ in DTA micelle (50 mM). ●, [MB] = 10 μM, [DMA] = $1{\cdot}0 \times 10^{-4}$ M, [DTA] = 0 M, $\chi_{EtOH} = 0{\cdot}3$; △, [MB] = 10 μM, [DMA/DTA] = $1{\cdot}0 \times 10^{-4}$ M; ○, [MB] = 5 μM, [DPO/DTA] = 25 μM; □, [EOS/DTA] = 10 μM, [DMA/DTA] = 37 μM

In the case of water and ethanol mixtures, the effect on the quantum yield for MB–O_2–DMA/DTA and EOS–O_2–DMA/SDS is anomalous. The increase in the molar fraction of ethanol produces a maximum $\Phi_{-\mathrm{DMA/DTA}}$ at $\chi_{\mathrm{EtOH}} \approx 0{\cdot}2$ (about four times of the value in pure water). $\Phi_{-\mathrm{DMA/DTA}}$ decreases in the region of $\chi_{\mathrm{EtOH}} > 0{\cdot}3$, the extent of the decrease being similar to the results obtained without the DTA micelle (Figure 5). The β values at $\chi_{\mathrm{EtOH}} = 0$ and $0{\cdot}2$ are approximately the same, being $6{\cdot}6 \times 10^{-4}$ and $5{\cdot}2 \times 10^{-4}$ M, respectively. Although more detailed investigations are required for a quantitative assessment, the increase in $\Phi_{-\mathrm{DMA/DTA}}$ might be accompanied by swelling of the micelle due to the addition of ethanol, whereas the decrease might be caused by the decomposition of the micelle in the region of $\chi_{\mathrm{EtOH}} > 0{\cdot}3$.

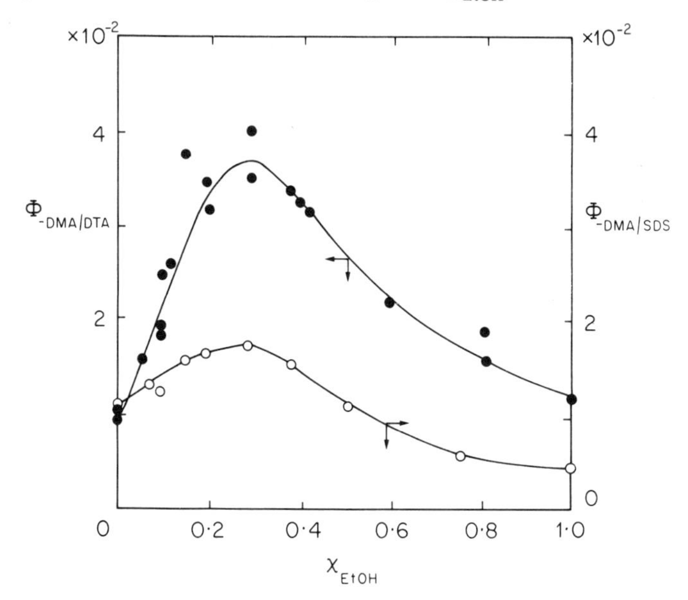

Figure 5 Effect of mixed H_2O–EtOH solvent on $\Phi_{-\mathrm{DMA/DTA}}$ and $\Phi_{-\mathrm{DMA/SDS}}$. ●, [MB] = 5 μM, [DMA/DTA] = 26 μM, [DTA] = 50 mM; ○, [EOS] = 20 μM, [DMA/DTA] = 14 μM, [SDS] = 20 mM

Rånby and Rabek reported that sensitized oxidation of polybutadiene in solution may occur when the dye is deposited on the surface of silica gel and the coated particles are suspended in the polymer solution.[13] A kinetic study of this dye-sensitized photo-oxidation is useful in order to establish the reaction mechanism in the investigated field of photo-oxidative degradation and photostabilization of polymers, for which the role of singlet oxygen in the initial step has been suggested.[14]

REFERENCES

1. J. P. Spikes and B. W. Glad, *Photochem. Photobiol.*, **3**, 907 (1969).
2. A. G. Kepka and L. I. Grossweiner, *Photochem. Photobiol.*, **10**, 49 (1973).
3. K. Kobayashi and T. Ito, *Photochem. Photobiol.*, **23**, 21 (1976).

4. Y. Usui and S. Kodera, *32nd Annual Meeting Chem. Soc. Japan, Abstr.*, **1975**, 311.
5. A. A. Gorman, G. Lovering, and M. A. J. Rodgers, *Photochem. Photobiol.*, **23**, 399 (1976).
6. K. Gollnick and G. O. Schenck, *Pure Appl. Chem.*, **9**, 507 (1964).
7. C. S. Foote, *Accounts Chem. Res.*, **1**, 104 (1968).
8. P. B. Merkel and D. R. Kearns, *J. Amer. Chem. Soc.*, **94**, 7244 (1972).
9. Y. Usui, *Chem. Lett.*, **1973**, 743.
10. Y. Usui and K. Kamogawa, *Photochem. Photobiol.*, **19**, 245 (1974).
11. C. A. Long and D. R. Kearns, *J. Amer. Chem. Soc.*, **97**, 2018 (1975).
12. C. Long and D. R. Kearns, *J. Chem. Phys.*, **59**, 5729 (1973).
13. J. F. Rabek and B. Rånby, *J. Polym. Sci. A1*, **14**, 1463 (1976).
14. B. Rånby and J. F. Rabek, *Photodegradation, Photo-oxidation and Photostabilization*, Wiley, New York, London, 1975.

21

Singlet Oxygen Reactions with Synthetic Polymers

B. RÅNBY and J. F. RABEK

Department of Polymer Technology, The Royal Institute of Technology, Stockholm, Sweden

INTRODUCTION

Reactions of polymers with oxygen from air are very important processes from both the scientific and the industrial points of view. 'Autoxidation', i.e. oxidation of polymers in air, occurs also to various extents under mild conditions. Oxidation may result in modification, chain degradation or crosslinking, which change the physical and mechanical properties.

Oxidation processes of polymers due to reactions with molecular and singlet oxygen are difficult to distinguish. Both types of oxidation may occur simultaneously.

SOME ASPECTS OF MOLECULAR OXYGEN OXIDATION OF POLYMERS

Oxidation with a free-radical mechanism proceeds in three fundamental steps:[1–3] initiation, propagation, and termination, and it involves the formation of peroxy radicals. 'Self-initiation' is a process which may involve different mechanisms in various kinds of polymers. At present no interpretation of the initiation step of the molecular oxidation reactions of polymers is generally accepted. A direct reaction of molecular oxygen with the polymer, resulting in abstraction of hydrogen, is improbable, because it is an endothermic reaction requiring about 30–40 kcal mol^{-1}. Several workers have proposed that small amounts of hydroperoxides present in polymers are responsible for the initiation step. UV irradiation or heating can decompose hydroperoxide groups, to free radicals, and subsequent reactions of the macroradicals may cause further degradation and oxidation of the polymer chain.

The formation of radicals from hydroperoxides can proceed not only as a monomolecular decomposition of hydroperoxides:

$$ROOH \longrightarrow RO\cdot + \cdot OH \qquad (1)$$

but also as a bimolecular interaction of two saturated molecules:

$$ROOH + RH \longrightarrow RO\cdot + R\cdot + H_2O \qquad (2)$$

The presence of carbonyl groups in polymer molecules can be responsible for the initiation process. Under UV irradiation, carbonyl groups absorb light are excited to the triplet state. In this form carbonyl groups are biradicals which can abstract hydrogen from the same or a neighbouring molecule (PH):

$$>C{=}O \xrightarrow{+h\nu} >\dot{C}{-}\dot{O} + PH \longrightarrow >\dot{C}{-}OH + P\cdot \quad (3)$$

One initiation event may produce a series of radicals in subsequent reactions, e.g. a rapid reaction of alkyl polymer radicals with molecular oxygen to form polymer peroxy radicals (POO·):

$$P\cdot + O_2 \longrightarrow POO\cdot \ t \quad (4)$$

$$POO\cdot + PH \longrightarrow POOH + P\cdot \quad (5)$$

The abstraction of hydrogen may occur from the same or another polymer molecule. The rate constant of this abstraction depends on the activation energy, and the reaction can proceed rapidly only if the activation energy is small. The rate of abstraction of hydrogen from hydrocarbon chains increases in the order primary$<$secondary$<$tertiary bonded hydrogen, depending upon the strength of the C—H bonds being broken. Thus, tertiary C—H bonds are weaker and tertiary hydrogen atoms are abstracted at faster rates than primary and secondary hydrogens.

It is evident from reaction (5) that each propagation step produces a new polymer radical (P·). Hence this radical may react with molecular oxygen in a similar way, producing peroxy polymer radicals (POO·). and the reaction can be repeated several times.

The unimolecular decomposition of hydroperoxy groups yields alkoxy polymer radicals and hydroxyl radicals:

$$POOH \longrightarrow PO\cdot + \cdot OH \quad (6)$$

Each of the radicals formed may react with other polymer molecules and the initiation of new oxidative chains may result:

$$PO\cdot + PH \longrightarrow POH + P\cdot \quad (7)$$

$$PH + HO\cdot \longrightarrow P\cdot + H_2O \quad (8)$$

The accumulation and decomposition of hydroperoxides into radicals during the oxidation process by molecular oxygen accelerate the rate of oxidation, and this effect is known as the 'autocatalytic stage'. This reaction can be stopped by free-radical termination processes.

Free-radical oxidation can easily be detected by ESR spectroscopy.[4] Peroxy radicals have g-values in the range 2·014–2·019 and no hyperfine structure. This observation serves to distinguish them from alkyl radicals commonly present in oxidizing systems which have g-values near that of the free electron ($g = 2{\cdot}0023$).

The effect of structure on the rate of autoxidation of polymers has been a topic of concern for some time: however, only recently has much effort been devoted to the separation of the structural effects that influence the termination and propagation. This separation requires measurements of the absolute rates of termination and propagation or the determination of relative rates of propagation.[1–3] These reactions obviously occur by a free-radical mechanism.

SOME ASPECTS OF SINGLET OXYGEN OXIDATION OF POLYMERS

The reactions of singlet oxygen in polymer oxidation should be considered in terms of the following questions:

1. How can singlet oxygen be generated in polymers?
2. What is the role of singlet oxygen in the initiation step of the oxidative degradation of polymers?
3. What is the mechanism of the singlet oxygen oxidation of polymers?
4. How can polymers be stabilized against singlet oxygen oxidation?

1. Singlet oxygen in a polymer matrix can be formed in two ways:

(i) In photosensitizing reactions due to traces of impurities present. It has been demonstrated that polyolefins exposed to air may absorb polynuclear aromatics released into the atmosphere from combustion processes.[5–7] Many polynuclear aromatic hydrocarbons are able to generate singlet oxygen efficiently.[8–10]

(ii) In an energy-transfer reaction between excited triplet states of chromophoric groups (e.g. carbonyl groups) embedded in a polymer backbone chain and molecular oxygen.[11]

The surface of a polymer can also be affected by singlet oxygen formed in photochemical reactions which occur, for example, in a polluted atmosphere or water in the environment of the polymer sample. Gaseous singlet oxygen, $^1O_2(^1\Delta_g)$, has a sufficiently long lifetime to cause oxidation of a solid polymer surface at some distance from the site of formation of the active agent. The lifetime of singlet oxygen in the polymer matrix is important. Singlet oxygen is reported to survive $3{\cdot}5 \times 10^4$ collisions with solid walls and more than 10^8 molecular collisions in the vapour phase.[12] It has been found that singlet oxygen can diffuse a significant distance (>100Å) through a hydrocarbon polymer film before appreciable deactivation occurs.[13]

2. Assuming that singlet oxygen is responsible for the initiation mechanism of polymer oxidation, it should be remembered that the rate constant of hydrogen abstraction depends primarily on the activation energy of the reaction, which cannot be lower than the heat of reaction. A reaction can proceed rapidly only if the activation energy is small. Furthermore, an oxidation reaction can be rapid only if the bond which is formed is at least as strong as that which is broken. From the comparison of bond strength in saturated hydrocarbon polymers (e.g. polyethylene, where P–H is ca. 80 kcal mol^{-1}) with that of the

new bond formed (P–OOH, ca. 70 kcal mol^{-1}), it is obvious that this type of reaction is not favoured. Kaplan and Kelleher[14] have reported that singlet oxygen can abstract hydrogen atoms from saturated hydrocarbons such as *n*-tetracosane. On the other hand, Carlsson and Wiles[15] did not find detectable amounts of hydroperoxide in hexadecane after exposure to singlet oxygen from a microwave source for 5 h. The resonance-weakened O—H, S—H, and N—H bonds of phenols (about 40 kcal mol^{-1}), thiophenols, and amines provide more readily abstractable hydrogen atoms by singlet oxygen. Matsuura and co-workers[16,17] have carried out valuable work on abstraction of hydrogen from hydroxyl groups in substituted phenols by singlet oxygen:

$$ArOH + {}^1O_2 \longrightarrow [ArOH\ldots O_2] \longrightarrow ArO\cdot + HO_2\cdot \quad (9)$$

It can be assumed that this reaction may occur with polymers that contain hydroxyl groups.

The assumption that singlet oxygen is responsible for the formation of hydroperoxy groups in polymers elucidates the initiation step of polymer oxidation. This initiation step, however, has not been fully proved for saturated polymers. The very few hydroperoxy groups assumed to be formed are not detectable by common analytical methods. Still they may further initiate a rapid free-radical oxidation.

3. Studies of the singlet oxygen oxidation mechanism of polymers are often highly complex, for the following reasons:

(i) experimental methods for singlet oxygen oxidation, separation of the oxidized products, analytical characterization, etc., of low molecular weight compounds are much simpler than those for polymers;

(ii) it is very difficult to separate and distinguish free-radical oxidation from pure singlet oxygen oxidation;

(iii) in any case unreacted polymer must be oxidized in the presence of the oxidized reaction products formed;

(iv) ESR spectroscopy, which is a very sensitive detection method for free radicals present in the polymer, does not provide evidence that radicals are present in the polymer treated with singlet oxygen;[18]

(v) low concentrations of compounds such as initial impurities, additives, and reaction products are capable of considerably changing the singlet oxygen oxidation reaction products and their rates of formation. We can even expect a complete quenching of singlet oxygen in two ways:

(a) chemical deactivation:

$${}^1O_2 + Q \longrightarrow QO_2 \quad (10)$$

(b) physical deactivation:

$$^1O_2 + Q \longrightarrow Q + {^3O_2} \tag{11}$$

where Q is the quencher. It is well known that the synthesis of pure polymers is experimentally impossible. External impurities such as initiator residues and internal impurities such as chromophoric groups are present even in a carefully purified polymer. Some impurities may cause rapid and complete deactivation of singlet oxygen, and this would prevent any effect of singlet oxygen on the polymer.

4. Another important aspect of singlet oxygen in polymer chemistry is the role of photostabilizers and antioxidants in singlet oxygen quenching.[19–22] This problem is discussed by Wiles in Chapter 34.

REACTIONS OF SINGLET OXYGEN WITH POLYMERS

The possible role of singlet oxygen in polymer oxidation has been reviewed in several recent publications.[1,2,23–29] The purpose of this paper is to summarize the present experimental evidence for reactions of singlet oxygen with polymers.

Polymers with Saturated Backbone Chains

Polyolefins, poly(vinyl chloride), polystyrene, and poly(methyl methacrylate) belong to this group of polymers. It should be pointed out, however, that all of these polymers may contain small amounts of unsaturated bonds and other chromophoric groups which may not even be detected analytically. These groups may, however, still be important for the singlet oxygen oxidation mechanism.

Trozzolo and Winslow[11] were the first to propose a singlet oxygen mechanism for the photo-oxidation of polyolefins. They suggested that the reaction proceeds in the following stages:

1. Carbonyl groups, e.g. in commercial polyethylene, are excited to the triplet state by UV irradiation.
2. The excitation of carbonyl groups to the triplet state may cause a cleavage of carbon–carbon bonds by the Norrish Type II reaction and result in the degradation of the polymer chains:

$$-CH_2-CH_2-{^3\left[\overset{\overset{\displaystyle O}{\|}}{C}-\right]}-CH_2-CH_2-CH_2-CH_2- \longrightarrow$$

$$-CH_2-CH_2-\overset{\overset{\displaystyle O}{\|}}{C}-CH_3 + CH_2{=}CH-CH_2-CH_2- \tag{12}$$

3. Some of the excited carbonyl groups in the triplet state may transfer their excitation energy to molecular oxygen, forming singlet oxygen $^1O_2(^1\Delta_g)$.

4. Singlet oxygen $^1O_2(^1\Delta_g)$ can react with vinyl groups and form hydroperoxides:

$$^1O_2(^1\Delta_g) + CH_2{=}CH{-}CH_2{-}CH_2{-} \longrightarrow HOO{-}CH_2{-}CH{=}CH{-}CH_2{-} \qquad (13)$$

5. UV irradiation or heating can decompose hydroperoxide groups to free radicals, and subsequent disproportionation of the macroradical may cause further degradation of the polymer chain.

Carlsson and Wiles[15] examined the oxidation of polypropylene by singlet oxygen from a microwave discharge unit. After exposure for 51 h, the concentration of hydroperoxides was found to be $5{\cdot}0 \times 10^{-4}$ M. This is the result of the reaction of singlet oxygen with unsaturated double bonds present in most polypropylene chains. Pre-treatment of polypropylene film with singlet oxygen decreased its photostability. It eliminated the usual induction period and greatly accelerated the destruction of the film. On the other hand, Russell *et al.* did not find any significant amount of peroxides in polyethylene foam after exposing it to singlet oxygen for 6 h. A similar result was obtained for polypropylene in solution by Mill *et al.*[30] They suggested that the concentration of unsaturated bonds is very low and the peroxides formed cannot be detected.

Exposure of polystyrene film to a stream of singlet oxygen produced by a microwave discharge did not give any evidence of the formation of detectable amounts of hydroperoxides.[18, 31]

Rabek and Rånby[32] proposed that the initial stage of the photo-oxidative degradation of polystyrene may involve reactions of singlet oxygen with polystyrene molecules. Singlet oxygen may be formed in a reaction between excited phenyl groups in polystyrene molecules and molecular oxygen:

$$-CH_2-CH(C_6H_5)- \xrightarrow{h\nu} -CH_2-CH(^3C_6H_5)- \xrightarrow{^3O_2} -CH_2-CH(C_6H_5)- + {}^1O_2 \qquad (14)$$

$$-CH_2-CH(C_6H_5)- \xrightarrow{^3O_2} [-CH_2-CH(C_6H_5 \cdots O_2)-]_{CTC} \xrightarrow{h\nu} -CH_2-CH(C_6H_5)- + {}^1O_2$$

This assumption was made on the basis of the tested reaction that singlet oxygen can be formed by energy transfer from the excited singlet (^1S) or triplet (^1T) state of benzene molecules to the ground state of molecular oxygen:[33, 34]

$$\text{Benzene } (S_0) + h\nu \longrightarrow \text{benzene } (^1S) \qquad (15)$$

$$\text{Benzene } (^1\text{S}) \longrightarrow \text{benzene } (^3\text{T}) \tag{16}$$

$$\text{Benzene } (^1\text{S or } ^3\text{S}) + {}^3\text{O}_2 \longrightarrow \text{benzene } (\text{S}_0) + {}^1\text{O}_2 \tag{17}$$

As we reported previously,[32] it has not been established whether singlet oxygen can abstract a tertiary hydrogen atom from a polystyrene molecule:

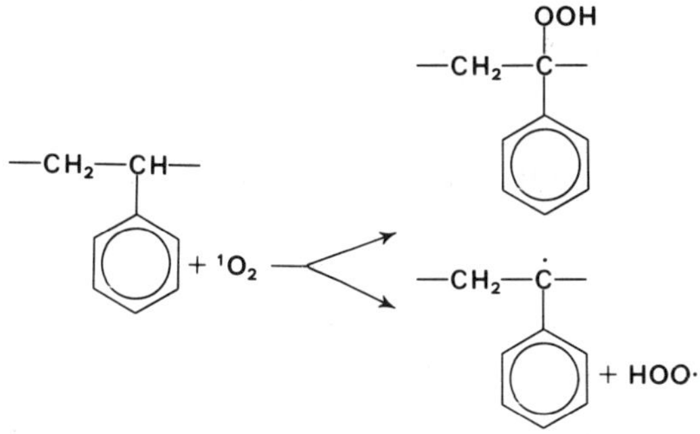

$—CH_2—CH(C_6H_5)— + {}^1O_2 \longrightarrow —CH_2—C(OOH)(C_6H_5)—$ (18)

$\longrightarrow —CH_2—\dot{C}(C_6H_5)— + HOO\cdot$ (19)

It is more probable that singlet oxygen reacts with phenyl groups in polystyrene molecules. As a result of this reaction, a ring-opening reaction is expected:

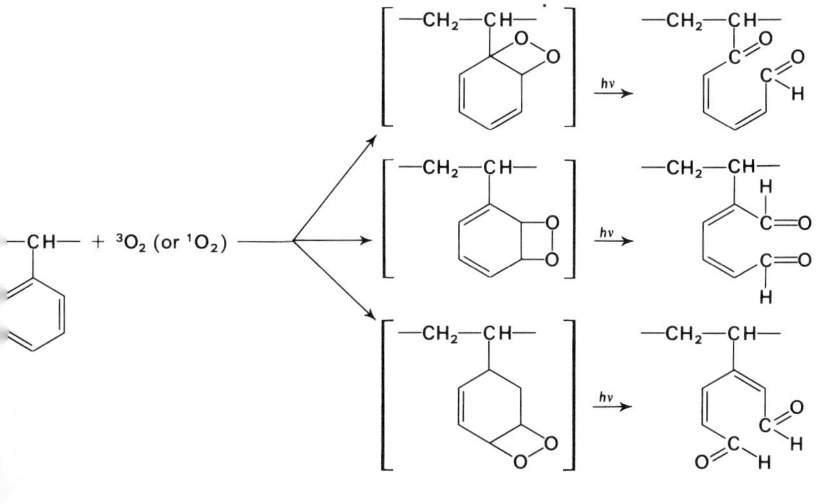

$—CH(C_6H_5)— + {}^3O_2 \text{ (or } {}^1O_2) \longrightarrow [\text{cyclic peroxides}] \xrightarrow{h\nu} \text{unsaturated dialdehydes}$ (20)

The formation of unsaturated dialdehydes as a result of ring opening photo-oxidation of benzene molecules has previously been reported by Wei *et al.*[35]

The role of singlet oxygen in the photo-oxidation of polystyrene is indicated by the observation that addition of β-carotene (an effective singlet oxygen quencher) decreases the photo-oxidative degradation of polystyrene.[32] However, this result cannot be used as definitive proof of a singlet oxygen mechanism of polystyrene oxidation. β-Carotene may also quench the triplet state of polystyrene molecules, and furthermore act as a UV absorber which reduces the amount of UV energy absorbed by the polystyrene. The experimental verification of the various possible processes in the photo-oxidative degradation of polystyrene is difficult and the interpretations remain uncertain.

The effects of singlet oxygen in the photo-oxidation of poly(vinyl chloride) were reported by Kwei[36] and Gibb and MacCallum.[37] The conjugated double bonds formed during photo-oxidative degradation of PVC play a special role in this case.[38, 39] The following mechanism is proposed for the photoreaction of molecular oxygen with conjugated double bonds in partially dehydrochlorinated PVC:[37]

1. Polyene double bonds are excited to the triplet state by UV irradiation:

$$\text{—(CH=CH)—} + h\nu \longrightarrow \text{—}^{3}\text{(CH=CH)*—} \tag{21}$$

2. Some of the double bonds excited to the triplet state may transfer their excitation energy to molecular oxygen, forming singlet oxygen $^{1}O_2(^{1}\Delta_g)$:

$$\text{—}^{3}\text{(CH=CH)*—} + {}^{3}O_2 \longrightarrow \text{—(CH=CH)—} + {}^{1}O_2 \tag{22}$$

3. The singlet oxygen $^{1}O_2(^{1}\Delta_g)$ formed may react with vinyl groups present in the backbone chain:

$$\text{—CH=CH—CH}_2\text{—} + {}^{1}O_2 \longrightarrow \text{—}\underset{}{\overset{\text{OOH}}{\overset{|}{\text{C}}\text{H}}}\text{—CH=CH—} \tag{23}$$

4. UV irradiation is reported[37] to decompose hydroperoxide groups to carbonyl groups and radical species which may accelerate further the dehydrochlorination of PVC:

$$\text{—}\overset{\text{OOH}}{\overset{|}{\text{C}}\text{H}}\text{—CH=CH—} \longrightarrow \text{—}\overset{\text{O}}{\overset{|}{\text{C}}}\text{—CH=CH—} + \begin{matrix}\text{unidentified}\\ \text{free radicals}\end{matrix} \tag{24}$$

Details of this mechanism were not given.

Exposure of PVC film to singlet oxygen generated by microwave excitation does not give any detectable results such as a loss in weight or a change in contact angle with distilled water.[31]

Polymers with Unsaturated Backbone Chains

Unsaturated polymers such as *cis*- and *trans*-polybutadienes and *cis*-polyisoprene were found to react readily with singlet oxygen produced by a microwave discharge.[18, 40–44] During this reaction hydroperoxides are formed at the surface of the sample and are detected by internal reflection infrared spectroscopy (ATR) as a band at 3400 cm^{-1}. The following mechanism for singlet oxygen oxidation of *cis*-polybutadiene has been proposed by Kaplan and Kelleher:[42]

```
—CH2    H2C—CH2    H2C—CH2    H2C—
   \    /      \    /      \    /
   HC=CH       HC=CH       HC=CH

            | 1O2
            ↓

—CH2    HC—CH     H2C—CH2    H2C—
   \   //    \\   /      \    /
   HC—CH      HC—CH       HC=CH
   |             |
   OOH           OOH                              (25)

            | 1O2
            ↓

—CH2    HC=CH     H2C—CH2    H2C—
   \   /     \    /      \    /
   HC—CH      HC—CH       HC=CH
   |    \    /   |
   OOH   O—O     OOH
```

It is interesting that a vinyl-polybutadiene sample containing 91·5% vinyl units, 7% *cis*-, and 1·5% trans-unsaturated units was not oxidized under these conditions.[42]

PHOTOSENSITIZED SINGLET OXYGEN OXIDATION

The photosensitized singlet oxygen oxidation of low molecular weight organic compounds is well known and has been studied extensively during the last decade. In this type of reaction, singlet oxygen is formed by an energy-transfer process from the excited singlet and/or triplet state of a sensitizer (S) to molecular oxygen, which can be represented as follows:

$$S_0 + h\nu \longrightarrow {}^1S \qquad (26)$$

$${}^1S \longrightarrow {}^3S \qquad (27)$$

$${}^1S + {}^3O_2 \longrightarrow S_0 + {}^1O_2 \qquad (28)$$

$${}^3S + {}^3O_2 \longrightarrow S_0 + {}^1O_2 \qquad (29)$$

A theoretical examination of singlet oxygen formation in photosensitized reactions was carried out by Kearns and co-workers.[45–47] When the excitation energy of a sensitizer molecule exceeds 22·5 kcal mol^{-1}, singlet oxygen ${}^1O_2({}^1\Delta_g)$ may be formed. For the formation of singlet oxygen ${}^1O_2({}^1\Sigma_g^+)$, the excitation energy of the sensitizer molecule should be higher than 37·5 kcal mol^{-1}.

The singlet oxygen formed can participate in two reactions:

1. deactivation (quenching):

$${}^1O_2 \xrightarrow{k_1} {}^3O_2 \qquad (30)$$

2. oxidation:

$${}^1O_2 + A \xrightarrow{k_2} AO_2 \qquad (31)$$

The ratio of the rate constants, $\beta = k_1/k_2$, is well known for various low molecular weight compounds.[48–50] For polymers, the determination of β-values becomes more complicated, because it is difficult to determine the amount of oxidized product formed (AO_2). As yet, no β-values for polymers have been published.

It has been reported by several workers that polynuclear aromatic hydrocarbons accelerate the photo-oxidative degradation of polyolefins,[15,19,20] polystyrene,[25,51,52] poly(methyl methacrylate),[53–61] and polydienes.[25,62] Polynuclear aromatic hydrocarbons photosensitize the generation of singlet oxygen, which may probably then react with these polymers, but the detailed mechanism of these reactions has not yet been elucidated.

Italian workers[63] reported that 1,1,4,4-tetraphenylbutadiene is a sensitizer for the oxidative photodegradation of polyolefins and that this reaction occurs by a singlet oxygen mechanism.

Rabek and Rånby[64] found that *p*-quinones sensitize a rapid photo-oxidative degradation of polystyrene in solution and the solid state, and this reaction can be inhibited by addition of β-carotene.

Dye-sensitized oxidative photodegradation of polydienes has been reported by several workers.[22,44,65,66] Small amounts (10^{-3}–10^{-4} M) of different dyes (e.g. methylene blue, fluorescein, Rose Bengal, acridine orange) added to a polymer solution and even present in a solid film (cast from solution) may efficiently accelerate the photodegradation (in solution) and the crosslinking (in solid films). Rabek and Rånby[44] reported that sensitized oxidation of polybutadiene in solution may also occur when the dye is deposited on the surface of silica gel and the coated particles are suspended in the polymer solution during UV irradiation.

Treatment of polybutadienes with singlet oxygen generated by a physical method or by a dye-photosensitizing method results in the formation of an IR absorption band at 3400 cm^{-1}, which can be attributed to hydroxyl and hydroperoxide groups. It is characteristic that under these conditions the formation of carbonyl groups is not observed and no absorption in the 1700–1760 cm^{-1} region is recorded. A plausible complete mechanism for the observed oxidative degradation of polydienes by singlet oxygen has not yet been formulated.

Terpolymers of ethylene, propylene, and ethylidenenorbornene react rapidly with singlet oxygen generated by dye sensitization.[67] Singlet oxygen attacks the ethylidene double bonds, forming at least two different hydroperoxides:

$^1O_2$ → OOH (44%, (*exo* + *endo*)) + OOH (24%) (3

The instability of these hydroperoxides causes degradation of the photo-oxidized terpolymers at relatively low temperatures. Terpolymers of ethylene, propylene, and dicyclopentadiene are inert to singlet oxygen.

Egerton and co-workers[68–70] found that anthraquinone dyes such as 1-piperidino-anthraquinone, 1,4-bismethylaminoanthraquinone, and 1-anilino-anthraquinone are able to sensitize the photo-oxidation of nylons and cellulose.

Acriflavine and eosin are two photosensitizing dyes which have been shown to sensitize the photochemical degradation of textile fibres.[71,72]

Davies and Dixon[73] reported that addition of eosin increased the sensitivity of hydroxypropylmethoxycellulose to visible light. Zweig and Henderson[67] have also reported dye-sensitized photo-oxidation of cellulose acetate film.

CONCLUSIONS

The role of singlet oxygen in the oxidation of synthetic polymers and the possible importance of singlet oxygen quenching in photostabilization processes can now be considered for industrial applications. Controlled stability of commercial polymers to oxygen is a key problem for the long-term properties of polymeric materials. Fundamental research in this field is the basis for technological advances in stabilization problems.

ACKNOWLEDGEMENTS

The authors thank the Swedish Board for Technical Development (STU) and the Swedish Polymer Research Foundation (SSP) for their generous support of this research.

REFERENCES

1. J. F. Rabek, in *Degradation of Polymers, Comprehensive Chemical Kinetics*, (Ed. C. H. Bamford and C. F. Tipper), Elsevier, Oxford, Vol. 14, 1975, p. 265.
2. B. Rånby and J. F. Rabek, *Photodegradation, Photo-oxidation and Photostabilization of Polymers*, Wiley, London, 1975.
3. L. Reich and S. Stivala, *Elements of Polymer Degradation*, McGraw-Hill, New York, 1971.
4. B. Rånby and J. F. Rabek *ESR Spectroscopy in Polymer Research*, Springer Verlag, Heidelberg, 1977.
5. R. H. Partridge, *J. Chem. Phys.*, **45**, 1679 (1966).
6. I. Boustead and A. Charlesby, *Eur. Polym. J.*, **3**, 459 (1967).
7. A. P. Pivovarov, Y. V. Gak, and A. F. Lukovnikov, *Vysokomol. Soedin.*, **A13**, 2110 (1971).
8. K. Gollnick, *Adv. Photochem.*, **6**, 1 (1968).
9. K. Gollnick, T. Franken, G. Schader, and G. Dorhofer, *Ann. N.Y. Acad. Sci.*, **171**, 89 (1970).
10. B. Stevens and B. E. Algar, *Ann. N.Y. Acad. Sci.*, **171**, 50 (1970).
11. A. M. Trozzolo and F. H. Winslow, *Macromolecules*, **1**, 98 (1968).
12. A. M. Winer and K. D. Bayes, *J. Phys. Chem.*, **70**, 302 (1966).
13. B. A. Schnuriger, J. Bourdon, and J. Bedu, *Photochem. Photobiol.*, **8**, 361 (1968).

14. M. L. Kaplan and P. G. Kelleher, *J. Polym. Sci. B*, **9**, 565 (1971).
15. D. J. Carlsson and D. M. Wiles, *J. Polym. Sci. B*, **11**, 759 (1973).
16. T. Matsuura, K. Omura, and R. Nakashima, *Bull. Chem. Soc. Japan*, **38**, 1358 (1965).
17. T. Matsuura, N. Yoshimura, A. Nishinaga, and I. Saito, *Tetrahedron Lett.*, **1972**, 4933.
18. A. K. Breck, C. L. Taylor, K. E. Russell, and J. K. S. Wan, *J. Polym. Sci. A*, **12**, 1505 (1974).
19. D. J. Carlsson, T. Suprunchuk, and D. M. Wiles, *J. Polym. Sci. B*, **11**, 61 (1973).
20. D. J. Carlsson, G. D. Mendenhall, T. Suprunchuk, and D. M. Wiles, *J. Amer. Chem. Soc.*, **94**, 8960 (1972).
21. J. Flood, K. E. Russell, and J. K. S. Wan, *Macromolecules*, **6**, 669 (1973).
22. A. Zweig and W. A. Henderson, Jr., *J. Polym. Sci. A1*, **13**, 717, 993 (1975).
23. J. F. Rabek, *Polimery*, **16**, 257 (1971).
24. J. F. Rabek, *Wiad. Chem.*, **25**, 293, 365, 435 (1971).
25. J. F. Rabek, in *XXIII IUPAC Congress, Boston, Mass.*, Butterworths, London, Vol. 8, 1971, p. 29.
26. J. F. Rabek and B. Rånby, *International Symposium on Degradation and Stabilization of Polymers, Brussels, Sept. 11–13, 1974*, p. 257.
27. J. F. Rabek and B. Rånby, *Polym. Eng. Sci.*, **15**, 40 (1975).
28. B. Rånby and J. F. Rabek, *Ultraviolet Light Induced Reactions in Polymers* (Ed. A. A. Labana), *ACS Symp. Ser.*, No. 25, 1976, p. 391.
29. D. J. Carlsson and D. M. Wiles, *Rubb. Chem. Technol.*, **49**, 991 (1976).
30. T. Mill, H. R. Richardson, and F. R. Mayo, *J. Polym. Sci. A*, **11**, 2899 (1973).
31. J. R. MacCallum and C. T. Rankin, *Makromol. Chem.*, **175**, 2477 (1974).
32. J. F. Rabek and B. Rånby, *J. Polym. Sci. A1*, **12**, 273 (1974).
33. D. R. Snelling, *Chem. Phys. Lett.*, **2**, 346 (1968).
34. J. A. Davidson and E. W. Abrahamson, *Photochem. Photobiol.*, **15**, 403 (1972).
35. K. Wei, J. C. Mani and J. N. Pitts, *J. Amer. Chem. Soc.*, **89**, 4225 (1967).
36. K. P. S. Kwei, *J. Polym. Sci. A1*, **7**, 1075 (1969).
37. W. H. Gibb and J. R. MacCallum, *Eur. Polym. J.*, **10**, 533 (1974).
38. J. F. Rabek, G. Canbäck, J. Lucki, and B. Rånby, *J. Polym. Sci. A1*, **14**, 1447 (1976).
39. B. Rånby and J. F. Rabek, *Second International Symposium on Polyvinylchloride, Lyon, France, 5–9 July, 1976.*
40. M. L. Kaplan and P. G. Kelleher, *Science, N.Y.*, **169**, 1206 (1970).
41. M. L. Kaplan and P. G. Kelleher, *Rubb. Chem. Technol.*, **45**, 423 (1972).
42. M. L. Kaplan and P. G. Kelleher, *J. Polym. Sci. A1*, **8**, 3163 (1970).
43. J. P. Dalle, R. Magous, and M. Mousseron-Canet, *Photochem. Photobiol.*, **15**, 141 (1972).
44. J. F. Rabek and B. Rånby, *J. Polym. Sci. A1*, **14**, 1463 (1976).
45. D. R. Kearns, *Chem. Rev.*, **71**, 395 (1971).
46. A. U. Khan and D. R. Kearns, *Adv. Chem. Ser.*, No. 76, 143 (1968'.
47. D. R. Kearns, *Amer. Chem. Soc. Div. Pet. Chem. Prepr.*, **16**, A9 (1971).
48. B. Stevens, *Ann. N.Y. Acad. Sci.*, **171**, 50 (1970).
49. T. Wilson, *J. Amer. Chem. Soc.*, **88**, 2898 (1966).
50. R. H. Young, R. Martin, K. Wehrly, and D. Feriozi, *Amer. Chem. Soc. Div. Pet. Chem. Prepr.*, **16**, A89 (1971).
51. T. Takeshita, K. Tsuji and T. Seiki, *J. Polym. Sci. A1*, **10**, 2315 (1972).
52. J. F. Rabek, *Scientific Report*, Uppsala University, 1968.
53. H. Mönig, *Naturwissenschaften*, **45**, 12 (1958).
54. H. Mönig in *Probleme und Ergebnisse aus Biophysik und Strahlenbiologie*, Akademie Verlag, Berlin, 1960, Vol. 2, p. 174.
55. H. Mönig and H. Kriegel, in *Progress in Photobiology* (Ed. B. C. Chriestensen and B. Buchman), Elsevier, Amsterdam, 1961, p. 618.

56. H. Mönig in *Probleme und Ergebnisse aus Biophysik und Strahlebiologie*, Akademie Verlag, Berlin, 1962, Vol. 3, p. 55.
57. H. Mönig and H. Kriegel, *Z. Naturforsch.*, **15b**, 333 (1960).
58. H. Mönig and H. Kriegel, *Proc. Int. Congr. Photobiol., Copenhagen*, 1960, p. 618.
59. H. Mönig and H. Kriegel, *Biophysik*, **2**, 22 (1964).
60. H. Siewert, *Z. Naturforsch.*, **19b**, 806 (1964).
61. R. B. Fox and T. R. Price, *J. Appl. Polym. Sci.*, **11**, 2373 (1967).
62. J. F. Rabek, Y. J. Shur, and B. Rånby, in preparation (cf. Chapter 26).
63. E. Cernia, E. Mantovani, W. Marconi, M. Mazei, N. Palladino, and A. Zanobi, *J. Appl. Polym. Sci.*, **19**, 15 (1975).
64. J. F. Rabek and B. Rånby, *J. Polym. Sci. A1*, **12**, 295 (1974).
65. T. Mill, K. C. Irvin, and F. R. Mayo, *Rubb. Chem. Technol.*, **41**, 296 (1968).
66. J. F. Rabek and L. Skorupa, *Scientific Report*, Institute of Organic Chemistry and Plastics, Technical University of Wroclaw, 1970.
67. E. E. Duynstee and M. E. A. H. Mevis, *Eur. Polym. J.*, **8**, 1375 (1972).
68. G. S. Egerton, *Brit. Polym. J.*, **3**, 63 (1971).
69. G. S. Egerton, N. E. N. Assaad and N. D. Uffindell, *Chem. Ind., Lond.*, **1967**, 1172.
70. G. S. Egerton, J. M. Gleadle, and A. G. Roach, *Nature, Lond.*, **202**, 345 (1964).
71. G. S. Egerton, *J. Soc. Dyers Colour.*, **65**, 764 (1949).
72. G. S. Egerton, *Nature, Lond.*, **204**, 1153 (1964).
73. D. H. Davies and D. Dixon, *J. Appl. Polym. Sci.*, **16**, 2449 (1972).

22

Theoretical Investigations on the Degradation of Polymers by Singlet Oxygen

J. J. LINDBERG and P. PYYKKÖ

Department of Wood and Polymer Chemistry, University of Helsinki, Helsinki, Finland

Department of Physical Chemistry, Åbo Akademi, Turku, Finland

INTRODUCTION

The calculation of reasonably accurate wavefunctions for moderately large sized molecules is now completely routine (Figure 1). The primary aim of such calcu-

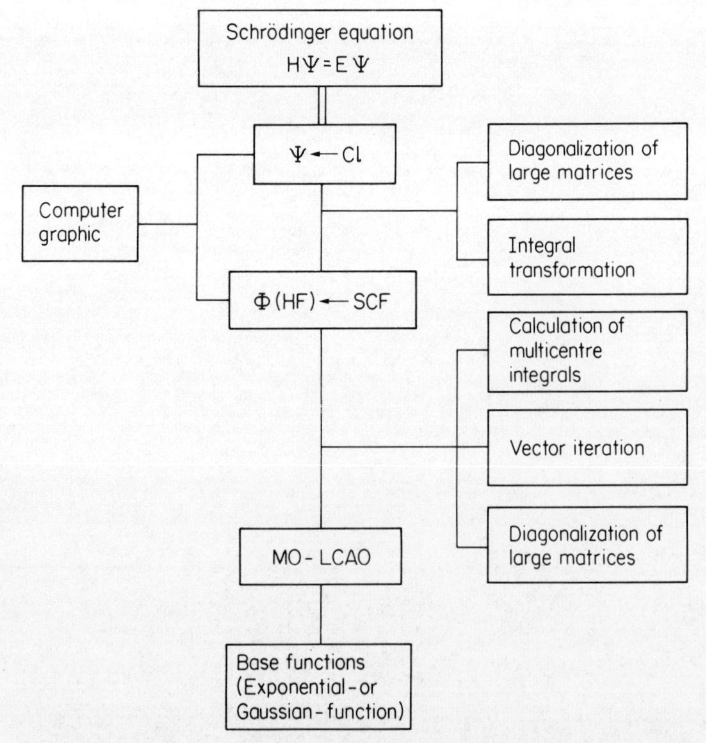

Figure 1 Use of computer for carrying out numerical calculations in quantum chemistry

lations is to evaluate physical properties, such as the dipole moment of a molecule, the electric field gradient at a nucleus, spectroscopic constants, etc. Also the much more complex question of chemical reactivity and reaction paths has been tackled by wave-mechanical calculations. Today there are many different methods, with various degrees of accuracy and sophistication, available for solving the above problems. Some of them are as follows:

Semi-empirical methods: HMO;[1,2] Free electron theory;[3–5] π-electron self-consistent field theory;[6,7] EHT;[8] CNDO, INDO;[9] MINDO;[9,10] PCILO.[11]
Non-empirical calculations: Ab initio methods.[24]

The simple Hückel molecular orbital method (HMO), like other empirical methods, rely too heavily on experimental data in the selection of adjustable parameters to be of use for more serious calculations of chemical reactivity and conformations. The semi-empirical methods, especially the extended Hückel molecular orbital method (EHT), described by Hoffman,[8] together with other methods taking in consideration the various amounts of differential overlap functions (CNDO, INDO, MINDO), have been used extensively for the calculation of ground-state properties of molecules. The attempts to use these for open-shell structures and excited states have indicated that, at best, rough approximations of true values may be obtained (Table 1). However, the properties calculated are primarily related to the gaseous or unperturbed state. Moreover, by suitable choice of empirical parameters, other states of aggregation may be considered.

Recently, Bingham *et al.*[10] reported that the differential orbitals version (MINDO/3) of the self-consistent field molecular orbital treatment comes reasonably close to meeting the requirements of accuracy. The same applies to open-shell structures and excited states. However, for exact results, the consideration of all orbitals is necessary as in the *ab initio* methods. Owing to excessive requirements of computer space and time methods of this type are, however, restricted to systems with high symmetry or comparatively few orbitals compared with the other methods mentioned. Recently, in some cases a simulated *ab initio* molecular orbital (SAMO) technique has proved to be an effective but less expensive means for producing wavefunctions of near *ab initio* accuracy for large molecules.[12]

SINGLET OXYGEN ($^1\Delta_g$, $^1\Sigma_g^+$)

The three lowest states of the oxygen molecule may be represented by the following electronic configurations:

$^3O_2$ ($1\sigma_g^2 1\sigma_u^2 2\sigma_g^2 2\sigma_u^2 1\pi_u^4 3\sigma_g^2 1\pi_g^2$, $^3\Sigma_g^-$: stable)

$^1O_2$ ($^1\Delta_g$: lifetime 2700 s)

($^1\Sigma_g^+$: lifetime 12 s)

All three systems are simple enough for possible use in very elaborate and detailed *ab initio* calculations, and much work of this type has been performed. Investigations by Langhoff[14] have shown that about two-thirds of splitting effects may be described as spin–orbit effects. Dewar *et al.*[13] have also tackled the problem of their MINDO/3 method. They obtained results of reasonable accuracy (Table 1) compared with other calculations and experimental results. The best results for O_2 have been obtained by Moss *et al.*[26,27] using a generalized valence bond method.

From Table 1 it is evident that an *ab initio* calculation (SCF-CI) using 128 configurations can only give as good, usually less accurate, results than the much discussed method of Dewar *et al.* It should, however, be mentioned that the MINDO/3 method was adjusted by reference to heats of atomization (binding energy) and molecular geometries in the ground state. This is also emphasized by the increasing error with increasing distance from the ground state.

Table 1 Theoretical calculations on the three lowest states of the oxygen molecules

State	Parameter	Experimental	Method of calculation			
			CNDO[9]	INDO[9]	MINDO/3[13]	*Ab initio*
$^3O_2\ (^3\Sigma_g^-)$	Equilibrium bond length (Å)	1·207	1·132	1·140	1·206	1·30[a] (1·220)[b]
	Binding energy (eV)	5·21	17·44	15·37	5·25	3·72[a] (4·72)[b]
$^1O_2\ (^1\Delta_g)$	Equilibrium bond length (Å)	1·216	—	—	1·206	1·33
	Binding energy (eV)	4·10	—	—	4·45	2·72
$^1O_2\ (^1\Sigma_g^+)$	Equilibrium bond length (Å)	1·227	—	—	1·206	1·34
	Binding energy (eV)	3·438	—	—	3·71	2·35
	Energy differences:					
	$^3\Sigma_g^- - {}^1\Delta_g$	0·95	—	—	0·77	1·00
	$^3\Sigma_g^- - {}^1\Sigma_g^+$	1·65	—	—	1·54	1·37

[a] 64 configurations.[15]
[b] 128 configurations.[23]

REACTIONS BETWEEN SINGLET OXYGEN ($^1\Delta_g$) AND POLYMERS

Rånby and Rabek,[16] Gollnick,[17] Belluš,[18] and Jefford[19] have extensively reviewed the possible models for the reaction of polymers with singlet oxygen in its $^1\Delta_g$ state. The practical and technological aspects of these reactions have been pointed out by Wiles.[20]

From the above studies, it can be concluded that a simple model for the reaction of singlet oxygen with the active centres of solid-state polyolefins can be represented by the following simple scheme:

These reactions may be calculated within the framework of *ab initio* methods. Dewar and Thiel[21] have recently studied closely related reaction models by their MINDO/3 method. (See also Ref. 25.)

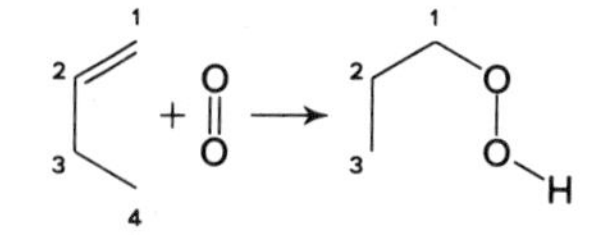

(1)

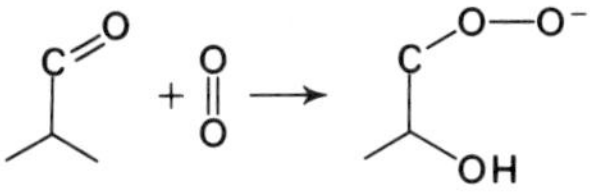

(2)

According to the MINDO/3 method, the cycloaddition of singlet oxygen to ethylene proceeds exclusively by the formation of a peroxirane (1) intermediate, which rearranges to 1,2-dioxetane (**2**). In the region of interaction, structures which are geometrically and electronically similar to peroxirane were always more stable than dioxetane-like structures.

(3)†

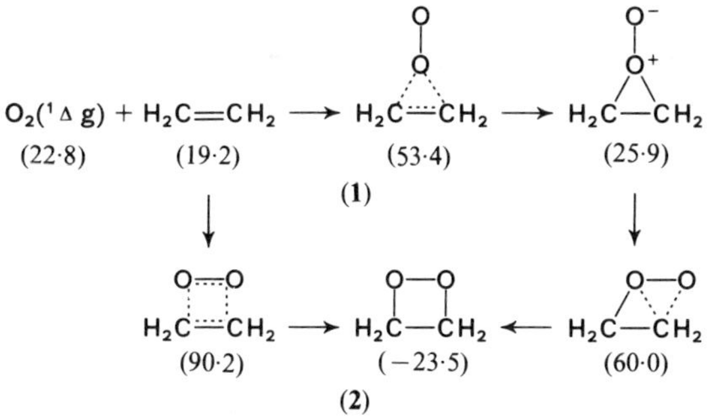

$O_2(^1\Delta g) + H_2C{=}CH_2 \rightarrow H_2C{\cdots}CH_2 \rightarrow H_2C{-}CH_2$

(22·8) (19·2) (53·4) (25·9)

(**1**)

$H_2C{\cdots}CH_2 \rightarrow H_2C{-}CH_2 \leftarrow H_2C{-}CH_2$

(90·2) (−23·5) (60·0)

(**2**)

As with ethylene, the reaction of singlet oxygen with propylene yields methylperoxirane as the first product. It can be formed in *cis*- (**3**) or *trans*- (**4**) geometries whose energies are similar. Both (**3**) and (**4**) can rearrange to the corresponding dioxetane (**5**) via very unsymmetrical and highly polar transition states. The presence of the methyl group in *cis*-methylperoxirane allows an alternative mode of rearrangement via the cyclic transition state to form propylene 3-hydroperoxide (**6**). A comparison of the various transition states indicates that (**7**) has the lowest energy. Thus MINDO/3 predicts that the intermediate should rearrange more easily to the hydroperoxide (ene reaction) rather than decompose or rearrange to the dioxetane (1,2-cycloaddition). The result is evidently in accordance with general views.

Furthermore, in attempts to find a competing one-step process, a two-dimensional potential surface for the reaction was calculated. The bond lengths, R_{CO} and R_{OH}, of the forming CO and OH bonds were taken as independent variables, all the other coordinates being optimized. In this grid search, the workers mentioned found no path for a one-step ene reaction.

The rate-determining transition state for the ene reaction was predicted to be

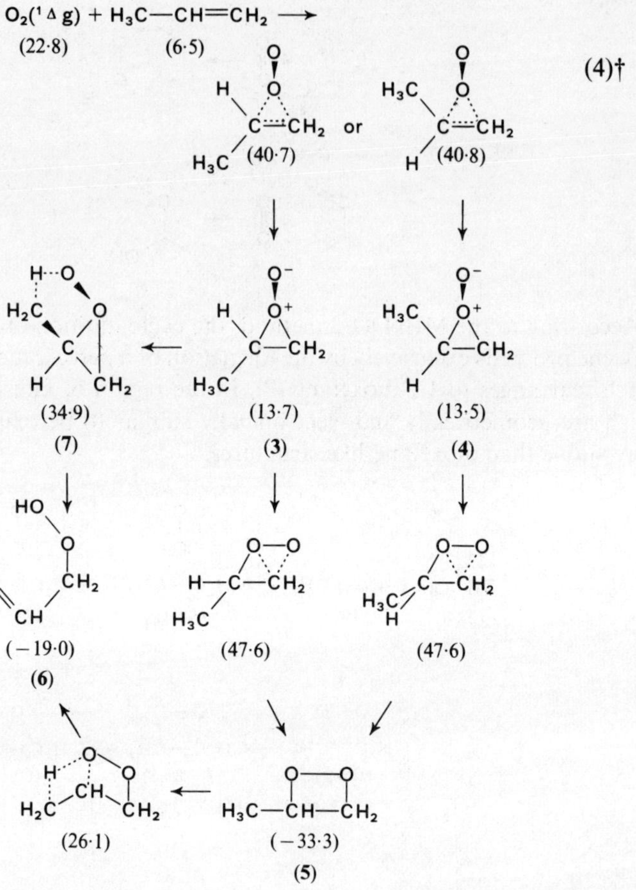

$O_2(^1\Delta g) + H_3C{-}CH{=}CH_2 \longrightarrow$ (4)†

(22·8) (6·5)

$H(H_3C)C{=}CH_2{\cdots}O{-}O$ (40·7) or $H_3C(H)C{=}CH_2{\cdots}O{-}O$ (40·8)

(34·9) (7) ← (13·7) (3); (13·5) (4)

(−19·0) (6); (47·6); (47·6)

(26·1) ← $H_3C{-}CH{-}CH_2$ (−33·3) (5)

† Calculated heats of formation (kcal/mol^{-1}) in parentheses.

11·5 kcal mol^{-1} higher in energy than the energy of the reactants. For comparison, it may be mentioned that the experimental value for *cis*-butene is 10·0 ± 1 kcal mol^{-1},[22] which is in excellent agreement with the above, considering that the substitution of one methyl group by a hydrogen atom should slightly increase the activation energy.

CONCLUSIONS

The above discussion indicates clearly that MINDO/3 can provide useful preliminary information concerning the mechanisms of chemical reactions. However, in spite of the contrary views of some workers, time-consuming *ab initio* calculations should be performed in order to obtain more accurate predictions. Furthermore, in the field of the singlet oxygen chemistry, the reactions of carbonyl groups, the problem of charge transfer, the formation of charge–transfer complexes, and the stabilization of polymers may also be profitably studied by the theoretical methods indicated. However, theoretical deductions without the strong support of experimental knowledge are still not possible, owing to the difficulties in predicting the influence of impurities and the surrounding reaction medium. In the near future we hope to be able to present more theoretical material, calculated by *ab initio* methods, to support and to complete the experimental evidence.

REFERENCES

1. A. Streitwieser, Jr., *Molecular Orbital Theory for Organic Chemists*, Wiley, New York, 1961.
2. E. Heilbronner and H. Bock, *Das HMO-Modell und seine Anwendung*, Verlag Chemie, Weinheim, 1968.
3. K. Ruedenberg, *J. Chem. Phys.*, **22**, 1878 (1954).
4. N. S. Ham and Ruedenberg, *J. Chem. Phys.*, **29**, 1199 (1958).
5. H. Kuhn, *J. Chem. Phys.*, **16**, 840 (1948).
6. L. Salem, *The Molecular Orbital Theory of Conjugated Systems*, Benjamin, New York, 1966.
7. J. N. Murrell, *The Theory of the Electronic Spectra of Organic Molecules*, Wiley, New York, 1968.
8. R. Hoffman, *J. Chem. Phys.*, **39**, 1397 (1963).
9. J. A. Pople and D. L. Beveridge, *Approximate Molecular Orbital Theory*, McGraw-Hill, New York, 1970.
10. R. C. Bingham, M. J. S. Dewar, and D. H. Lo, *J. Amer. Chem. Soc.*, **97**, 1285, 1294, 1302, 1311 (1975).
11. M. Giacomini, B. Pullman, and B. Maigret, *Theor. Chim. Acta*, **19**, 347 (1970).
12. J. E. Eilers and D. R. Whitman, *J. Amer. Chem. Soc.*, **95**, 2067 (1973).
13. M. J. S. Dewar, R. C. Haddon, W.-K. Li, W. Thiel, and P. K. Weiner, *J. Amer. Chem. Soc.*, **97**, 4540 (1975).
14. S. R. Langhoff, *J. Chem. Phys.*, **61**, 1708 (1974).
15. H. F. Schaefer, III, and F. E. Harris, *J. Chem. Phys.*, **48**, 4946 (1968).
16. B. Rånby and J. Rabek, *Photodegradation, Photo-oxidation and Photostabilization of Polymers*, Wiley, New York, 1975.
17. K. Gollnick, *Mechanism and Kinetics of Chemical Reaction of Singlet Oxygen with Organic Compounds*, First EUCHEM Conference on Singlet Oxygen Reactions with Polymers, Stockholm, 1976.
18. D. Belluš, *Quenchers of Singlet Oxygen—A Critical Review*, First EUCHEM Conference on Singlet Oxygen Reactions with Polymers, Stockholm, 1976.
19. C. W. Jefford, *Reaction of Singlet Oxygen with α-Keto-carboxylic Acids of Biological Interest*, First EUCHEM Conference on Singlet Oxygen Reactions with Polymers, Stockholm, 1976.
20. D. M. Wiles, *Photo-oxidative Reactions of Polymers*, IUPAC Macromolecular Symposium, Stockholm, 1976.
21. M. J. S. Dewar and W. Thiel, *J. Amer. Chem. Soc.*, **97**, 3978 (1975).
22. E. Koch, *Tetrahedron Lett.*, **1968**, 3271.
23. H. F. Schaefer, III, *J. Chem. Phys.*, **54**, 2207 (1971).
24. L. Radom and J. A. Pople, *MTP International Review of Science, Physical Chem., Ser. One*, **1**, 71 (1972).
25. S. Inagaki and K. Fukui, *J. Am. Chem. Soc.*, **97**, 7480 (1975).
26. B. J. Moss and W. A. Goddard III, *J. Chem. Phys.*, **63**, 3523 (1975).
27. B. J. Moss, F. W. Bobrowicz and W. A. Goddard III, *J. Chem. Phys.*, **63**, 4632 (1975).

23

The Role of Singlet Oxygen in the Photo-oxidation of Polymers—Some Practical Considerations

G. Scott

University of Aston in Birmingham, Gosta Green, Birmingham, B4 7ET, England

INTRODUCTION

The anomalous sensitivity of polyethylene to light [1] has led to much speculation about the nature of the chromophore responsible for the initiation step. Carbonyl groups have been widely discussed as being responsible for the UV absorption in polyolefins [2] which occurs at wavelengths above 300 nm and copolymerization of carbon monoxide during manufacture has been suggested as a source.[3] Hydroperoxide groups are known to be formed at an early stage in the thermal oxidation of polyethylene [4] and it has been suggested [1] that the photolability of these primary oxidation products is responsible for photo-initiation during the early stages of photo-oxidation. More recently there has been similar speculation about the function of UV stabilizers and in particular the class of metal complex stabilizers whose UV absorption characteristics do not seem adequately to explain their effectiveness. Some of the more effective of these (notably the dithiocarbamates and dithiophosphates) had been shown to be effective catalysts for the decomposition of hydroperoxides [5] and it was suggested that this was an important mechanistic role which contributed to their UV-stabilizing function in polymers.[6]

It was later found that many metal complexes, some of which were effective and others ineffective as UV stabilizers, were efficient quenching agents for excited states of molecules [7–14] (in particular carbonyl triplet and oxygen singlet) and it is now common to invoke this mechanism to explain the function of any UV stabilizer whose mechanism is not unequivocably established.[15]

There is no question that singlet oxygen reacts readily with polyenic polymers such as polybutadiene,[14] nor is there much doubt that the hydroperoxides so produced must be effective photoactivators.[1] Furthermore, it seems likely that sensitizing groups such as carbonyl, if present in polymers, should be readily quenched by ground-state oxygen to give singlet oxygen. However, subsequent reactions of singlet oxygen will depend on its ability to react with chemical groups present in the polymer in competition with physical quenching by small molecules present.

The purpose of this paper is to examine the possible role of singlet oxygen in photo-oxidation in three stages. Firstly, evidence will be sought to show that singlet oxygen produced by quenching of carbonyl compounds can compete with other sensitizers as initiators of photo-oxidation of aliphatic polymers. The second question that will be examined is whether quenching of singlet oxygen can compete with other photo-inhibition processes in photo-oxidizing media. Finally, the likely relevance of these processes in the light of the known sequence of chemical steps that occur during the thermal oxidation and photo-oxidation of polymers will be discussed.

EFFECTS OF CARBONYL COMPOUNDS IN THE PHOTODEGRADATION OF POLYMERS

Carbonyl impurities seem to be the only type which need be considered as a potential general cause of photo-sensitization in polymers. Polycyclic hydrocarbons [8] and dye sensitizers may be present in certain instances (e.g. in polymers exposed to an industrial atmosphere) but there is no evidence that they are present in newly fabricated polymers.

Ketonic groups can be conveniently introduced into hydrocarbon polymers by thermal oxidation. Figure 1(c) shows the effect of a substantial concentration

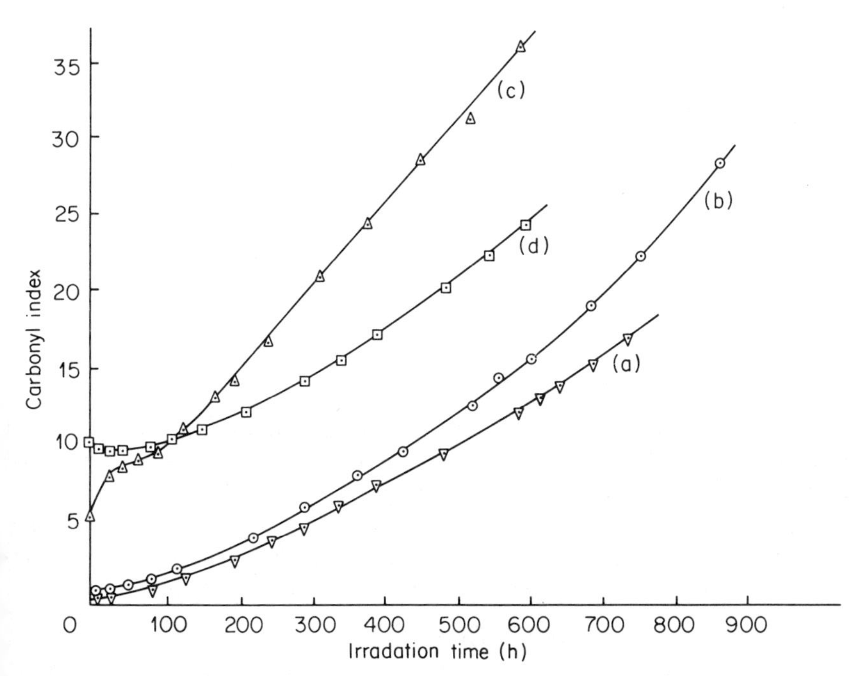

Figure 1 Effect of processing and subsequent heat treatment in argon on the photo-oxidation of LDPE. Samples: (a) control, compression-moulded without processing; (b) processed in an atmosphere of argon; (c) processed in closed mixer for 30 min; (d) as (c) but heat-treated in argon at 110 °C for 16 h

of carbonyl, introduced by a severe processing operation, on the subsequent photo-oxidative behaviour of low-density polyethylene (LDPE). Curve (d) represents a similar sample which in addition was subjected to a high-temperature anealing process in argon to remove hydroperoxides by thermolysis. It can be seen that the latter, although it contains a higher concentration of ketone than sample (c), photo-oxidizes at a slower rate. Furthermore, the rate of photo-oxidation of (d) is very similar to that of a control sample (a) which contains only a fraction of the carbonyl concentration of (d) and it actually photo-oxidizes more slowly than (b), which was processed in a small amount of oxygen.

Only sample (c) contained a measurable amount of hydroperoxide and it is evident that hydroperoxide photolysis is a much more important initiating process than is carbonyl sensitization. Nevertheless, the carbonyl groups present undergo photolysis since there is a rapid formation of vinyl groups by the Norrish Type II process (1) in the polymer:

$$\mathrm{RCOCH_2CH_2CH_2R'} \longrightarrow \mathrm{RCOCH_3 + CH_2{=}CHR'} \tag{1}$$

This process is known to be favoured over the alternative Norrish Type I photolysis (2)[16] in the initial stages of photolysis of solid polymers:

$$\mathrm{RCOCH_2CH_2CH_2R'} \longrightarrow \mathrm{R\dot{C}O + \cdot CH_2CH_2CH_2R} \tag{2}$$

There is no evidence that the increased concentration of carbonyl in polymer (d) leads to a higher rate of initiation of photo-oxidation than that observed in polymers (a) and (b) by either direct abstraction of hydrogen by the triplet state [reaction (3a)], or by the formation of singlet oxygen [reaction (3b)], and its subsequent reaction with vinyl [reaction (4)].[17]

$$\mathrm{R\underset{\bullet}{\overset{\overset{\displaystyle \dot{O}}{|}}{C}}CH_2CH_2CH_2R'} \xrightarrow[\text{(a)}]{\text{RH}} \mathrm{R\underset{\bullet}{\overset{\overset{\displaystyle OH}{|}}{C}}CH_2CH_2CH_2R' + R\cdot} \tag{3}$$

$$\xrightarrow[\text{(b)}]{^3\mathrm{O_2}} \mathrm{RCOCH_2CH_2CH_2R' + {}^1O_2}$$

$$\mathrm{R''CH_2CH{=}CH_2 + {}^1O_2} \longrightarrow \mathrm{RCH{=}CHCH_2OOH} \tag{4}$$

There is now considerable evidence confirming that the presence of initial unsaturation (vinylidene groups) in polyethylene is at least partly responsible for its lack of photostability. It is, therefore, of interest to see how the two possible photoinitiators, *tert*-butyl hydroperoxide (TBH) and diamyl ketone (**1**), affect the rate of photo-oxidation of 1,1′-diamylethylene (**2**), a low molecular weight model of the vinylidene function in polyethylene.

$$\mathrm{CH_3CH_2CH_2CH_2CH_2\underset{\underset{\displaystyle O}{\|}}{C}CH_2CH_2CH_2CH_2CH_3}$$

(1)

$$\mathrm{CH_3CH_2CH_2CH_2CH_2\underset{\underset{\displaystyle CH_2}{\|}}{C}CH_2CH_2CH_2CH_2CH_3}$$

(2)

The results are shown in Figure 2, from which it can be seen that TBH, curve (b), removes the induction period associated with the photo-oxidation of the vinylidene compound [curve (a)] completely, whereas diamyl ketone (**1**) at the same molar concentration has virtually no effect on the rate of oxidation. The initial reduction in carbonyl concentration is again associated with carbonyl photolysis and clearly indicates that triplet carbonyl was formed. The presence of an allylic methylene group is a particularly favourable situation for hydrogen abstraction by reaction (3a) since it has been shown that the reaction of triplet carbonyl with olefins is two orders of magnitude faster than that with saturated hydrocarbons [18] and yet this reaction does not appear to take place. This suggests that Norrish Type II photolysis is the preferred process.

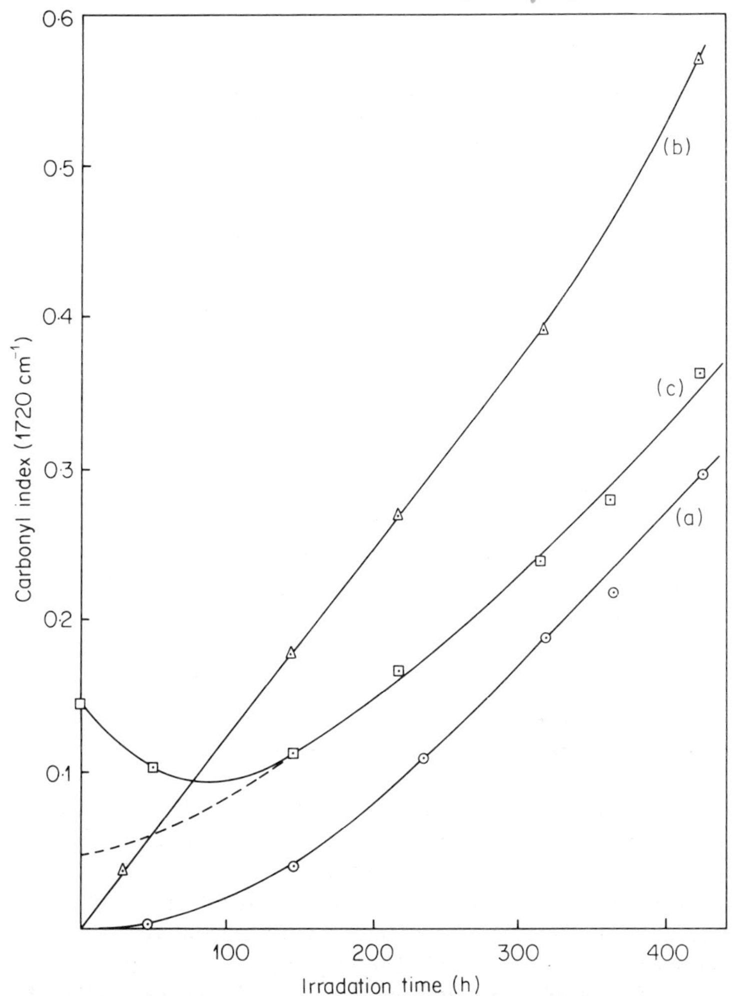

Figure 2 Effect of photoactivators on the photo-oxidation of 1,1-diamylethylene: (a) control (no additive); (b) *tert*-butyl hydroperoxide (2×10^{-4} mol cm^{-3}); (c) diamyl ketone (2×10^{-4} mol cm^{-3})

The absence of hydroperoxide formation through attack of singlet oxygen on the olefin groups suggests that this species, if it is formed under the conditions of the experiment, must be quenched by other molecules present. Table 1 lists the singlet oxygen quenching constants for some simple chemicals[19] which are present either in the surface of the polymer or in the gas phase with which it is in contact. It also gives the quenching constants for typical 1-alkyl-and 1,2-dialkylethylenes. It is clear that several of the gaseous contaminants (particularly water) are able to compete effectively with the olefins for singlet oxygen. Since they are all present in a large molar excess over the unsaturation in the polymer, this is a likely explanation for the fact that reactions of singlet oxygen with the polymer or with 1,1′-diamylethylene do not appear to take place.

Table 1 Singlet oxygen ($^1\Delta_g$) quenching by other molecules[19]

Molecule	Rate constant, k_Q (l mol^{-1} s^{-1} $\times 10^{-3}$)
O_2	1·40
N_2	0·06
CO_2	2·30
H_2O	9·00
$CH_3CH{=}CHCH_3$	25·00
$C_4H_9CH{=}CH_2$	6·70

Breck *et al.*[14] found that singlet oxygen did not react with polyethylene foams exposed to singlet oxygen over a 6 h period. Similar studies in these laboratories on polyethylene film have confirmed this conclusion.[20] No hydroperoxide or carbonyl could be detected by the methods described previously[21] and which show the presence of hydroperoxide in processed polymers. Moreover, the oxidation curves of such singlet oxygen-treated samples were identical with the control. This contrasted with the effect of an ozone-containing atmosphere over the same period of time which led to the presence of a substantially increased carbonyl concentration but again did not increase the rate of photo-oxidation.[20]

The low reactivity of vinyl and vinylidene groups in polymers towards singlet oxygen is in accord with the order of activity found in simple olefins in homogeneous solution.[22] Studies carried out using sensitizers (Rose Bengal or chlorophyll) in the presence of oxygen gave the order of decreasing activity of some relevant model compounds as dimethylcyclohexene > methylcyclohexene > cyclohexene ≫ oct-1-ene.[23] The very low reactivity of 1-olefins under even the most favourable conditions is relevant to the present findings in polymers since this is the olefinic structure which has been postulated to be involved in the singlet oxygen photosensitization process.[17]

Polybutadiene, which contains the *cis*-1,2-dialkylethylene structure, should be more reactive towards singlet oxygen than are the vinylidene and vinyl groups in polyethylene (see Table 1). Previous studies have shown that singlet oxygen does indeed react with polybutadiene-containing polymers.[14,24] The possibility of carbonyl photosensitized initiation of this polymer and of PB-modified polystyrene (high-impact polystyrene, HIPS) has, therefore, been examined. Dinonyl ketone was added to HIPS at a concentration which corresponded to an extensively oxidized HIPS sample (25h). It was found that this concentration of carbonyl had almost no effect on the photo-oxidation process (see Figure 3).[25]

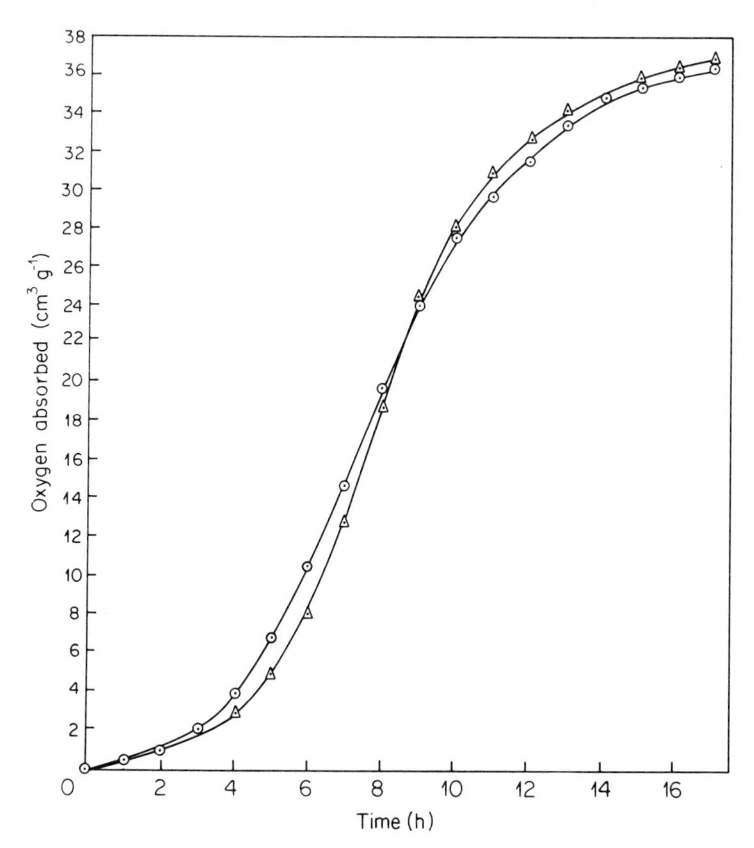

Figure 3 Photo-oxidation of HIPS film containing dinonyl ketone. △, Control; ○, 10×10^{-3} mol per 100 g

The initial ketone product formed by photolysis of an allylic hydroperoxide is, however, not a saturated but a conjugated ketone:[15]

$$\text{—CH}_2\text{CH=CH}\underset{}{\overset{\text{OOH}}{\overset{|}{\text{C}}}}\text{HCH}_2\text{—} \longrightarrow \text{—CH}_2\text{CH=CH}\overset{\text{O}}{\overset{\|}{\text{C}}}\text{CH}_2\text{—} \qquad (5)$$

Mesityl oxide was, therefore, added to polybutadiene in chlorobenzene solution and photo-oxidized. This proved to be equally ineffective as a photoactivator at low concentrations and at higher concentrations it behaved as a retarder of photo-oxidation (see Figure 4).[31] Although mesityl oxide absorbs just above 300 nm, the extinction coefficient is very low and this stabilizing behaviour is not thought to be a UV-screening effect but is more probably due to initial interference with the radical chain reaction. However, it is clear that even under these favourable conditions there is no acceleration of photo-oxidation even after the induction period, and hence singlet oxygen activation on the polymer cannot have occurred.

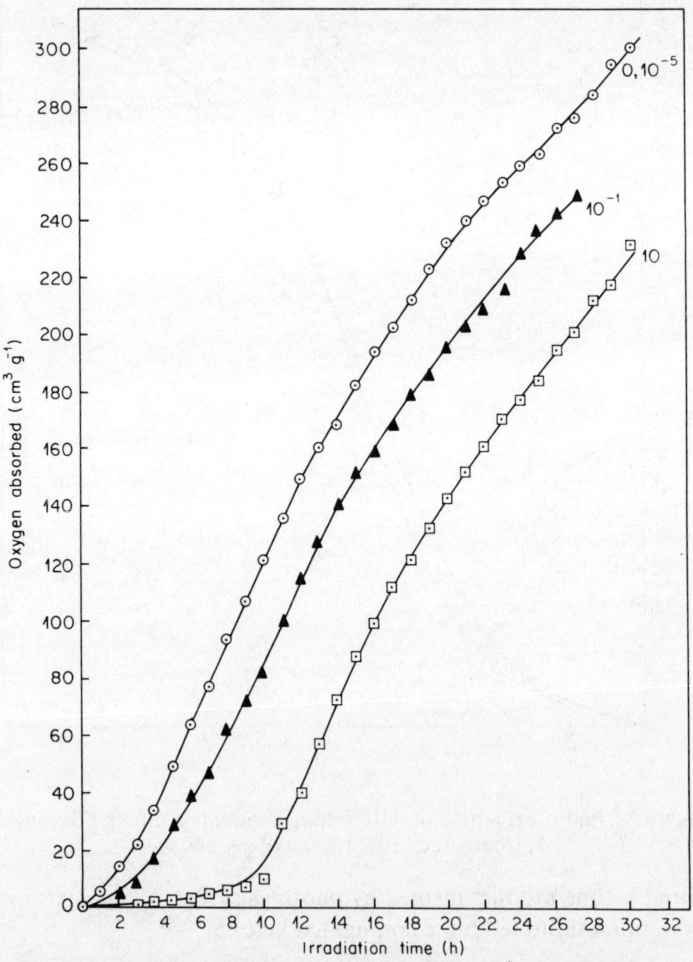

Figure 4 Photo-oxidation of polybutadiene in chlorobenzene containing mesityl oxide (numbers on curves are moles of mesityl oxide per 100 g of rubber)

UV-STABILIZING MECHANISMS

Although there is no doubt that many nickel complex UV stabilizers can quench singlet oxygen (with rate constants up to 10^6 times as fast as those listed in Table 1),[26] the evidence presented above casts considerable doubt upon the relevance of quenching to their UV-stabilizing behaviour. Recent studies have shown that they also have the ability to react rapidly with hydroperoxides both thermally and in light-catalyzed reactions.[27]

Nickel dibutyldithiocarbamate (NiDBC) (**3**), whose effectiveness as a UV stabilizer was suggested by Guillory and Becker[13] to be due to its ability to quench singlet oxygen, is one of the most powerful light-activated catalysts for the ionic destruction of hydroperoxides known. The products are those expected on the basis of a non-radical decomposition [for example, cumene hydroperoxide (CHP) gives almost entirely phenol and acetone in a Lewis acid catalysed process].[27]

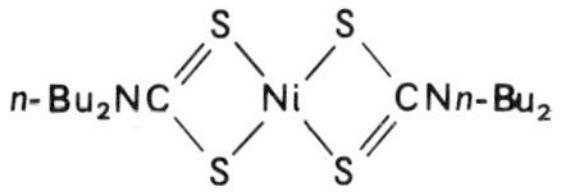

(**3**) 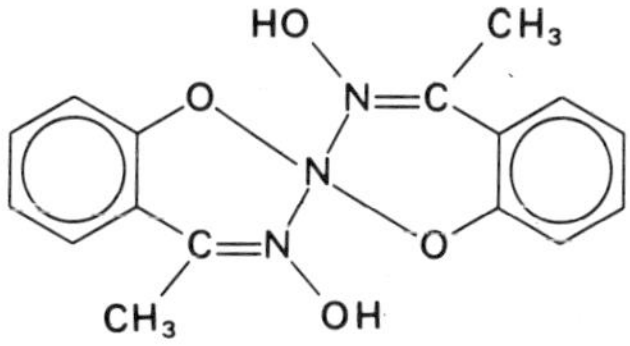 (**4**)

Derivatives of nickel 2-hydroxyacetophenone oxime (NiOx) (**4**) have been extensively investigated by Briggs and McKellar[8] and by Adamczyk and Wilkinson[28] and, on the basis of quenching experiments, the latter concluded[26] that quenching of singlet oxygen is probably involved in their action. However, it has been shown that in a benzophenone-initiated photo-oxidation of cumene, NiOx behaves entirely as a UV-screening agent.[29,30] In a hydroperoxide-initiated photo-oxidation, on the other hand, it has an additional function which has been shown[27] to be a light-catalyzed stoichiometric ionic reaction with the hydroperoxide in which one molecule of NiOx destroys up to eight molecules of hydroperoxide:

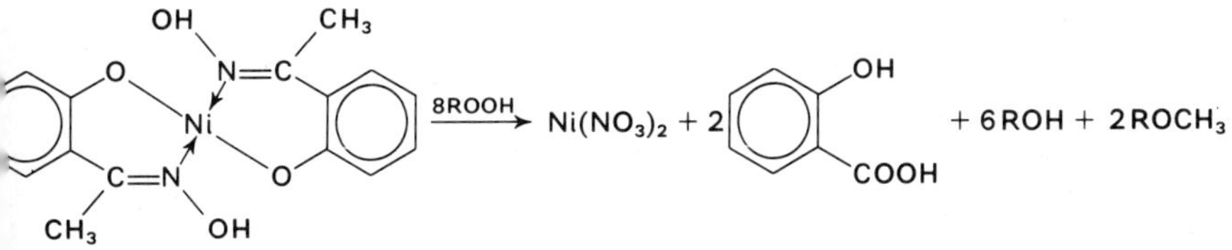

$$\xrightarrow{8ROOH} Ni(NO_3)_2 + 2 \quad + 6ROH + 2ROCH_3$$

This process occurs at high temperatures during the processing of the polymer as well as at ambient temperatures during environmental exposure of the polymer.

RELEVANCE OF INITIATION AND STABILIZATION MECHANISMS TO THE PHOTODEGRADATION OF COMMERCIAL POLYMERS

It has been shown in separate studies that hydroperoxides are the primary initiators formed in several polymers as a result of thermal processing operations. The rate of photo-oxidation of these polymers has been shown to be a function of the hydroperoxide concentration in the polymer both before and during UV exposure.

HIPS, when heated at 98 °C in air, is rapidly oxidized with the formation of hydroperoxide in the polymer (see Figure 5).[31] Hydroperoxide can be measured by conventional chemical techniques even after a few minutes of thermal treatment (there appears to be no induction period under these conditions) and conjugated carbonyl, the expected primary breakdown product, can also be

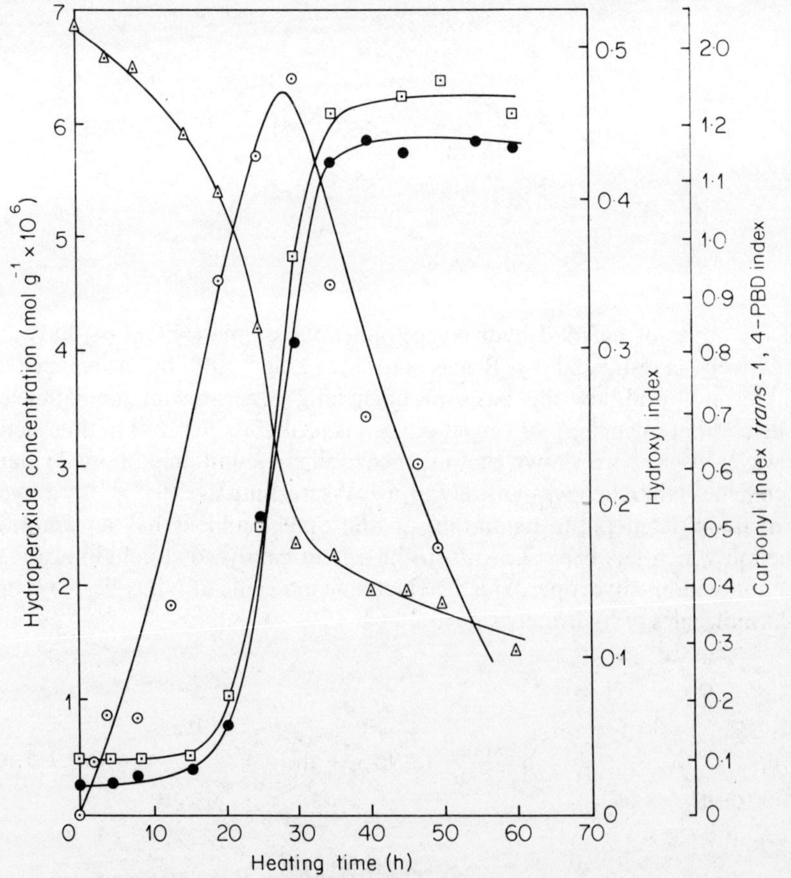

Figure 5 Change in concentration of functional groups in HIPS film on heating in air at 98 °C. ○, Hydroperoxide; □, hydroxyl; ● carbonyl; △, *trans*-1,4-polybutadiene

detected by IR spectroscopy at almost the same time. A similar change in functional group concentration occurs during processing of HIPS at 200 °C (see Figure 6), although at this temperature hydroperoxides thermolyse rapidly to give carbonyl compounds. Hydroperoxide can be detected in HIPS even after a normal commercial extrusion process[31] since it is impossible to exclude oxygen completely in any commercial operation. Soluble oxygen present in the polymer and occluded between the polymer particles is primarily responsible for this behaviour and even a compression-moulded sample without mechanical shearing leads to some oxidation of the polymer.[32]

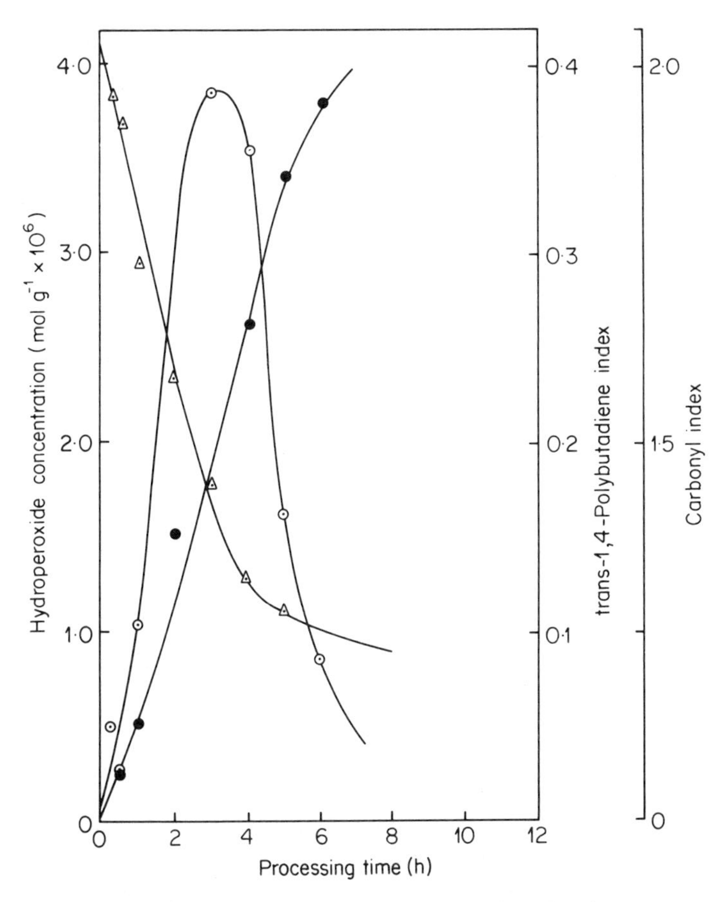

Figure 6 Effect of processing at 200 °C on functional group concentration in HIPS. ○, Hydroperoxide; ●, conjugated carbonyl; △, *trans*-1,4-polybutadiene

The effect of prior thermal treatment on the kinetics of the photo-oxidation process is shown in Figure 7. It is evident that the autocatalytic period which precedes the linear rate of photo-oxidation is removed by the prior thermal treatment. The autoacceleration disappears in samples which have been

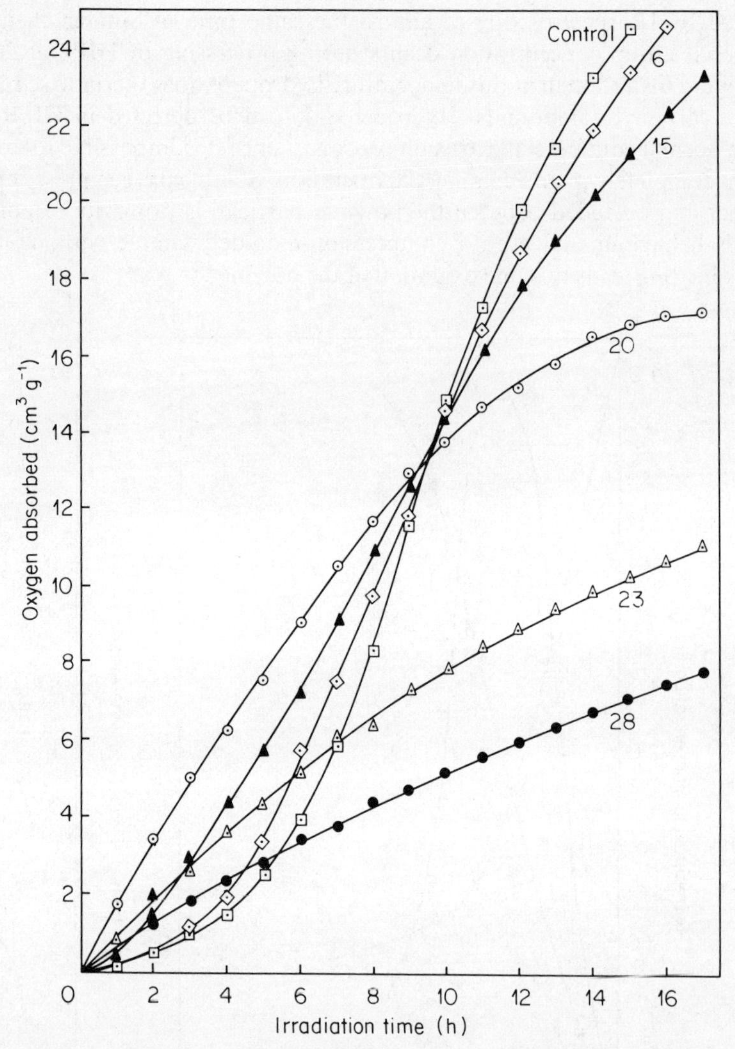

Figure 7 Photo-oxidation of HIPS extruded film heated at 98 °C in air (numbers on curves indicate heating time in hours)

thermally oxidized for 20–25 h and in which, according to Figure 5, the hydro-peroxide concentration is approaching the maximum value. This closel identifies the photoinitiation step with the hydroperoxide photolysis and confirmation is found for this by following the hydroperoxide concentratio during UV irradiation of an extruded film (see Figure 8). As in thermal oxidatio of HIPS, the hydroperoxide concentration rises to a maximum (10 h of UV exposure) and this is the time of UV exposure at which the photo-oxidation rat reaches a maximum in the unoxidized extruded sample of HIPS (see Figure 7) In contrast, the UV absorbance (286 nm) due to conjugated carbonyl of th extruded sample does not appear until about 4 h of UV exposure (see Figure 8

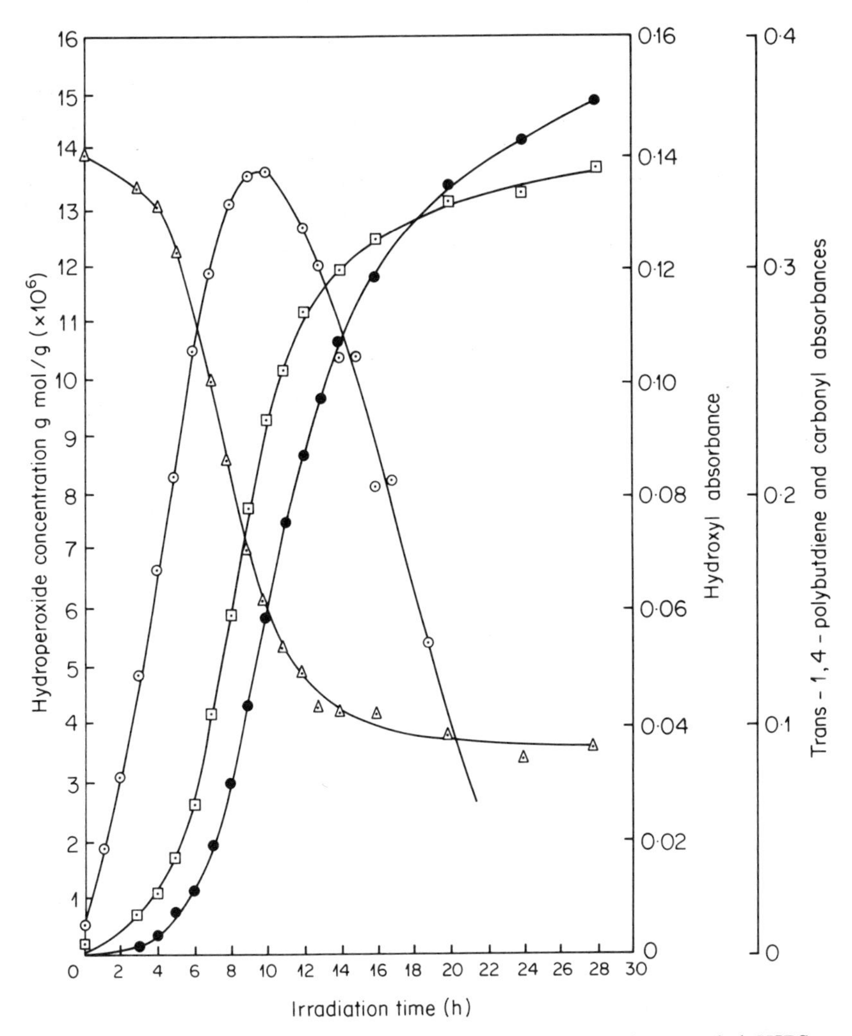

Figure 8 Change in concentration of functional groups in extruded HIPS film on UV irradiation. ○, Hydroperoxide; □, hydroxyl; ●, carbonyl; △, *trans*-1,4-polybutadiene

and, by the time the maximum concentration of conjugated carbonyl is approached (15 h), the rate of photo-oxidation of the HIPS has slowed considerably owing to depletion of the unsaturation. This evidence confirms that presented earlier that hydroperoxide rather than carbonyl compounds are responsible for the photoinitiation step.

A very similar sequence of chemical reactions has been observed during thermal processing of polypropylene.[33] The photo-oxidation rate correlates well with the hydroperoxide concentration initially present in the polymer before photo-oxidation and, as in the case of LDPE (Figure 1), there is no correlation with initial carbonyl concentration.

CONCLUSIONS

It is concluded that neither triplet carbonyl nor derived singlet oxygen is initially important as a photoinitiator in aliphatic saturated and olefinically unsaturated polymers. It is further concluded that quenching of these excited states is not an important stabilizing mechanism in practice but that the preventive non-radical decomposition of hydroperoxides is the most important of the conventional antioxidant mechanisms involved in the action of two nickel complexes.

ACKNOWLEDGEMENT

The contribution of Dr. M. Humphrey to this investigation is gratefully acknowledged.

REFERENCES

1. G. Scott, *Atmospheric Oxidation and Antioxidants*, Elsevier, London and New York, 1965, p. 280.
2. B. S. Biggs, *Polymer Degradation Mechanisms*, NBS Circ., 1953, p. 137.
3. A. R. Burgess, *Nat. Bur. Stand. Circ. No. 525*, 149 (1953).
4. Ref. 1, p. 276.
5. Ref. 1, p. 193.
6. Ref. 1, p. 216.
7. J. C. W. Chien and W. P. Connor, *J. Amer. Chem. Soc.*, **90**, 1001 (1968).
8. P. J. Briggs and J. F. McKellar, *J. Appl. Polym. Sci.*, **12**, 1825 (1968).
9. J. P. Guillory and C. F. Cook, *J. Polym. Sci. A1*, **11**, 1927 (1973).
10. J. P. Guillory and C. F. Cook, *J. Amer. Chem. Soc.*, **95**, 4885 (1973).
11. D. J. Carlsson, G. D. Medenhall, T. Suprunchuk, and D. M. Wiles, *J. Polym. Sci.*, **94**, 8960 (1972).
12. D. J. Carlsson, T. Suprunchuk, and D. M. Wiles, *J. Polym. Sci. B*, **11**, 61 (1973).
13. J. P. Guillory and R. S. Becker, *J. Polym. Sci. A1*, **12**, 993 (1974).
14. A. K. Breck, C. L. Taylor, K. E. Russell, and J. K. S. Wan, *J. Polym. Sci. A1*, **12**, 1505 (1974).
15. S. W. Beaven, P. A. Hackett, and D. Phillips, *Eur. Polym. J.*, **10**, 925 (1974).
16. G. H. Hartley and J. E. Guillet, *Macromolecules*, **1**, 169 (1968); F. J. Golemba and J. E. Guillet, *Macromolecules*, **5**, 63 (1972).
17. A. M. Trozzolo and F. H. Winslow, *Macromolecules*, **1**, 98 (1968).
18. G. Porter, S. K. Dogra, R. O. Loulfy, S. F. Sugimore, and R. W. Yip, *J. Chem. Soc. Faraday Trans. I*, **69**, 1462 (1973).
19. D. R. Kearns, *Chem. Rev.*, **71**, 395 (1971).
20. G. Scott and L. M. K. Tillekeratne, unpublished work.
21. M. U. Amin, G. Scott, and L. M. K. Tillekeratne, *Eur. Polym. J.*, **11**, 85 (1975).
22. N. A. Khan and G. Scott, unpublished work.
23. K. R. Kopecky and H. J. Reich, *Can. J. Chem.*, **43**, 2265 (1965).
24. A. Zweig and W. A. Henderson, *J. Polym. Sci. A1*, **13**, 993 (1975).
25. A. Ghaffar, A. Scott, and G. Scott, *Eur. Polym. J.*, **12**, 615 (1976).
26. A. Adamczyk and F. Wilkinson, *J. Appl. Polym. Sci.*, **18**, 1225 (1974).
27. R. P. R. Ranaweera and G. Scott, *Eur. Polym. J.*, **12**, 825 (1976).
28. A. Adamczyk and F. Wilkinson, *J. Chem. Soc. Faraday Trans. II*, **68**, 2031 (1972)
29. R. P. R. Ranaweera and G. Scott, *Chem. Ind., Lond.*, **1974**, 774.

30. R. P. R. Ranaweera and G. Scott, *Eur. Polym. J.*, **12**, 591 (1976).
31. A. Ghaffar, A. Scott and G. Scott, *Eur. Polym. J.*, **13**, 83 (1977).
32. A. Ghaffar, A. Scott and G. Scott, *Eur. Polym. J.*, unpubl. work.
33. K. B. Chakraborty and G. Scott, *Polymer*, **18**, 98 (1977).

24

The Degradation of Anthracene-doped Polymers by Singlet Oxygen

M. KRYSZEWSKI and B. NADOLSKI

Centre of Molecular and Macromolecular Studies, Polish Academy of Sciences, 90–362 Łódź, Poland

INTRODUCTION

The effect of the quenching of excited states, especially excited triplet states, of aromatics by oxygen has recently attracted experimental[1–4] and theoretical[5–9] interest. Investigations in solutions[2,3,10,11] and in the vapour phase[12,13] have been carried out since it was found that excited oxygen molecules may be very important in photosensitized oxygenation reactions.[14,15]

A number of papers have been published that describe the photophysical processes related to the generation and physical properties of singlet oxygen as well as to its quenching mechanism. On the other hand, the mechanism and kinetics of chemical reactions of singlet oxygen with organic compounds and polymers have also been extensively studied.

It is not within the scope of this paper to discuss all results of the studies of singlet oxygen generation. Many direct methods[16–20] of singlet oxygen formation are known, including radiationless transitions between molecules excited by UV light (sensitizers) and oxygen, which generate singlet oxygen. Some important problems of singlet oxygen generation in radiationless transition have been discussed by Kearns and co-workers.[21,22] It was shown that the singlet oxygen molecule $^1O_2$ (Δ^1_g) in its ground state may be formed when the excitation energy of the sensitizer molecule exceeds 22·5 kcal mol^{-1}, whereas for the excited singlet oxygen molecule $^1O_2$ $(^1\Sigma_g^+)$ the energy of the sensitizer molecule must be higher than 37·5 kcal mol^{-1}. Many organic molecules satisfy these conditions and are effective sensitizers for singlet oxygen generation. Most important among these compounds are aromatics and dyes.

Careful studies of oxygen quenching by aromatics in the singlet state and especially in triplet states[23,24] leading to singlet oxygen molecules $^1O_2(^1\Delta_g)$ or $^1O_2(^1\Sigma_g)$ have shown that the intermolecular enhancement of spin-forbidden excited state decay by $^3O_2$ is due to electronic interactions between the excited molecule and a molecule of the quencher within the collision complex. Gijzeman *et al.*[23] have shown by detailed kinetic and energetic analysis that the oxygen quenching rates vary by an order of magnitude, being dependent on the polarity and viscosity of the solvent. These results were analysed in terms of non radiative transitions within the collision complex between the ground-state oxygen and the aromatic molecule in its triplet state. This quenching reaction occurs much faster in the case of low triplet energies (10 000 cm^{-1} < E_t < 16 000 cm^{-1}) and depends on restrictive Franck–Condon factors which are determined by the structure of the hydrocarbon. An important result for further consideration is that anthracene is quenched with a very high probability on account of its low triplet energy leading mostly to singlet oxygen $^1O_2(^1\Delta_g)$.[24]

Singlet oxygen may be formed directly during UV irradiation of polymers containing specific groups and internal or external impurities. Internal impurities, e.g. in polyolefins or vinyl polymers, are formed during the synthesis, processing, and storage of polymers. The most important are carbonyl groups which, in the excited triplet state (absorption at 260–340 nm), may react with molecular oxygen leading to the formation of singlet oxygen.[10,25–27] External impurities, mostly aromatics and substituted aromatics, which are present as traces of compounds originating during the synthesis or from additives and related to different processing methods, are effective sensitizers for singlet oxygen generation,[28–34] and they are therefore very important in the initiation step of oxidation reactions. Singlet oxygen may react with vinyl groups present in polymers resulting in the formation of hydroperoxides. It was clearly shown that on polyolefin surfaces treated with singlet oxygen generated by microwaves, hydroperoxides are created which promote degradation processes.[28,35] Analogous reactions were postulated for polydienes.[36]

Singlet oxygen may abstract hydrogen atoms with the formation of hydroperoxides. An important reaction which leads to photo-oxidation of polymers containing hydroxyl groups is the abstraction of hydrogen atoms from hydroxyl groups with the formation of alkoxy radicals and hydroperoxides.[37] Although at present there is no clear interpretation of all of the steps involved in polymer degradation by singlet oxygen, it seems that the formation of hydroperoxides is very important. These groups can decompose (on heating or irradiation) leading to macro-radicals which are important in further degradation processes in polymers.[38]

A special case is the degradation of polymers that contain phenyl or aromatic groups. In this case singlet oxygen can be formed by an intermolecular energy transfer between benzene rings in polystyrene excited by UV light (irradiation by UV light below 280 nm) and molecular oxygen.[39] It was assumed that, in addition to the direct energy transfer, a charge-transfer (CT) complex may be formed which, after excitation, produces singlet oxygen:

$$-CH_2CH(C_6H_5)- \xrightarrow{h\nu} -CH_2-CH(C_6H_5^{*})- \xrightarrow{^3O_2} -CH_2-CH(C_6H_5)- + {}^1O_2 \quad (1)$$

$$-CH_2CH(C_6H_5)- \xrightarrow{^3O_2} \left[-CH_2-CH(C_6H_5 \cdots O_2)-\right] \xrightarrow{h\nu} -CH_2-CH(C_6H_5)- + {}^1O_2$$

Both reactions lead to the degradation of polystyrene because radicals and singlet oxygen molecules may react with the polymer chains, inducing degradation processes.

When polystyrene (PS) with an added aromatic compound, e.g. anthracene (ANT), in liquid or solid solution is irradiated with UV light both of the above reactions of singlet oxygen formation must be considered, the first reaction related to the presence of excited phenyl groups of the matrix and the second reaction related to the oxygen quenching of the excited states of the added aromatic compound. Both can occur by the pathway in which a CT complex with $^3O_2$ may be produced when the samples are irradiated with UV light. Thus, in order to separate (at least partially) the influence of aromatic additives (known impurities in commercial polymers) on the formation of singlet oxygen in the system, a third component, which is a stronger acceptor than $^3O_2$ and a quencher for both excited species, may be added. In this way acceptor molecules would influence in a controlled manner the generation of singlet oxygen, which in turn may change (reduce) the oxidative degradation of polymers.

The aim of this preliminary report is to describe the changes in molecular weight of PS with added ANT, chloranil (CA), and trinitrobenzene (TNB) in the process of photo-oxidative degradation. Both CA and TNB act as acceptors for benzene rings in PS and aromatic additives. The systems PS–ANT–CA and PS–ANT–TNB were chosen for these studies because their photophysical properties (absorption, fluorescence, and its quenching) had been carefully studied by us.[40] We report some results of studies of solid solutions (films), not only because they are interesting model systems, but also because they may have some practical importance.

As our interest in these investigations was related to the oxidative degradation of the PS matrix, no flash photolysis techniques were used and steady illumination conditions were applied.

EXPERIMENTAL

The purification of anthracene, chloranil, trinitrobenzene, and solvents and the preparation of polystyrene have been described previously.[40] The same applies to the preparation of films, which have been the subject of studies on the fact that CA may undergo some degradation reactions in liquid solution.[41,42] The maximum concentration of ANT, TNB, and CA in solid PS solutions was about 3 mol-% of monomer unit, i.e. 0·36 mol dm^{-3}. At higher concentrations these compounds precipitate in the form of crystals during the evaporation of the solvent.

The absorption spectra of all of the compounds used were in agreement with the literature data.[43]

Polystyrene films (pure polymer or with a known amount of added ANT–CA and TNB) were cast from 2% methylene chloride solutions that had been filtered through sintered-glass crucibles. Films were formed on clean, carefully polished quartz plates or on clean mercury surfaces in Petri dishes. The volume of solution used for casting was such that the film thickness was of the order of 10 μm. After controlled evaporation of the solvent, the films were placed in a drying chamber connected to a vacuum apparatus and dried for 48 h at 60 °C to constant weight. The residual methylene chloride content in the films investigated did not exceed 1%, as tested by IR spectroscopy. The films were kept in dry air for a long period in the dark. Taking into account the diffusion coefficient of oxygen in pure PS,[44] it can be assumed that these films were saturated with oxygen. The same applies to the films with small amount of low molecular weight aromatic additives which do not influence the diffusion or solubility of oxygen appreciably. Polymer films prepared in this way were irradiated by using two types of UV lamps: a low-pressure mercury lamp (maximum emission at 254 nm) and a high-pressure mercury lamp with suitable filters of known transmittance. The current was stabilized by a constant voltage transformer and, after about 5 min, the light intensity became constant. Heat effects from the light source (high-pressure lamp) were made negligible by cooling it with an air blower. The intensity of radiation incident upon the solid film surface was varied both by varying the distance between the lamp and the film and by using intensity screens for the beam. These screens were made of blackened copper gauze of various mesh sizes and known transmission (the quantum flux of the low-pressure mercury lamp is approximately six times lower than that of the high-pressure lamp). The light intensity was determined using potassium ferrioxalate.[45] Each polymer received approximately $3{\cdot}6 \times 10^6$ quanta cm^{-2} s^{-1}. After exposure at room temperature in air for a known time, the films were weighed and dissolved in benzene, and the solutions were filtered through sintered-glass crucibles. The gel fraction was determined by drying to a constant weight and weighing. The viscosities of the solutions were measured for four concentrations and the values were extrapolated to zero concentration in order to determine the intrinsic viscosities, $[\eta]$. Molecular weights (M) of the initial PS sample and of degraded samples were calculated from the Mark–Houvink equation:

$$[\eta] = KM^{\alpha} \tag{2}$$

here $K = 1{\cdot}06 \times 10^4$ and $\alpha = 0{\cdot}735$, for polystyrene in benzene.[46] For samples ith random distributions of molecular weights it may be assumed that $[\eta]$ does ot change with moderate degrees of degradation. The relationship between scosity and number-average molecular weights M_v and M_n in such a case is ven by the equation

$$\overline{M}_n = \frac{\overline{M}_v}{(\alpha+1)\Gamma(\alpha+1)^{1/\alpha}} \tag{3}$$

where Γ is the gamma function and $\alpha = 0{\cdot}735$. The values of $\overline{M}_{n,o}$ and $\overline{M}_{n,t}$ can be calculated at any stage of the degradation reaction; thus, the degree of degradation, a, is given by

$$a = \frac{1}{\overline{M}_{n,t}} - \frac{1}{\overline{M}_{n,o}} \tag{4}$$

RESULTS AND DISCUSSION

The photoinduced oxidative degradation of PS in the solid state (films) is very complex. In the first stages of irradiation the initiation reaction depends on light absorption by macromolecules or by intrinsic chromophors (which can be generated during the reaction) and on the formation of CT complexes, energy transfer, etc. It depends upon the mobility of free radicals created in the polymer matrix as well as upon the concentration of oxygen in the sample and upon its diffusion into the polymer. The rate of oxidation depends upon the reactivity of oxygen molecules with radicals or other species, and on the competitive bimolecular recombination of two macro-radicals leading to crosslinking. Oxygen molecules may form CT complexes between phenyl rings in the polymer and/or with impurities.[33,38,47] These species may be active in the generation of singlet oxygen, which is an important species in subsequent processes.[33] In subsequent stages of PS oxidative degradation in the solid state, the reactions occur as in solutions and may be interpreted with the aid of hydroperoxidation theory.[48–50]

These general remarks show that the detailed mechanism of these reactions is very complex. The aim of this study was to show a probable role of singlet oxygen in the decrease in molecular weight of PS in the absence and presence of aromatic additives. It was possible to follow this reaction by plotting the change in viscosity number (η_{sp}/c) for solutions obtained from films irradiated for various periods or by discussing the degree of degradation (a) as a function of irradiation time. It seems that the latter method may give more information. The change in degree of degradation with time is given by the equation

$$a = \frac{1}{\overline{M}_{n,t}} - \frac{1}{\overline{M}_{n,o}} = K_{exp}t \tag{5}$$

Equation (5) has been used in many studies and was discussed by Jellinek and co-workers,[51,52] taking into consideration a detailed analysis of isotactic polystyrene degradation in the presence of SO_2 and O_2 under irradiation with UV light at wavelengths above 280 nm. In the systems examined we do not have sufficient data to propose a detailed kinetic scheme or to derive a kinetic equation which accounts for all steps in the photo-oxidative degradation. Thus equation (5) has only an approximate value but it provides facilities for comparison of different systems in a better way than the η_{sp}/c diagrams. The results obtained are presented in Figure 1.

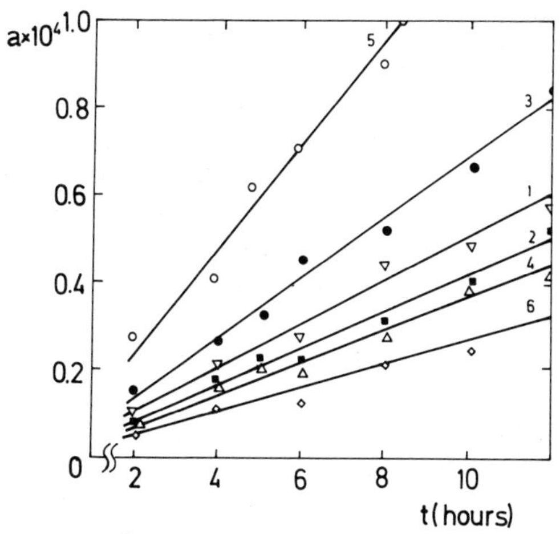

Figure 1 Degree of photo-oxidative degradation of pure PS films and PS with donor–acceptor additives: 1 = PS + O_2 (wavelength 254 nm); 2 = PS + O_2 (>280 nm); 3 = PS + ANT + O_2 (>280 nm); 4 = PS + TNB + O_2 (>280 nm); 5 = PS + CA + O_2 (>280 nm; 6 = PS + ANT + CA (>280 nm)

K values, given in Table 1, were calculated for irradiation times exceeding 2 h; linear relationships were obtained in this case. For shorter times the determination of molecular weight changes is difficult but distinct deviations from linearity could be detected; these effects will be not discussed here. From Figure 1 and Table 1 it follows that the presence of aromatic additives in PS (ANT, CA, and TNB) influences the photodegradation. Pure PS films 'saturated' with O_2 from air undergo slow photodegradation when irradiated with light of wavelength 254 nm. Absorption of PS occurs between 190 and 273 nm. From an analysis of the energy diagram for the photolysis of PS,[53] it follows that the photoelimination of H_2 proceeds via the triplet state. In the solid state the degradation leads to some crosslinking and formation of conjugated double bonds, but chain scission is also important.[49] PS usually contains a small

Table 1 Rate constants (K_{exp}) of photo-oxidative degradation of PS in absence and presence of donor–acceptor additives[a] (30 °C)

No.	System	Wavelength (nm)	$K_{exp} \times 10^6$ (h^{-1})
1	PS + O_2	254	4·8
2	PS + O_2	>280	4·2
3	PS + ANT + O_2	>280	8·0
4	PS + TNB + O_2	>280	3·6
5	PS + AC + O_2	>280	11·2
6	PS + ANT + CA	>280	3·2

[a] Aromatic additives concentration: 0·18 mol dm^{-3}.

amount of carbonyl groups[54] which may take part in the degradation reaction, but their concentration in the initial samples is not high enough to account for the observed effects. Thus it is probable that they are created as a result of the generation of singlet oxygen and its further reaction with PS chain elements. It is possible that singlet oxygen is formed by intermolecular energy transfer between excited phenyl rings of PS and $^3O_2$, or by photodissociation of CT complexes with $^3O_2$, as was mentioned before and discussed by Rånby and Rabek.[55] On irradiation at longer wavelengths (above 280 nm) absorption in styrene is small and consequently only the $\gt C{=}O$ groups or CT complexes between the phenyl rings of PS and $^3O_2$ may be active. They absorb at a wavelength of ca. 300 nm, in analogy with benzene–$^3O_2$ complexes.[56,57] It seems that they are the main source of the singlet oxygen which leads to degradation reactions. If singlet oxygen is indeed generated and if its reaction with the PS matrix is an important intermediate step then the other compounds added to the system examined, which are also active in $^1O_2$ generation, should enhance the degradation process. It is known that excited triplet states of the ANT molecule may be quenched by oxygen molecules, leading to singlet oxygen,[58,59] and therefore the solid solution of ANT in PS should undergo a faster degradation than pure PS when irradiated with light of wavelength above 280 nm (especially when irradiated in the absorption band of ANT). This was confirmed experimentally (curve 3 in Figure 1 and Table 1). It is also in accordance with the results of Rabek[60] obtained for benzene solutions of PS.

Irradiated quinones abstract hydrogen from several donors and can react over the carbonyl group and the ring double bonds.[61,62] It was shown that triplet states of quinone are reactive intermediates and that in the presence of oxygen singlet oxygen may be formed by an energy transfer between the excited triplet of quinone and molecular oxygen. It was shown that the addition of several quinones, e.g. chloranil (CA), causes as the first reaction a crosslinking and then a rapid degradation of PS in solid films and in benzene solutions.[33] The same result was obtained in this study. Curve 5 in Figure 1 provides evidence of fast PS degradation. The presence of typical oxygen quenchers in the systems investigated results in a decrease in the photodegradation of PS.[33] It was therefore concluded that $^1O_2$ molecules are important in this reaction.

Another argument for the role of singlet oxygen in the photodegradation of PS was that the addition of ANT (donor) and CA (acceptor) causes a decrease in the rate of PS degradation (curve 6 in Figure 1). This effect may be related to complex formation between the polycyclic compound (ANT) and quinone (CA), and with energy transfer from all complexed and non-complexed excited components, impurities and PS phenyl rings, to the complex. On account of the low energy of the first excited state of the CT level of the ANT–CA complex[40] it is probably inactive in singlet oxygen generation. In the system PS–ANT–TNB a similar decrease in photo-oxidative degradation of PS was found and therefore the $a = f(t)$ curves are not given in Figure 1.

These results provide indirect proof of the role of singlet oxygen in photodegradation reactions of PS. This reaction has, however, not yet been fully elucidated. The explanation of the observed decrease in PS photo-oxidation cannot be achieved by considerating the filtering effects of added donor and acceptor molecules. The addition of ANT and CA alone causes an increase in the rate of PS degradation, and therefore they do not exhibit a strong filtering activity. The absorption of CT complexes in the active range of light is comparable to that of their components. As has been mentioned above, when the CT complex is electronically excited in the UV range its energy is intramolecularly transferred to the lowest energy state visible CT band.

CONCLUSIONS

Polystyrene irradiated with UV light of wavelength above 280 nm undergoes chain scission. This reaction is enhanced by polycyclic aromatic compounds, e.g. ANT, and quinones, e.g. CA, both of which may be very important in singlet oxygen generation reactions. Singlet oxygen seems to be a very efficient initiator in the first stages of photo-oxidative reactions of PS. The presence of two additives which can form donor–acceptor complexes reduces the rate of PS photodegradation following the quenching of excited species which may be active in singlet oxygen generation. This study seems to show that in the search for antioxidants not only their reactivity with radicals or with hydroperoxydes should be considered but also their ability to form CT complexes with species that are active in singlet oxygen generation.

Further research is needed to elucidate several points in the proposed general scheme of PS degradation in the system examined, e.g. the dependence between wavelength and additive concentration and the effect of various components of CT complexes with different ionization potentials and electron affinities, etc. Some of these problems are being investigated in our laboratory.

REFERENCES

1. J. B. Birks, *Photophysics of Aromatic Molecules*, Wiley-Interscience, New York, 1970.
2. A. Adamczyk and F. Wilkinson, in *Organic Scintillators and Liquid Scintillation Counting* (Ed. D. L. Horrocks and C. T. Peng), Academic Press, New York, 1971, p. 223.
3. L. K. Patterson, G. Porter, and M. R. Topp, *Chem. Phys. Lett.*, **7**, 612 (1970).
4. P. B. Merkel and D. R. Kearns, *Chem. Phys. Lett.*, **12**, 120 (1972).
5. G. W. Robinson, *J. Chem. Phys.*, **46**, 572 (1967).
6. K. Kawaoka, A. U. Khan, and D. R. Kearns, *J. Chem. Phys.*, **46**, 1842 (1967).
7. A. U. Khan and D. R. Kearns, *J. Chem. Phys.*, **48**, 3272 (1968).
8. D. R. Kearns and A. J. Stone, *J. Chem. Phys.*, **55**, 3383 (1971).
9. N. E. Geacintov and C. E. Swenberg, *J. Chem. Phys.*, **57**, 378 (1972).
10. D. R. Adams and F. Wilkinson, *J. Chem. Soc. Faraday Trans. II*, **68**, 586 (1972).
11. B. E. Algar and B. Stevens, *J. Phys. Chem.*, **74**, 3029 (1970).
12. L. J. Andrews and E. W. Abrahamson, *Chem. Phys. Lett.*, **10**, 113 (1971).

13. C. K. Duncan and D. R. Kearns, *J. Chem. Phys.*, **54**, 5822 (1971).
14. C. S. Foote and S. Wexter, *J. Amer. Chem. Soc.*, **86**, 3879 (1964).
15. E. J. Corey and W. C. Taylor, *J. Amer. Chem. Soc.*, **86**, 3881 (1964).
16. K. Furukawa, E. W. Gray, and E. A. Ogryzlo, *Ann. N.Y. Acad. Sci.*, **171**, 175 (1975).
17. I. B. C. Matheson and J. Lee, *J. Amer. Chem. Soc.*, **94**, 3310 (1972).
18. S. J. Arnold, R. J. Brown, and E. A. Ogryzlo, *Photochem. Photobiol.*, **4**, 963 (1965).
19. E. McKeown and W. A. Waters, *J. Chem. Soc.*, **1966**, 1040.
20. R. W. Murray and M. L. Kaplan, *J. Amer. Chem. Soc.*, **90**, 537 (1968).
21. D. R. Kearns, *Chem. Rev.*, **71**, 395 (1971).
22. A. U. Khan and D. R. Kearns, *Adv. Chem. Ser.*, No. 76, 143 (1968).
23. O. L. J. Gijzeman, F. Kaufman and G. Porter, *J. Chem. Soc. Faraday Trans. II*, **69**, 708 (1973).
24. O. L. J. Gijzeman and F. Kaufman, *J. Chem. Soc. Faraday Trans. II*, **69**, 721 (1973).
25. A. M. Trozzolo and F. H. Winslow, *Macromolecules*, **1**, 98 (1968).
26. A. M. Trozzolo, *Adv. Chem. Ser.*, No. 77, 167 (1968).
27. J. F. Rabek, *Wiad. Chem.*, **25**, 293, 365, 435 (1971).
28. D. J. Carlsson and D. M. Wiles, *J. Polym. Sci. B*, **11**, 759 (1973).
29. H. Mönig, *Naturwissenschaften*, **45**, 12 (1958).
30. R. B. Fox and T. R. Price, *J. Appl. Polym. Sci.*, **11**, 2373 (1976).
31. E. D. Owen and R. J. Bailey, *J. Polym. Sci. A1*, **10**, 113 (1972).
32. J. F. Rabek, *Chem. Stosow.*, **11**, 53 (1967).
33. J. F. Rabek and B. Rånby, *J. Polym. Sci. A1*, **12**, 295 (1974).
34. G. S. Egerton, *Brit. Polym. J.*, **3**, 63 (1971).
35. M. L. Kaplan and P. G. Kelleher, *J. Polym. Sci. B*, **11**, 565 (1971).
36. M. L. Kaplan and P. G. Kelleher, *J. Polym. Sci. A1*, **8**, 3163 (1970).
37. T. Matsuura, N. Yoshimura, A. Nishinaga, and I. Saito, *Tetrahedron Lett.*, **1972**, 4933.
38. J. F. Rabek and B. Rånby, *Polym. Eng. Sci.*, **15**, No. 1, 41 (1975).
39. J. R. Rabek and B. Rånby, *J. Polym. Sci. A1*, **12**, 273 (1974).
40. A. Degórski and M. Kryszewski, *J. Chem. Soc. Faraday Trans. II*, **71**, 1503, 1513 (1975).
41. P. J. Reucroft, O. N. Rudyj, R. E. Salomon, and M. M. Labes, *J. Chem. Phys.*, **43**, 767 (1965).
42. M. Fisch and W. Hammerlin, *Tetrahedron Lett.*, **31**, 3125 (1972).
43. *DNF–UV Atlas of Organic Compounds*, Verlag Chemie, Weinheim, and Butterworths, London, 1968.
44. V. Stannett, in *Diffusion in Polymers* (Ed. J. Crank and G. S. Park), Academic Press, London and New York, 1968, p. 47.
45. G. G. Hatchard and C. A. Parker, *Proc. Roy. Soc.*, *A*, **235**, 532 (1956).
46. G. Danusso and G. Moraglio, *J. Polym. Sci.*, **24**, 161 (1957).
47. K. Tsuji and T. Seiki, *Rep. Progr. Poly. Phys.*, *Japan*, **14**, 581 (1971).
48. J. L. Bolland, *Quart. Rev.*, **3**, 1 (1949).
49. N. Grassie and N. A. Wier, *J. Appl. Polym. Sci.*, **9**, 975, 987, 999 (1965).
50. N. A. Weir and J. B. Laurence, *IUPAC Int. Sym. Macromol. Chem.*, *Budapest*, 1969, Vol. 5, p. 323 (Ed. F. Tüdös).
51. H. H. G. Jellinek and J. F. Kryman, in *Photochemistry of Macromolecules* (Ed. R. F. Reinisch), Plenum Press, New York, 1970, p. 85.
52. H. H. Jellinek and J. Pavlinec, in *Photochemistry of Macromolecules* (Ed. R. F Reinisch), Plenum Press, New York, 1970, p. 91.
53. R. F. Reinisch, H. R. Gloria, and G. M. Androes, in *Photochemistry of Macromolecules* (Ed. R. F. Reinisch), Plenum Press, New York, 1970, p. 201.
54. O. I. Selivanov, V. L. Maksimov and E. I. Kirilova, *Vysokomol. Soedin. A*, **11**, 48[illegible] (1969).

55. B. Rånby and J. F. Rabek, *Photodegradation Photo-oxidation and Photostabilization of Polymers*, Wiley, New York, 1975.
56. J. C. Chien, *J. Phys. Chem.*, **69**, 4317 (1965).
57. D. E. Evans, *J. Chem. Soc.*, **1957**, 1351.
58. C. W. Cowell and J. N. Pitts, *J. Amer. Chem. Soc.*, **90**, 1106 (1968).
59. P. H. Bolton, R. D. Kenner, and A. U. Khan, *J. Chem. Phys.*, **57**, 5604 (1972).
60. J. F. Rabek, *XXIIIrd IUPAC Congress, Boston, Mass.*, 1971, Butterworths, London, Vol. 8, p. 29.
61. N. K. Bridge, *Trans. Faraday Soc.*, **56**, 1001 (1960).
62. C. F. Wells, *Trans. Faraday Soc.*, **57**, 1719 (1961).

25

The Use of β-Carotene for the Evaluation of the Role of Singlet Oxygen in Longwave UV Photo-oxidation of Polystyrene

M. NOWAKOWSKA

Department of Physical Chemistry, Institute of Chemistry, Jagiellonian University, 41 Krupnicza Street, 30–060 Kraków, Poland

INTRODUCTION

During recent years there has been increasing interest in the singlet oxygen mechanism of polymer photo-oxidation originally proposed by Trozzolo and Winslow.[1] The reaction of electronically excited oxygen molecules with a polymer is considered to be the most important process in this mechanism. The singlet oxygen molecules are formed when $n \rightarrow \pi^*$ triplet states of carbonyl compounds are quenched by molecular oxygen.[1–4]

This manner of singlet oxygen generation may be particularly important for industrial polymers in which a high concentration of carbonyl compounds is found. It is, however, less important in carefully purified polymers which contain only trace amounts of carbonyl compounds. This paper reports the results of investigations which indicate the possibility of singlet oxygen formation in polystyrene films by intramolecular energy transfer in the excited state of the polystyrene–oxygen complex.

EXPERIMENTAL

Polystyrene was obtained by thermal polymerization of styrene at 85 °C for 450 h in the absence of oxygen. The polymer was purified by triple precipitation from a chloroform solution into methanol and extraction in a Soxhlet apparatus for 210 h at the boiling point of methanol.

The polymer was dried in air at room temperature. The weight–average molecular weight of the polymer determined viscometrically was 462 000. The polystyrene was irradiated in the form of thin films, which were obtained by casting a chloroform solution of polystyrene (or of polystyrene and β-carotene) on a smooth glass surface and slowly evaporating the solvent.

The polymer films were irradiated with an HBO–200 high-pressure mercury lamp at wavelengths longer than 300 nm or using a monochromatic filter with a wavelength of 313 nm. The films were placed in a pressure cell with two quartz optical windows 6 mm thick. UV absorption spectra of the films and of the polystyrene–oxygen complex were obtained by using a Carl Zeiss Jena UV visible spectrophotometer. β-Carotene was obtained from BDH Chemicals, Poole, England.

RESULTS AND DISCUSSION

Oxygen under high pressure induces longwave absorption in polystyrene films in the region where polystyrene does not absorb. Figure 1 shows the absorption spectrum of a polystyrene film 165 μm thick over the range 35 000–28 000 cm^{-1} in air and its spectrum in the presence of oxygen at the pressure of 20 atm. The absorption band induced by oxygen begins at a wavenumber $\bar{\nu}=28\ 000$ cm^{-1}. The band is unstructured.

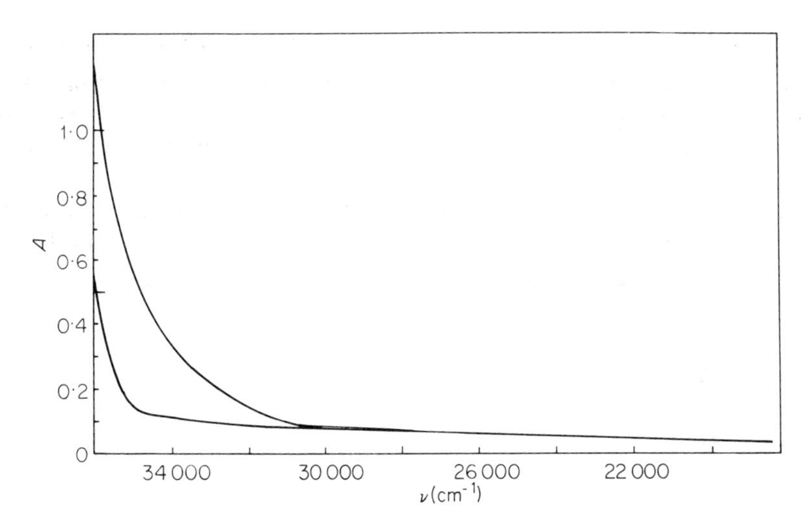

Figure 1 Absorption spectrum of PS film 165 μm thick: 1, in air ($p_{O_2}=$ 0·2 kg cm^{-2}), 2, in the presence of oxygen at a pressure of 20 kg cm^{-2}

The differential absorption spectrum (Figure 2) is similar in shape to the differential spectrum of the benzene–oxygen system at a pressure of $p_{O_2}=68$ atm which was reported by Lim and Kowalski.[5] The longwave part of induced absorption in benzene (wavelength 290–300 nm) was explained by Lim and Kowalski in terms of the influence of charge transfer of the benzene–oxygen collision complex.

The polystyrene film absorption induced by oxygen is completely reversible. The chemical oxidation of polystyrene does not occur under these conditions.

The increase in absorption of carefully purified PS film at wavenumbers higher than $\bar{\nu}=28\ 000$ cm^{-1} is observed during irradiation with light absorbed only by the polystyrene–oxygen complex.

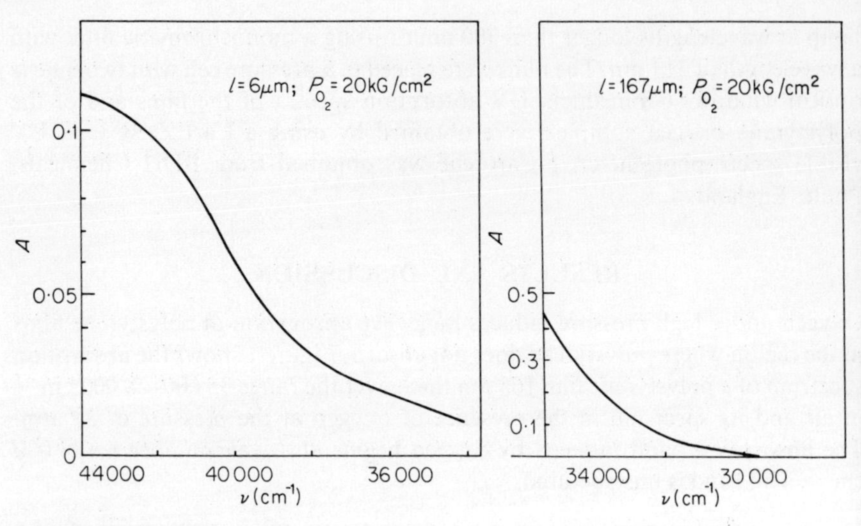

Figure 2 Differential absorption spectrum of the polystyrene–oxygen complex

Figure 3 shows the UV absorption spectra of polystyrene film 235 μm thick recorded after various irradiation times in the presence of oxygen at a pressure of 20 atm. The changes in the UV spectrum are greatest in the initial stage of this process.

Application of the light of wavelength $\lambda > 300$ nm instead of monochromatic light of wavelength $\lambda = 313$ nm increases the amount of energy absorbed by the

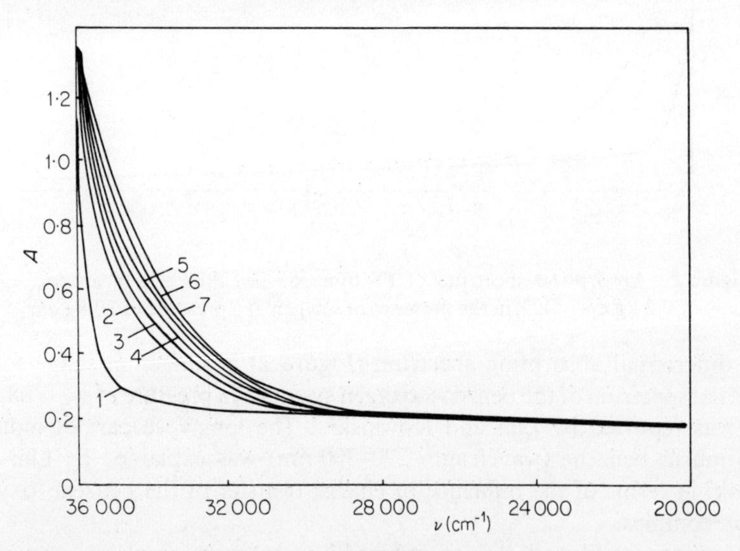

Figure 3 Absorption spectra of PS film 235 μm thick: 1, in air ($p_{O_2} =$ 0·2 kg cm^{-2}); 2, in the presence of oxygen ($p_{O_2} = 20$ kg cm^{-2}); 3, after irradiation in presence of oxygen ($p_{O_2} = 20$ kg cm^{-2}) for $t = 5$ min; 4, as (3), $t = 15$ min; 5, as (3), $t = 25$ min; 6, as (3), $t = 40$ min; 7, as (3), $t = 50$ min

polystyrene–oxygen complex. This change permits a more rapid achievement of the deeper stages of photo-oxidation than when monochromatic light is used.

The IR spectra of polystyrene films thus irradiated indicate that carbonyl compounds are formed in this process. In addition, alcohols are produced in polystyrene films after 25 h of irradiation. The presence of these products, however, may be observed after a long period of irradiation, and they should be considered as the secondary products of the photo-oxidation of PS.

Rabek and Rånby's suggestion [6] of the possibility of singlet oxygen formation by an intramolecular energy transfer in the excited state of the polystyrene–oxygen complex was applied to explain the PS photo-oxidation mechanism under these conditions.

Photochemical experiments with different thicknesses of polystyrene films containing various concentrations of β-carotene (a well known singlet oxygen quencher [7–12]) were carried out in the presence of oxygen at pressures of 0·2–20 atm. The use of monochromatic light of wavelength 313 nm as initiation radiation in these experiments was the most rational because polystyrene does not absorb this light, contrary to the polystyrene–oxygen complex. β-Carotene also absorbs this light, but its extinction coefficient is the lowest in this region $\varepsilon^{313} = 5{\cdot}2 \times 10^3\,\mathrm{l\,mol^{-1}\,cm^{-1}}$). First, the reactivity of the electronic excited states of β-carotene and the influence of oxygen on the rate and quantum yield of the consumption of β-carotene in n-hexane were determined. Consumption of β-carotene was observed during irradiation with light of wavelength 313 nm of β-carotene solutions in n-hexane in the presence of oxygen at any pressure within the range 0·2–20 atm. Figure 4 shows the UV–visible absorption spectra of β-carotene solution in n-hexane irradiated in the presence of oxygen at 20 atm for various times.

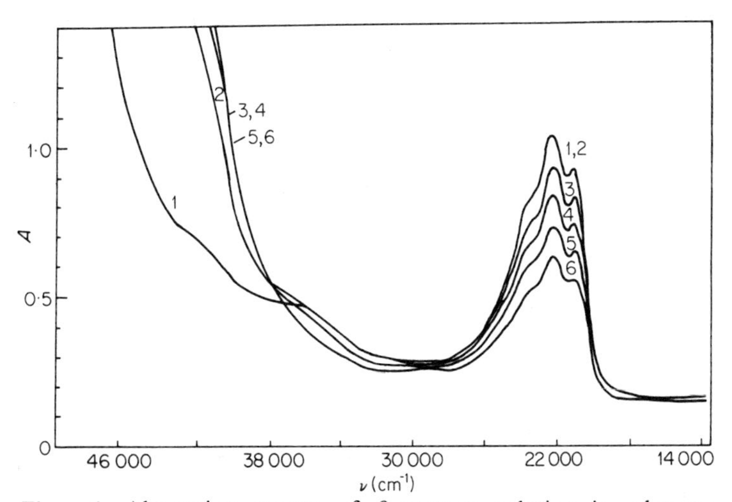

Figure 4 Absorption spectra of β-carotene solution in n-hexane: 1, in air ($p_{O_2} = 0{\cdot}2$ kg cm^{-2}); 2 in the presence of oxygen ($p_{O_2} = 20$ kg cm^{-2}); 3, after irradiation ($p_{O_2} = 20$ kg cm^{-2}) for $t = 15$ min; 4, as (3), $t = 30$ min; 5, as (3), $t = 45$ min; 6, as (3), $t = 60$ min

The consumption of β-carotene was considerably more rapid during the irradiation under these conditions of polystyrene films containing this compound.

Figure 5 shows the UV–visible absorption spectra of polystyrene film 150 μm thick with the addition of β-carotene at a concentration (c) of 9×10^{-4} mol l^{-1} recorded after various irradiation times in the presence of oxygen at 20 atm.

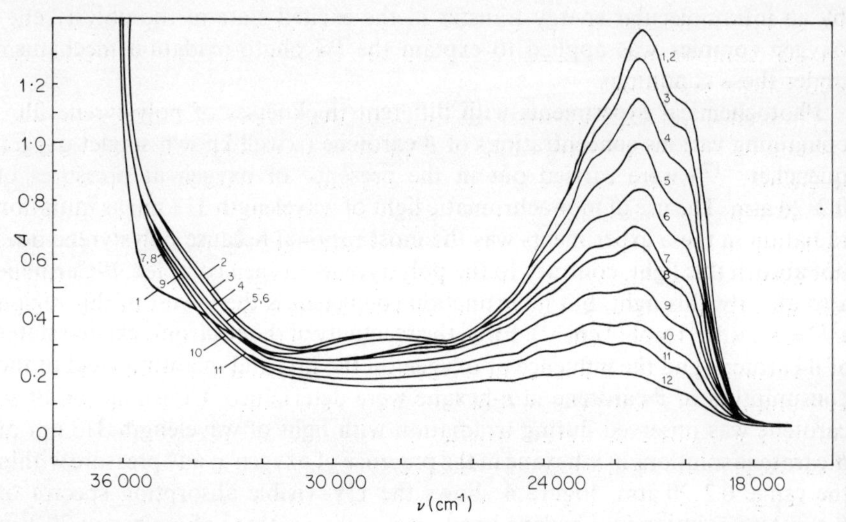

Figure 5 Absorption spectra of PS film + β-carotene: 1, in air ($p_{O_2}=0{\cdot}2$ kg cm^{-2}); 2, in the presence of oxygen ($p_{O_2}=20$ kg cm^{-2}); 3, after irradiation ($p_{O_2}=20$ kg cm^{-2}) for $t=1$ min; 4, as (3), $t=3$ min; 5, as (3), $t=6$ min; 6, as (3), $t=10$ min; 7, as (3), $t=15$ min; 8, as (3), $t=20$ min; 9, as (3), $t=30$ min; 10, as (3), $t=45$ min; 11, as (3), $t=60$ min; 12, as (3), $t=90$ min

The quantum yields of β-carotene consumption during the irradiation of its n-hexane solutions and polystyrene films are shown in Table 1. The quantum yields were estimated by using the expression

$$\phi=\frac{(\mathrm{d}A/\mathrm{d}t)}{10^3\rho'\varepsilon_{max}}$$

where

ϕ = quantum yield;

ε_{max} = extinction coefficient at wavenumber $\bar{\nu}=22\,000$ cm^{-1};

ρ = density of the incident radiation ($\lambda=313$ nm, $\rho=1\times10^{-6}$ einsteins min^{-1} cm^{-2});

$\rho'=\rho a$

$a=1-T/100$

T = transmission of sample at 313 nm.

Table 1 Quantum yields of β-carotene consumption (ϕ) during the irradiation of its n-hexane solutions and of polystyrene films in the presence of oxygen at pressures of 0·2, 10, 15, and 20 atm

p_{O_2}(kg cm^{-2})	ϕ in PS film 150 μm thick (mol einstein^{-1})	ϕ in n-hexane solution (mol einstein^{-1})
0·2	$1{\cdot}17\times10^{-3}$	$1{\cdot}43\times10^{-4}$
10	$1{\cdot}20\times10^{-3}$	$2{\cdot}55\times10^{-4}$
15	$1{\cdot}33\times10^{-3}$	$2{\cdot}73\times10^{-4}$
20	$1{\cdot}90\times10^{-3}$	$2{\cdot}97\times10^{-4}$

The results in Table 1 show that an increase in oxygen pressure causes an increase in the quantum yields of β-carotene consumption in n-hexane solution as well as in polystyrene films. However, the quantum yields of β-carotene consumption in PS films are considerably greater than in n-hexane solution. This fact may be explained by the interaction of β-carotene with singlet oxygen, the latter being formed by an intramolecular energy transfer in the excited state of the polystyrene–oxygen complex. A general kinetic reaction course for the generation and quenching of singlet oxygen in the PS$+O_2+\beta$-carotene system may be proposed as follows:

$$(PS+O_2)\ \text{complex}+h\nu \longrightarrow {}^1O_2^* \qquad \rho_o\varepsilon_k[O_2]_k \qquad (1)$$

$$^1\beta_o+h\nu \longrightarrow {}^1\beta^* \qquad \rho_o\varepsilon_\beta[\beta] \qquad (2)$$

$$^1\beta^* \longrightarrow {}^1\beta_o \qquad k_{nr}^{S}[^1\beta^*] \qquad (3)$$

$$^1\beta^*+{}^3O_2 \longrightarrow {}^3\beta^*+{}^3O_2 \qquad k_{ISC}^{-O_2}[^1\beta^*][O_2] \qquad (4)$$

$$^1\beta^* \longrightarrow {}^3\beta^* \qquad k_{ISC}[^1\beta^*] \qquad (5)$$

$$^{1,3}\beta^*+{}^1O_2^* \longrightarrow \beta_o+{}^3O_2 \qquad k_z[^{1,3}\beta^*][^1O_2^*] \qquad (6)$$

$$^1O_2^* \longrightarrow {}^3O_2 \qquad k_i[^1O_2^*] \qquad (7)$$

$$^1O_2^*+{}^1\beta_o \longrightarrow {}^3\beta^*+{}^3O_2 \qquad k_p[^1O_2^*][\beta_o] \qquad (8)$$

$$^1O_2^*+{}^1\beta_o \longrightarrow P \qquad k_{\beta_o}[\beta_o][^1O_2^*] \qquad (9)$$

$$^3\beta^* \longrightarrow {}^1\beta_o \qquad k_{nr}^{T}[^3\beta^*] \qquad (10)$$

$$^3\beta^*+{}^3O_2 \longrightarrow P' \qquad k_\beta[^3\beta^*][O_2] \qquad (11)$$

$$^3\beta^*+{}^3O_2 \longrightarrow {}^1\beta_o+{}^3O_2 \qquad k_q[^3\beta^*][O_2] \qquad (12)$$

$$^3\beta^* \longrightarrow P'' \qquad k_d[^3\beta^*] \qquad (13)$$

where $^1\beta_0$, $^1\beta^*$, and $^3\beta^*$, are the ground-state and the first excited singlet and triplet β-carotene states, respectively.

The rate of consumption (V) of β-carotene in this system is expressed by the equation

$$V=k_{\beta_o}[^1\beta_o][^1O_2^*]+k_\beta[^3\beta^*][O_2]+k_z[^{1,3}\beta^*][^1O_2^*]+k_d[^3\beta^*] \qquad (14)$$

The substitution of the expressions for $[^1O_2^*]$ and $[^{1,3}\beta^*]$ calculated from equations (1)–(13) in equation (14) gives a complicated expression which can be simplified on the assumption that in the presence of high oxygen concentrations: (a) radiationless processes of the deactivation of β-carotene electronic excited states [equations (3) and (10)] are less important in comparison with the competitive decay processes; $k_{nr}^S = k_{nr}^T = O$; (b) deactivation processes of the triplet excited state of β-carotene by oxygen [equations (11) and (12)] are more probable than the process represented by equation (13) under these conditions; $k_d = O$; and (c) *cis–trans* β-carotene isomerization [equation (6)] can be neglected in this system; $k_z = O$. Then V is given by the equation

$$V = \rho_o \varepsilon_k [O_2] \left\{ \frac{k_{\beta_0}[\beta]}{k_i + (k_p + k_{\beta_0})[\beta]} + \frac{k_\beta}{k_\beta + k_q} \cdot \frac{k_p[\beta]}{k_i + (k_p + k_{\beta_0})[\beta]} \right\} + \rho_o \varepsilon_\beta [\beta] \cdot \frac{k_\beta}{k_\beta + k_q} \tag{15}$$

Figure 6 shows as an example a plot of $V = -d\beta/dt$ *versus* β-carotene concentration for $P_{O_2} = 20$, 10, and 0·2 atm. The values of $dV/d[\beta]$ ($[\beta] \to O$) can be calculated from the slope of the tangent in the initial parts of these curves.

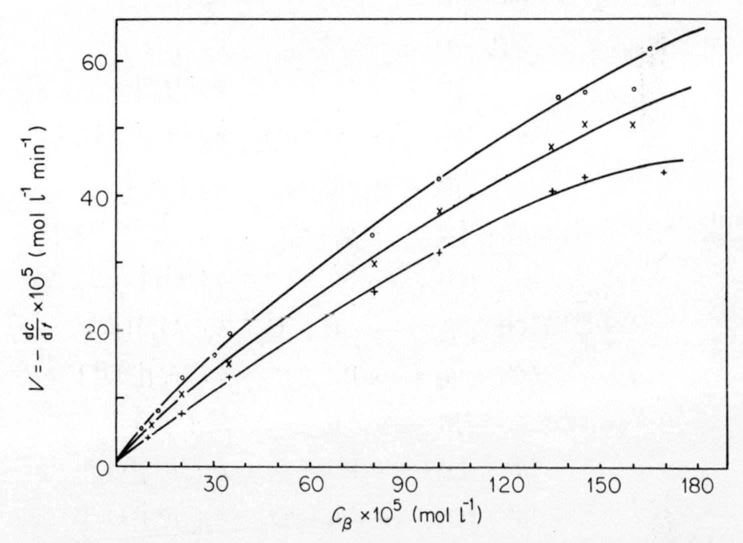

Figure 6 Plot of rate of β-carotene consumption *versus* β-carotene concentration in PS film for oxygen pressures of ○ 20 atm, × 10 atm, and + 0·2 atm

From equation (15) the following expression is obtained:

$$\frac{dV}{d[\beta]_{[\beta]\to O}} = \rho_o \varepsilon_k [O_2]_k \left[\frac{k_{\beta o}}{k_i} + \frac{k_p}{k_i} \cdot \frac{k_\beta}{k_q k_\beta} \right] + \rho_o \varepsilon_\beta \cdot \frac{k_\beta}{k_q + k_\beta} \tag{16}$$

The linear plot of $\mathrm{d}V/\mathrm{d}[\beta]$ ($[\beta]\rightarrow$ O) *versus* oxygen concentration gives

$$\frac{k_\beta}{k_\mathrm{q}+k_\beta}=6\times 10^3 \tag{17}$$

and

$$k_{\beta_0}+6\times 10^{-3}k_\mathrm{p}=6k_\mathrm{i} \tag{18}$$

when $\rho_\mathrm{o}=1\times 10^{-6}$ einsteins min^{-1} cm^{-2}, $\varepsilon_\mathrm{k}^{313}=45$ l mol^{-1} cm^{-1}, $\varepsilon_\beta^{313}=5{\cdot}2\times 10^3$ l mol^{-1} cm^{-1}, and oxygen solubility coefficient in polystyrene film[13] $\alpha=3{\cdot}2\times 10^{-3}$ M l^{-1} atm^{-1}.

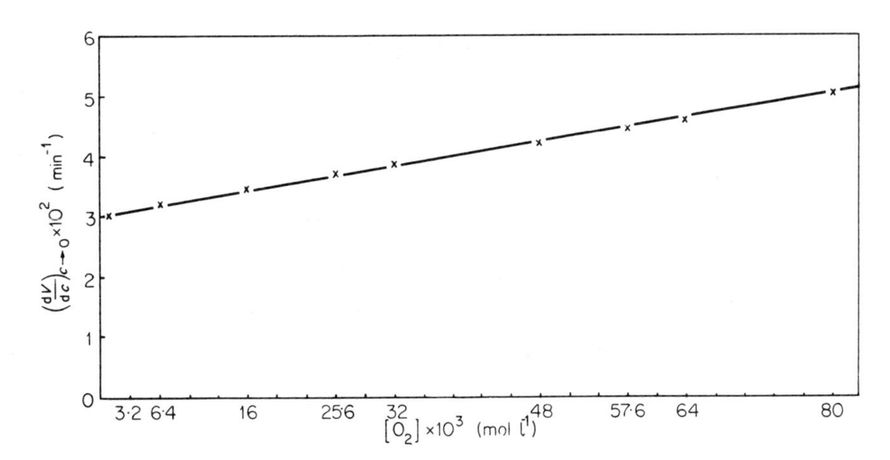

Figure 7 Linear plot of $\left(\frac{\mathrm{d}V}{\mathrm{d}c}\right)_{c\rightarrow 0}$ *versus* oxygen concentration in PS film

It was found that the process of singlet oxygen quenching in a liquid is controlled by diffusion.[7, 14]

The rate constant of the process controlled by diffusion is given by equation (15)

$$k_\mathrm{Q}=4\pi N'D_\mathrm{AB}R_\mathrm{AB}p\left(1+\frac{R_\mathrm{AB}}{\sqrt{D_\mathrm{AB}\tau_\mathrm{N}}}\right) \tag{19}$$

where

N' = Avogadro's number $\times 10^{-3}$;
R_AB = radius of interaction;
$D_\mathrm{AB}=D_\mathrm{A}+D_\mathrm{B}$ = the sum of the diffusion coefficients of the reagents;
P = probability of quenching in a collision;
τ_N = lifetime of excited state quenching molecule.

This rate constant for singlet oxygen in PS film deactivation is

$$K_Q = 2{\cdot}725 \times 10^8 \text{ l mol}^{-1} s^{-1} \tag{20}$$

when $p=1$, $R_{AB}=1$ nm, $D_{AB}=D_{O_2}=3{\cdot}2\times10^{-7}$ cm^2 s^{-1}, and $\tau_{^1O_2}=2$ μs. In the kinetic system proposed here singlet oxygen quenching processes controlled by diffusion are expressed by equations (8) and (9). Therefore:

$$k_{\beta_o} + k_p = 2{\cdot}725 \times 10^8 \text{ l mol}^{-1} \text{ s}^{-1} \tag{21}$$

The values of the rate constants k_{β_o}, k_p, and k_i were found from equations (18) and (21) and assuming that the lifetime of singlet oxygen in this system is $\tau_{^1O_2}=2$ μs:

$$k_p = 2{\cdot}711 \times 10^8 \text{ l mol}^{-1} \text{ s}^{-1} \tag{22}$$

$$k_{\beta_o} = 1{\cdot}4 \times 10^6 \text{ l mol}^{-1} \text{ s}^{-1} \tag{23}$$

$$k_i = 5 \times 10^5 \text{ s}^{-1} \tag{24}$$

The rate constant of the diffusion-controlled process of β-carotene excited triplet state quenching[11, 17] is

$$k^Q_{3\beta*} = 1/9 K_Q = 1/9 \times 2{\cdot}34 \times 10^8 = 2{\cdot}6 \times 10^7 \text{ l mol}^{-1} \text{ s}^{-1} \tag{25}$$

For the calculation of: K_Q:$p=1$, $R_{AB}=1$ nm, $D_{AB}=D_{O2}=3\times10^{-7}$ cm^2 s^{-1}, and $\tau=9$ μs. However, from the proposed kinetic system

$$k_\beta + k_q = 2{\cdot}6 \times 10^7 \text{ l mol}^{-1} \text{ s}^{-1} \tag{26}$$

Taking into account equations (17) and (26) it was calculated that

$$k_\beta = 1{\cdot}56 \times 10^5 \text{ l mol}^{-1} \text{ s}^{-1} \tag{27}$$

$$k_q = 2{\cdot}548 \times 10^7 \text{ l mol}^{-1} \text{ s}^{-1} \tag{28}$$

The consumption of β-carotene under vacuum or in the presence of very low oxygen concentrations in polystyrene films is described by equation (13) in the proposed kinetic system. The rate of consumption of β-carotene under these conditions is given by the equation

$$V = \rho_o \varepsilon_\beta [\beta] \left(\frac{k_{ISC}}{k_{ISC} + k^S_{nr}} \cdot \frac{k_r}{k_r + k^T_{nr}} \right) \tag{29}$$

All values estimated here have only approximate significance. They help to establish the role of these processes in the general course of the generation and quenching of singlet oxygen in polystyrene films during excitation by light which is absorbed only by the polystyrene–oxygen complex.

REFERENCES

1. A. M. Trozzolo and F. H. Winslow, *Macromolecules*, **1**, 98 (1968).
2. D. R. Adams and F. Wilkinson, *Trans. Faraday Soc.*, **68**, 586 (1972).
3. J. P. Guilory and C. F. Cook, *J. Polym. Sci. A1*, **11**, 1927 (1973).

4. M. L. Kaplan and P. G. Kelleher, *J. Polym. Sci. B*, **9**, 565 (1971).
5. E. C. Lim and V. L. Kowalski, *J. Chem. Phys.*, **36**, 1729 (1962).
6. J. F. Rabek and B. Rånby, *J. Polym. Sci. A1*, **12**, 273 (1974).
7. C. S. Foote and R. W. Denny, *J. Amer. Chem. Soc.*, **90**, 6233 (1968).
8. C. S. Foote, Y. C. Chang, and R. W. Denny, *J. Amer. Chem. Soc.*, **92**, 5216 (1970).
9. C. S. Foote, Y. C. Chang, and R. W. Denny, *J. Amer. Chem. Soc.*, **92**, 5218 (1970).
10. P. B. Merkel and D. R. Kearns, *J. Amer. Chem. Soc.*, **94**, 7244 (1972).
11. A. Farmilo and F. Wilkinson, *Photochem. Photobiol.*, **18**, 447 (1973).
12. I. B. C. Matheson and J. Lee, *Chem. Phys. Lett.*, **14**, 350 (1972).
13. M. Nowakowska, J. Najbar, and B. Waligóra, *Eur. Polym. J.*, **12**, 387 (1976).
14. W. G. Herkstroeter, *J. Amer. Chem. Soc.*, **97**, 4161 (1975).
15. A. M. Alwatter, in *Organic Molecular Photophysics* (Ed. J. B. Birks), Wiley-Interscience, London, New York, Toronto and Sydney, 1974, p. 403.
16. R. P. Wayne, in *Advances in Photochemistry* (Ed. J. N. Pitts, Jr., G. S. Hammond, and W. A. Noyes, Jr.), Interscience, New York, 1969, Vol. 7, p. 312.
17. O. L. J. Gijzeman, F. Kaufman, and G. Porter, *J. Chem. Soc., Faraday Trans. II*, **69**, 708 (1973).

26

Photosensitized Singlet Oxygen Oxidation of Polydienes

J. F. RABEK, Y. J. SHUR and B. RÅNBY

Department of Polymer Technology, The Royal Institute of Technology, Stockholm, Sweden

INTRODUCTION

It has been reported that polydienes are easily photo-oxidized by molecular and singlet oxygen,[1–3] and these reactions are important for the rubber industry and especially for the lifetime of tyres. Recently the effects of common pollutants present in the environment, e.g. in the atmosphere, have aroused a great interest. The energy–transfer process in which atmospheric pollutants capable of absorbing solar radiation transfer the energy to molecular oxygen, forming singlet oxygen in urban atmosphere,[4] present an important problem. Polycyclic hydrocarbons, which can be absorbed by rubbers and plastics from polluted air, form one of the most common atmospheric contaminants.

It has also been reported that certain dyes in the presence of oxygen sensitize the photodegradation of rubber,[1,5–7] polyamides, and cellulose.[8–11] The mechanism of singlet oxygen reactions with low molecular weight olefins and dienes has been the subject of a number of recent studies.[12–28] Singlet oxygen has been reported to attack unsaturated bonds in rubber to give hydroperoxides[29–32] and to react with unsaturated or even saturated groups in polyethylene[32] and polypropylene,[33] likewise giving hydroperoxides. The following mechanism for singlet oxygen oxidation with *cis*-polybutadiene has been proposed:[31]

$$\begin{array}{ccccccccc}
-CH_2 & & H_2C-CH_2 & & H_2C-CH_2 & & H_2C- \\
\backslash & / & & \backslash & / & & \backslash & / \\
 & HC{=}CH & & & HC{=}CH & & & HC{=}CH
\end{array}$$

$$\downarrow {}^1O_2$$

$$\begin{array}{ccccccccc}
-CH_2 & & HC-CH & & H_2C-CH_2 & & H_2C \\
\backslash & // & & \backslash\backslash & / & & \backslash & / \\
 & HC-CH & & HC-CH & & & HC{=}CH \\
 & | & & & | & & \\
 & OOH & & & OOH & &
\end{array}$$

$$\downarrow {}^{1}O_2$$

$$\text{—}CH_2 \quad HC{=}CH \quad H_2C\text{—}CH_2 \quad H_2C\text{—}$$

$$HC\text{—}CH \quad HC\text{—}CH \quad HC{=}CH$$

$$OOH \quad O\text{—}O \quad OOH$$

In this research we have carried out experiments of industrial interest which include mainly the study of changes in the structure of polybutadiene during sensitized photo-oxidation. Two series of experiments, with anthracene and dyes as sensitizers, are reported in this paper.

EXPERIMENTAL

Anthracene as sensitizer

An efficient method for studying the oxidation processes in rubber is to measure the oxygen consumption. The absorption of oxygen at equilibrium was obtained by monitoring gravimetrically the weight increase with time of UV irradiation of a rubber film sample exposed to oxygen (1 atm), using a Cahn electromicrobalance (Model RG, Figure 1). For measurement, a sample in a frame was placed in the balance and evacuated to 4×10^{-7} mmHg at 25 °C for 2 days (equilibrium weight, W_1). Oxygen was then introduced into the balance and the sample irradiated with UV light (mercury lamp, Type SP 500 W, Philips, The Netherlands). The weight increase of the sample after irradiation for t min was measured (weight W_2). In order to measure only the amount of oxygen consumed by photo-oxidation, the sample was then degassed under vacuum until

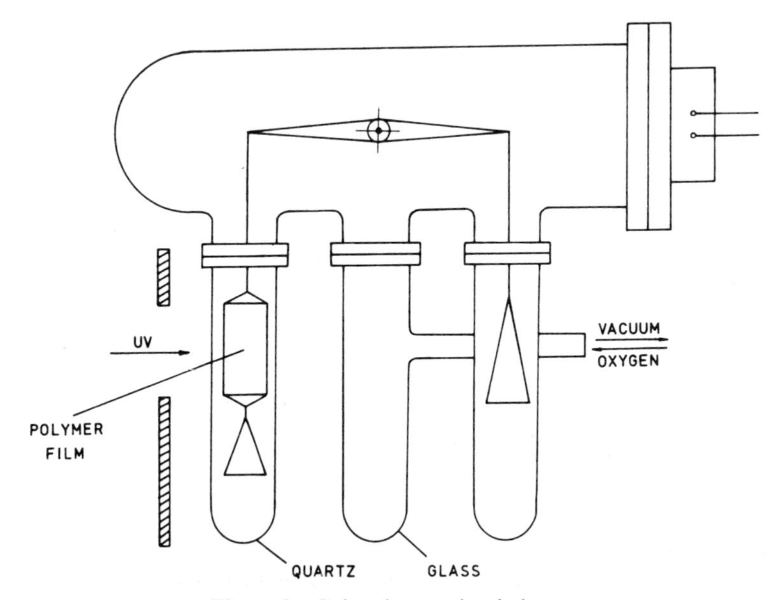

Figure 1 Cahn electromicrobalance

desorption equilibrium was reached (final weight W_3). The difference (ΔW) between the final weight (W_3) and original equilibrium weight (W_1) both in vacuum at equilibrium represents the consumption of oxygen by photo-oxidation of the rubber sample (Figure 2):

$$\Delta W = W_3 - W_1$$

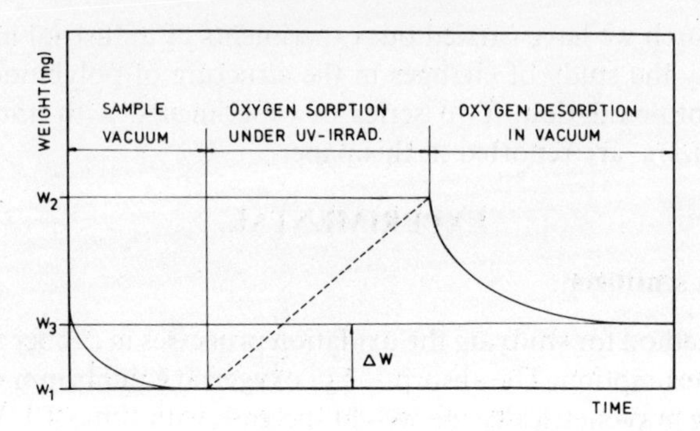

Figure 2 Procedure for measuring oxygen consumption by polymer sample

The rate of consumption of oxygen by *cis*-1,4-polybutadiene in the presence of anthracene (10^{-3} M) is much higher than that of the pure rubber sample (Figure 3). Samples of *cis*-1,4-polybutadiene photo-oxidized in the presence of anthracene (10^{-3} M) show an increased amount of hydroxy and/or hydroperoxy groups (Figure 4).

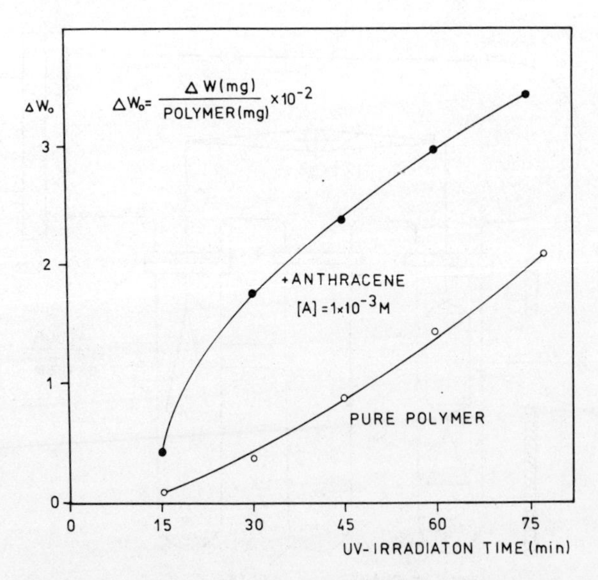

Figure 3 Kinetics of oxygen consumption of *cis*-1,4-polybutadiene during UV irradiation

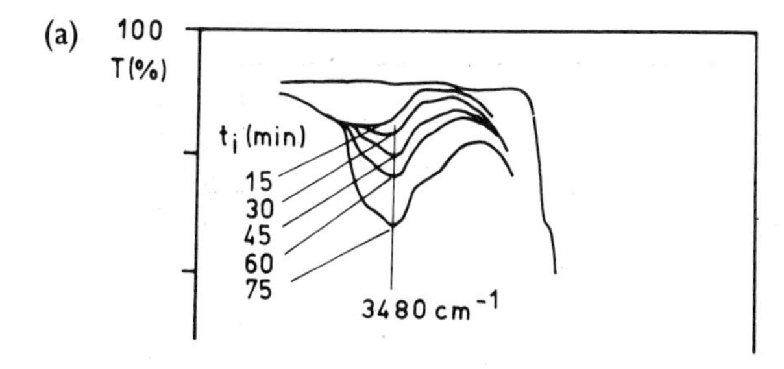

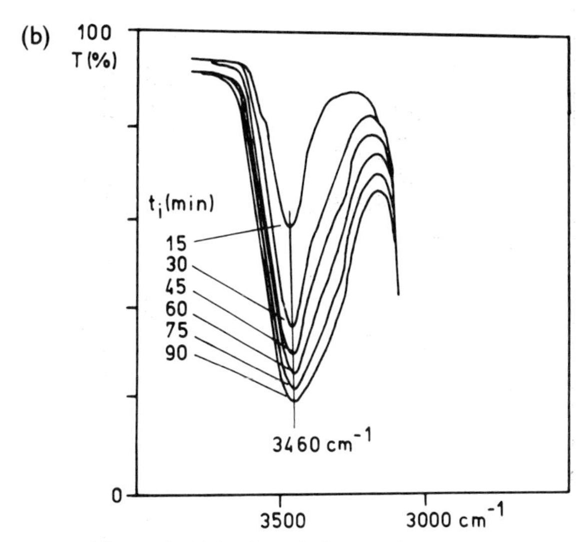

Figure 4 Kinetics of changes in IR absorption spectra of *cis*-1,4-polybutadiene during UV irradiation: (a) without and (b) with anthracene (1×10^{-3} M)

Solid polymer samples containing anthracene, after photo-oxidation for 1 h, are completely crosslinked and are insoluble in benzene and tetrahydrofuran. Photo-oxidation of polybutadiene in benzene solution (1 wt.-%) in the presence of anthracene (10^{-3} M) causes a rapid decrease in viscosity due to decreasing molecular weight and also to chain branching.

Dyes as sensitizers

Solutions for the experiments were prepared by dissolving purified *cis*-1,4-polybutadiene samples in benzene (spectral grade) to concentrations of about 1 g per 100 ml. The solutions were filtered through a fine cellulose fibre pad (4 mm thick) under nitrogen pressure. In this way gel particles over 20 μm were removed.

The following dyes were used as photosensitizers: methylene blue B, fluorescein and Rose Bengal. These dyes are insoluble in benzene but are soluble in polar solvents such as methanol. When the dyes are dissolved in methanol at

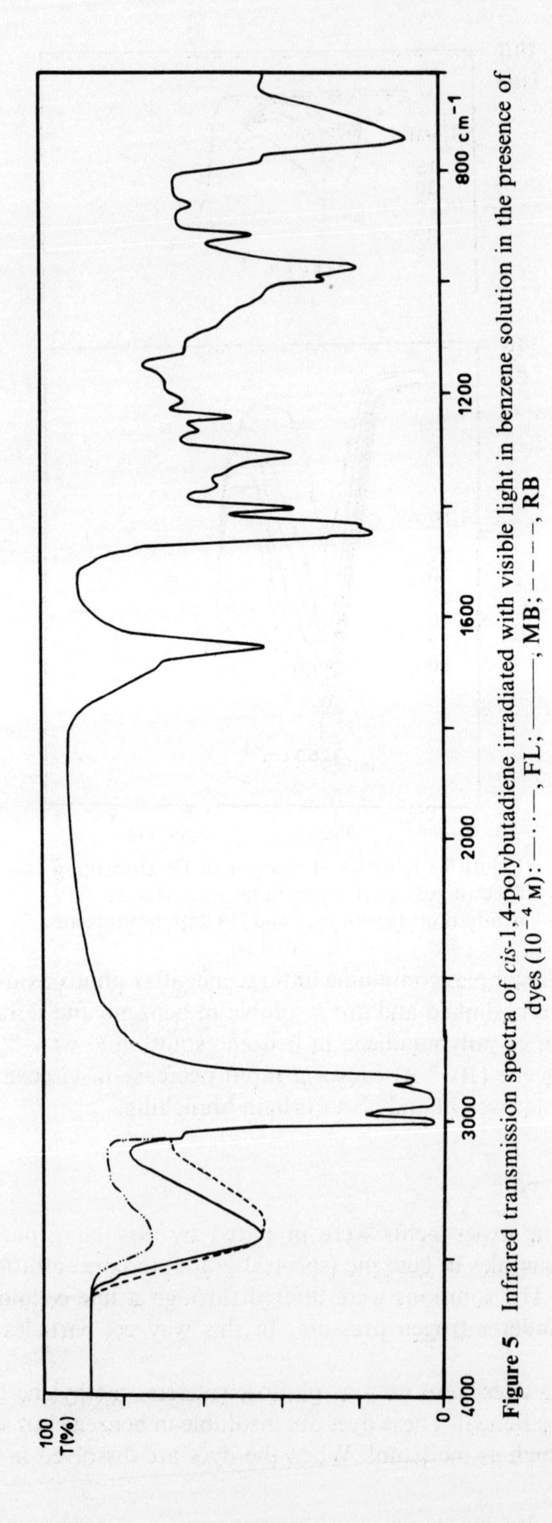

Figure 5 Infrared transmission spectra of *cis*-1,4-polybutadiene irradiated with visible light in benzene solution in the presence of dyes (10^{-4} M): —·—, FL; ——, MB; – – –, RB

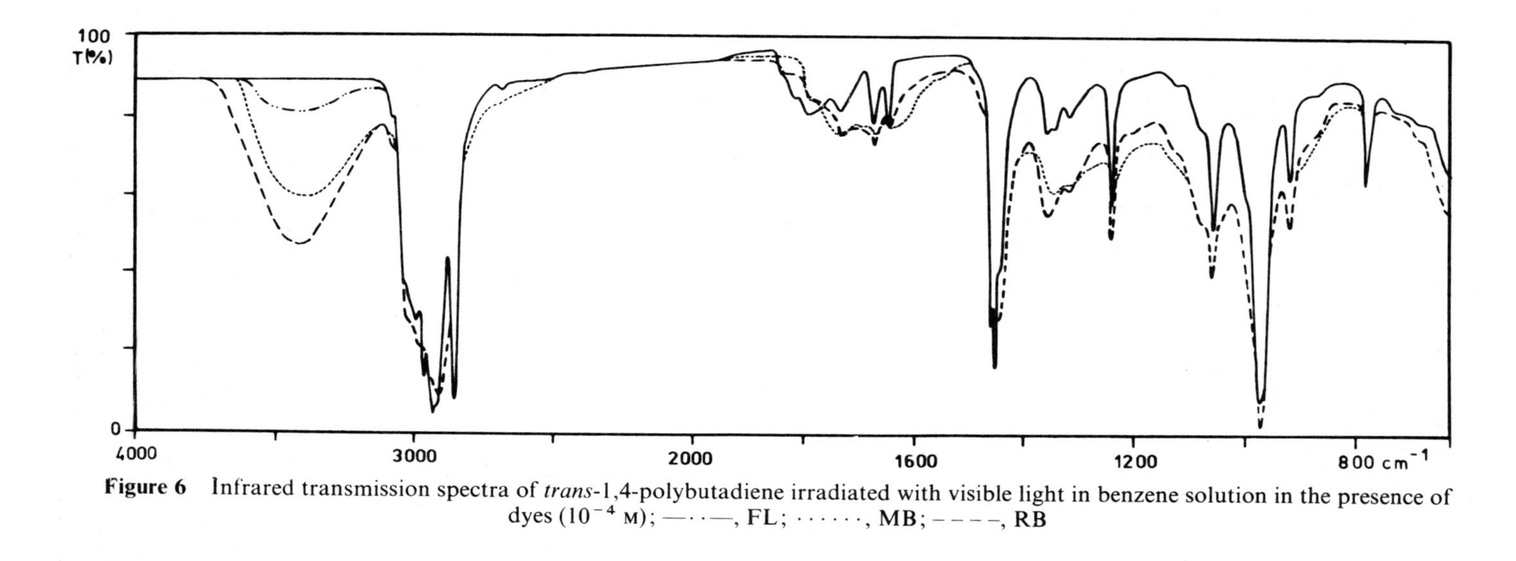

Figure 6 Infrared transmission spectra of *trans*-1,4-polybutadiene irradiated with visible light in benzene solution in the presence of dyes (10^{-4} M); —·—, FL; ······, MB; – – –, RB

concentrations of 10^{-2}–10^{-3} M and 1 volume of dye solution is added to 9 volumes of polybutadiene solution in benzene, neither the dyes nor the polymers are precipitated. Benzene–methanol solutions of *cis*-1,4-polybutadiene containing these sensitizing dyes (10^{-3}–10^{-4} M) rapidly consumed oxygen on irradiation with visible light. Photosensitized oxidation of *cis*-polybutadiene (Figure 5) and *trans*-polybutadiene (Figure 6) in benzene solution in the presence of different dyes under irradiation with visible light (500-W tungsten lamp), produces only a strong broad absorption band in the region 3600–3200 cm^{-1} which is assigned to hydroxyl groups and/or hydroperoxy groups. On heating the film sample to 100 °C the sharp band at 3420 cm^{-1} becomes much stronger while the broad spectrum decreases. This process is reversible. When the sample is cooled in air to room temperature, the original spectrum is completely restored. This result indicates that the observed IR absorption band at 3600–3200 cm^{-1} can be attributed to hydroperoxy groups only. The IR absorption band of hydroperoxy groups is probably overlapped by the strong absorption of hydroxyl groups. The hydroperoxide content is less than 0·02–0·04 mol-%. During heating of the sample to 100 °C the hydroperoxides present are thermally decomposed, whereas the hydroxyl groups are stable at that temperature. Formation of carbonyl groups in the range 1800–1600 cm^{-1} has not been observed. An additional band at 965 cm^{-1} is interpreted as being due to vinyl double bonds.

When *cis*-polybutadiene in benzene solution saturated with molecular oxygen is irradiated with UV light without a sensitizer (molecular oxygen oxidation), the formation of three absorption bands at 3400 cm^{-1} (–OH and/or –OOH), 1740 cm^{-1} ($>$C=O), and 965 cm^{-1} (vinyl double bonds) has been observed (Figure 7).

The presence of sensitizing dyes in *cis*- and *trans*-polybutadiene in benzene solutions produces an appreciable decrease in viscosity on exposure of these solutions to visible light (Figure 8). The strongest decrease in viscosity was observed for a dye concentration of 10^{-3} M. The viscosity data were additionally confirmed by gel permeation chromatographic measurements (Figure 9). The results obtained indicate a rapid degradation of the polymers during the singlet oxygen oxidation process.

DISCUSSION OF RESULTS

Basic studies and interpretation of singlet oxygen formation in photo-sensitized reactions have been made by Kearns.[17,34] When the excitation energy of a sensitizer molecule exceeds 22·5 kcal mol^{-1}, singlet oxygen $^1O_2(^1\Delta_g)$ may be formed. For the formation of singlet oxygen $^1O_2(^1\Delta_g^+)$ the excitation energy of the sensitizer molecule should exceed 37·5 kcal mol^{-1}. A schematic diagram for the energy transfer in dye-photosensitized reactions in which singlet oxygen

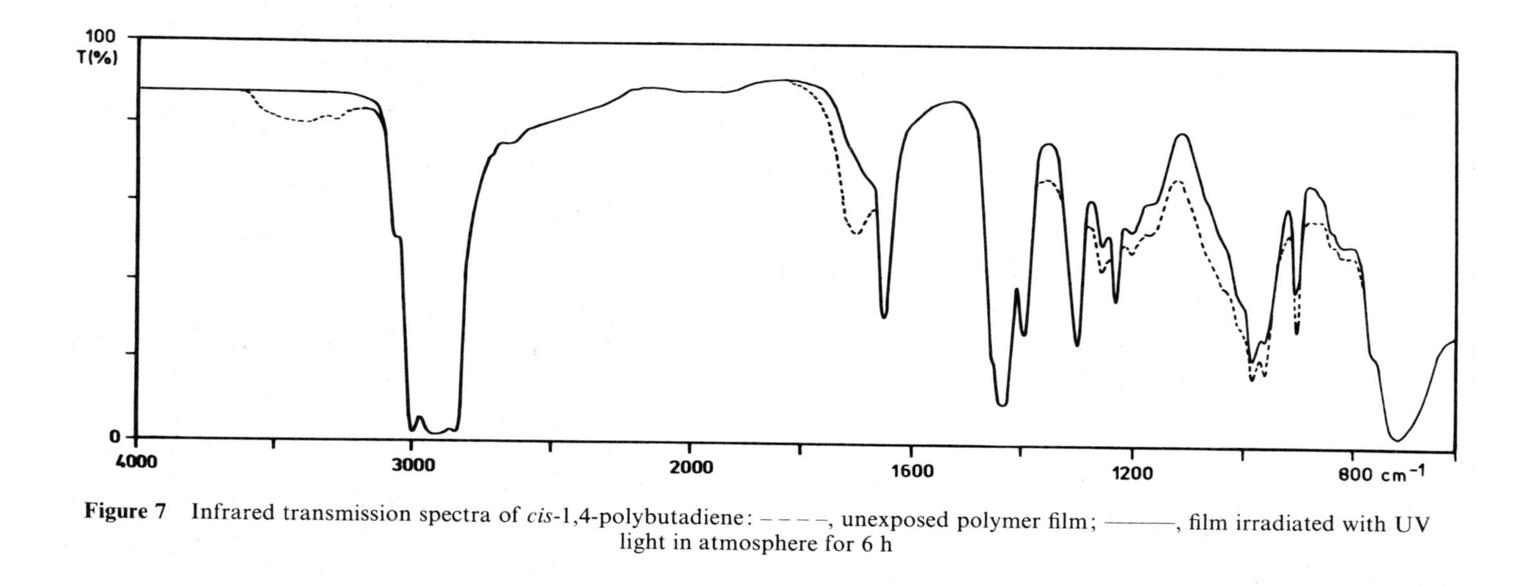

Figure 7 Infrared transmission spectra of *cis*-1,4-polybutadiene: – – – –, unexposed polymer film; ——, film irradiated with UV light in atmosphere for 6 h

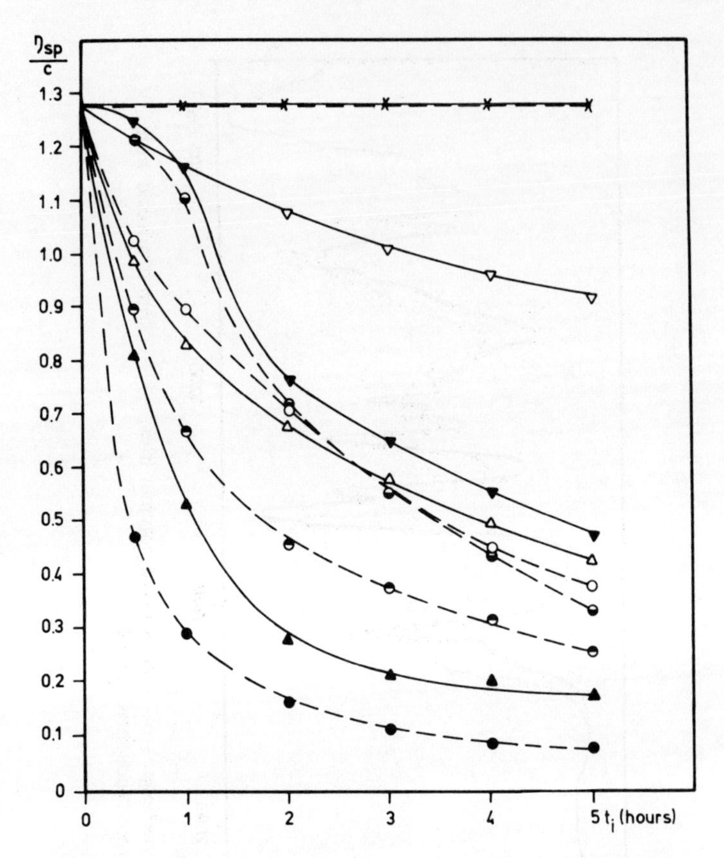

Figure 8 Change in viscosity of (— — —) *cis*-1,4-polybutadiene and (———) *trans*-1,4-polybutadiene in benzene solution (ca. 0·9 wt.-%) during irradiation with visible light in the presence of oxygen and with dyes used as sensitizers: ▽, ○, FL (10^{-4} M); ▲, ◒, RB (10^{-4} M); △, ◓, MB (10^{-4} M); ▲, ●, MB (10^{-3} M); ×, viscosity of solution irradiated without sensitizer

$^1O_2(^1\Delta_g$ and $^1\Sigma_g^+)$ may be formed is shown in Figure 10. The sensitizers used in our experiments have the following triplet energies (E_T): anthracene (A), 42 kcal mol^{-1}; fluoresceine (FL), 47·2 kcal mol^{-1}; Rose Bengal (RB), 42·0 kcal mol^{-1}; and methylene blue (MB), 32·0 kcal mol^{-1}. All of these triplet energies are sufficient for the formation of singlet oxygen $^1O_2(^1\Delta_g)$, but only anthracene, fluorescein and Rose Bengal have enough triplet energy (E_T) for the transfer forming the singlet oxygen $^1O_2(^1\Sigma_g^+)$. The sigma singlet oxygen is rapidly quenched in solution, indicating that $^1O_2(^1\Sigma_g^+)$ is not important as a reactant in these reactions. Singlet oxygen $^1O_2(^1\Delta_g)$ is presumably responsible for the initiation of oxidation of polydienes.

Singlet oxygen $^1O_2(^1\Delta_g)$ may attack the double bonds in polybutadienes to give an allylic hydroperoxide attached to the polymer chain:

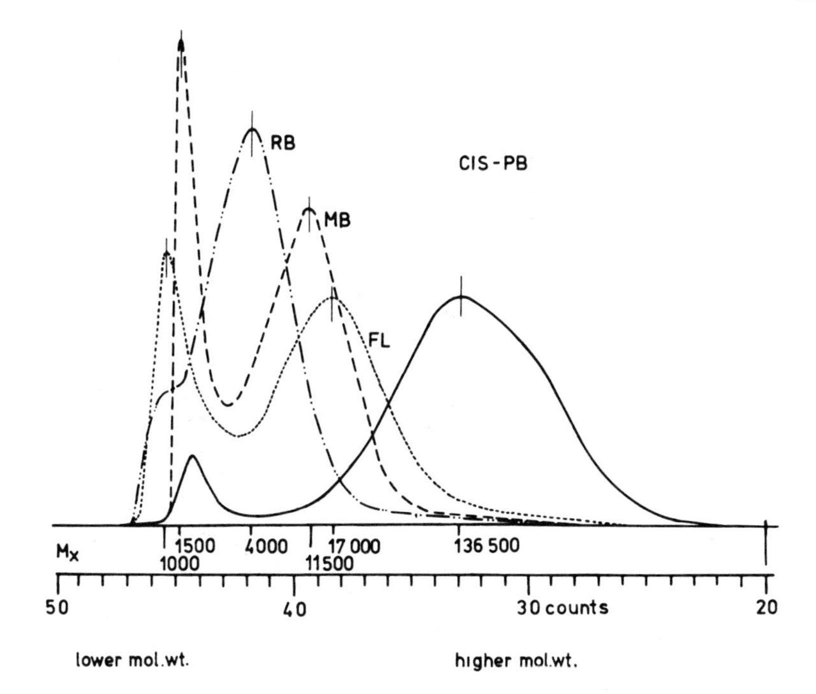

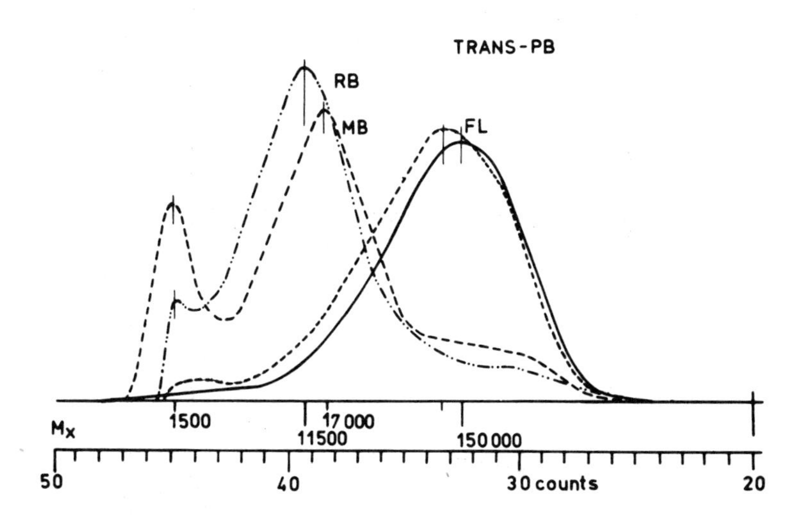

Figure 9 Gel permeation chromatograms of *cis*- and *trans*-1,4-polybutadienes: ———, undegraded samples and samples degraded in the presence of dye as sensitizers (10^{-4} M); ······, FL; – – – –, MB; —·—, BR under visible light

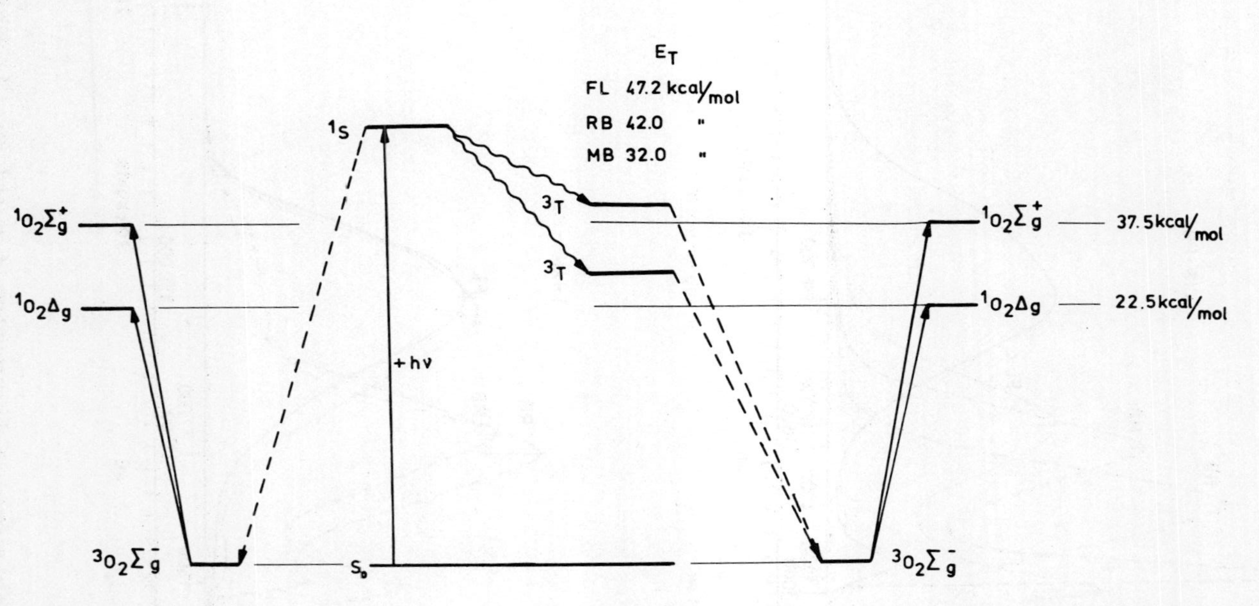

Figure 10 Diagram of energy transfer from the excited singlet (^1S) or triplet (^3T) of a sensitizer to molecular oxygen ($^3O_2$) from which singlet oxygen $^1O_2$ ($^1\Delta_g$) or $^1O_2$ ($^1\Sigma_g^+$) can be formed

$$-CH_2-CH{=}CH-CH_2- + {}^1O_2 \longrightarrow -CH_2-\underset{\displaystyle |}{\overset{\displaystyle OOH}{\overset{|}{C}H}}-CH{=}CH- \quad (1)$$

A small amount of hydroperoxide groups may be detected analytically.

Under UV irradiation, hydroperoxide groups can be decomposed to polymer oxy radical and hydroxyl radical according to the reaction

$$-CH_2-\overset{\displaystyle OOH}{\overset{|}{C}H}-CH{=}CH- \xrightarrow{+h\nu} -CH_2-\overset{\displaystyle \dot{O}}{\overset{|}{C}H}-CH{=}CH- + HO\cdot \quad (2)$$

The polymer oxy radicals are very reactive and can abstract hydrogen from a neighbouring polymer molecule, giving hydroxyl groups:

$$-CH_2-\overset{\displaystyle \dot{O}}{\overset{|}{C}H}-CH{=}CH- + -CH_2-CH{=}CH-CH_2- \longrightarrow$$

$$-CH_2-\overset{\displaystyle OH}{\overset{|}{C}H}-CH{=}CH- + -\dot{C}H-CH{=}CH-CH_2- \quad (3)$$

The mechanism, however, presented by reactions (2) and (3) cannot be completely accepted for the dye-sensitized photo-oxidation of polybutadienes under visible-light irradiation. Hydroperoxy groups have almost no absorption in the range 400–700 nm, and the light in this region cannot, therefore, decompose the hydroperoxide groups.

Our experimental observations lead us to propose a new mechanism for formation of hydroxyl groups in the dye-sensitized photo-oxidation of polybutadiene. The mechanism can be described by the following stages:

Stage 1: singlet oxygen reacts with a double bond by addition to give an endoperoxide:

$$-CH_2-CH{=}CH-CH_2- + {}^1O_2 \longrightarrow -CH_2-\overset{\displaystyle O}{\overset{|}{C}H}-\overset{\displaystyle O}{\overset{|}{C}H}-CH_2- \quad (4)$$

(with the two O atoms bonded to each other, O—O)

Stage 2: endoperoxides are unstable at room temperature and decompose with the formation of two reactive polymer oxy radicals:

$$-CH_2-\overset{\displaystyle O}{\overset{|}{C}H}-\overset{\displaystyle O}{\overset{|}{C}H}-CH_2- \xrightarrow{+\Delta} -CH_2-\overset{\displaystyle \dot{O}}{\overset{|}{C}H}-\underset{\displaystyle \dot{O}}{\underset{|}{C}H}-CH_2- \quad (5)$$

Stage 3: the polymer oxy radicals abstract hydrogen from methylene groups present in a neighbouring molecule and even in the same macromolecule:

$$-CH_2-\underset{\substack{|\\ \dot{O}}}{\overset{\substack{\dot{O}\\ |}}{CH}}-\underset{\substack{|\\ \dot{O}}}{CH}-CH_2- + -CH_2-CH{=}CH-CH- \longrightarrow$$

$$-CH_2-\overset{\substack{OH\\ |}}{CH}-\underset{\substack{|\\ \dot{O}}}{CH}-CH_2- + -\dot{C}H-CH{=}CH-CH_2- \quad (6)$$

Stage 4: the allyl polymer radical formed can easily be oxidized to a peroxy radical by addition of molecular oxygen which is present at the same time:

$$-\dot{C}H-CH{=}CH-CH_2- + {}^3O_2 \longrightarrow -\overset{\substack{\dot{O}\\ |\\ O\\ |}}{CH}-CH{=}CH-CH_2- \quad (7)$$

Stage 5: the peroxy polymer radicals may abstract hydrogen from neighbouring macromolecules (PH) and give hydroperoxides:

$$-\overset{\substack{\dot{O}\\ |\\ O\\ |}}{CH}-CH{=}CH-CH_2- + PH \longrightarrow -\overset{\substack{OOH\\ |}}{CH}-CH{=}CH-CH_2- + P\cdot \quad (8)$$

In the solid state, sensitized photo-oxidation of polybutadiene results in the crosslinking of the polymer, whereas in solution a rapid degradation is observed which obviously occurs by a free-radical mechanism. The reaction conditions determine which one predominates. In the solid state, the crosslinking of macro-radicals mainly occurs, whereas in solution the degradation is observed. Sometimes small amounts of microgel are also observed in solutions, especially at higher concentrations of polymers (>1 wt.-%).

CONCLUSION

Singlet oxygen takes part in the photosensitized oxidation of polydienes in the solid state and in solution. The reaction mechanism proposed involves the formation of polymer radicals as intermediates, which are expected to react with molecular oxygen present in the system. Therefore, under the experimental conditions used, it is not possible to separate the effects of singlet oxygen and molecular oxygen. It is possible that further information may be deduced from the analysis of the structure of oxidized polymers and other data.

ACKNOWLEDGEMENTS

The authors thank the Swedish Board for Technical Development (STU) and the Swedish Polymer Research Foundation (SSP) for their generous support of this research.

REFERENCES

1. B. Rånby and J. F. Rabek, *Photodegradation, Photo-oxidation and Photostabilization of Polymers*, Wiley, London, 1975.
2. D. J. Carlsson and D. M. Wiles, *Rubb. Chem. Technol.*, **49**, 991 (1976).
3. G. P. Canva and J. J. Canva, *Rubb. J.*, **1971**, 36.
4. J. N. Pitts, Jr., in *Chemical Reactions in Urban Atmospheres* (Ed. C. S. Tuesday), Americal Elsevier, New York, 1971, p. 29.
5. T. Mill, K. C. Irvin, and F. R. Mayo, *Rubb. Chem. Technol.*, **41**, 296 (1968).
6. J. F. Rabek, in *XXIII IUPAC Congress, Boston, Mass.*, Butterworths, London, 1971, Vol. 9, p. 29.
7. J. F. Rabek and B. Rånby, *J. Polym. Sci. A1*, **14**, 1643 (1976).
8. G. S. Egerton, *Brit. Polym. J.*, **3**, 63 (1971).
9. G. S. Egerton, J. M. Gleadle, and M. A. Roch, *Nature, Lond.*, **211**, 1087 (1966).
10. G. S. Egerton, N. E. N. Assaad, and N. D. Uffindel, *Chem. Ind., Lond.*, **1967**, 1172.
11. A. Zweig and W. A. Henderson, Jr., *J. Polym. Sci. A1*, **13**, 713, 993 (1975).
12. C. S. Foote, *Accounts Chem. Res.*, **1**, 104 (1968).
13. C. S. Foote, *Pure Appl. Chem.*, **27**, 635 (1971).
14. K. Gollnick and G. O. Schenck, *Pure Appl. Chem.*, **9**, 507 (1964).
15. K. Gollnick and G. O. Schenck, in *1,4-Cycloaddition Reactions* (Ed. J. Hamer), Academic Press, New York, 1967, p. 255.
16. K. Gollnick, *Adv. Photochem.*, **6**, 1 (1968).
17. D. R. Kearns, *Chem. Rev.*, **71**, 395 (1971).
18. A. U. Khan and D. R. Kearns, *Adv. Chem. Ser.*, No. 77, 143 (1968).
19. R. Higgins, C. S. Foote, and H. Cheng, *Adv. Chem. Ser.*, No. 77, 102 (1968).
20. C. S. Foote, R. W. Denny, L. Weaver, Y. Chang, and J. Peters, *Ann. N.Y. Acad. Sci.*, **171**, 139 (1970).
21. F. A. Litt and A. Nickon, *Adv. Chem. Ser.*, No. 77, 118 (1968).
22. W. Fenical, D. R. Kearns, and P. Radlick, *J. Amer. Chem. Soc.*, **91**, 3396 (1969).
23. D. R. Kearns, W. Fenical, and P. Radlick, *Ann. N.Y. Acad. Sci.*, **171**, 34 (1970).
24. W. Fenical, D. R. Kearns, and P. Radlick, *J. Amer. Chem. Soc.*, **91**, 7771 (1969).
25. N. Hasty, P. B. Merkel, P. Radlick, and D. R. Kearns, *Tetrahedron Lett.*, **1972**, 49.
26. K. Gollnick, D. Haisch, and G. Schade, *J. Amer. Chem. Soc.*, **94**, 1747 (1972).
27. C. S. Foote, T. T. Fujimoto, and Y. C. Chang, *Tetrahedron Lett.*, **1972**, 45.
28. N. A. Hasty and D. R. Kearns, *J. Amer. Chem. Soc.*, **95**, 3380 (1973).
29. M. L. Kaplan and P. G. Kelleher, *Rubb. Chem. Technol.*, **45**, 423 (1972).
30. M. L. Kaplan and P. G. Kelleher, *Science, N.Y.*, **169**, 1206 (1970).
31. M. L. Kaplan and P. G. Kelleher, *J. Polym. Sci. A1*, **8**, 3163 (1970).
32. M. L. Kaplan and P. G. Kelleher, *J. Polym. Sci. B*, **9**, 565 (1971).
33. D. J. Carlsson and D. M. Wiles, *J. Polym. Sci. B*, **11**, 759 (1973).
34. P. B. Merkel and D. R. Kearns, *J. Amer. Chem. Soc.*, **94**, 7244 (1972).

27

Photodegradation of Polyisoprene Catalysed by Singlet Oxygen

H. C. NG and J. E. GUILLET

Department of Chemistry, University of Toronto, Toronto, Canada M5S 1A1

INTRODUCTION

It is well known that singlet oxygen can attack the double bonds of unsaturated polymers to give an allylic hydroperoxide attached to the polymer chain:

$$R{-}CH_2{-}CH{=}CH{-}R' \xrightarrow{^1O_2} R{-}CH{=}\underset{\displaystyle H}{\underset{|}{C}}{-}\underset{\displaystyle H}{\underset{|}{\overset{\overset{\displaystyle H}{\overset{\displaystyle O}{\overset{\displaystyle O}{|}}}}{C}}}{-}R' \qquad (1)$$

This reaction has been postulated as a key initiating step in the photo-oxidation of both saturated and unsaturated polymers,[1,2] and for this reason it is important to understand the kinetics and mechanism of the degradation processes resulting from the introduction of such groups in polymer chains. We report here the results of experiments in which purified *cis*-polyisoprene was treated with singlet oxygen and then photolysed in solution with light of specific wavelengths. The reaction can be followed conveniently and with great precision by automatic viscometry.

EXPERIMENTAL

Generation of Singlet Oxygen

Singlet oxygen was generated by the microwave discharge method.[3,4] Oxygen at 5 torr was led over a mercury droplet and through an EMS 200 microwave discharge (2450 MHz, 100 W) in a quartz tube, where the energy from the waveguide was coupled with the gas stream. Singlet oxygen, oxygen atoms and ozone were generated. The last two species were removed by reaction with traces of mercury vapour and formed mercury oxide rings inside the quartz tube. The singlet oxygen stream was then passed through a cold trap (−78 °C) and led into the reaction vessel. The path length from the discharge zone to the reaction cell was about 1 m, which is short enough for the singlet oxygen to survive at the pumping rate used (30 l min^{-1}).

Treatment of Sample

A commercial sample of masticated *cis*-polyisoprene was purified by dissolving it three times in spectroscopic-grade benzene and precipitating it with methanol. No traces of antioxidants or other additives could be detected by IR or UV spectroscopy after this treatment. The sample was dissolved in benzene and stored in the dark.

Samples from the stock benzene solution (1·2 wt.-%) were freeze-dried overnight. The dried, porous films were introduced into the reaction vessel and treated with singlet oxygen in the dark. After treatment (2–20 h) the sample was dissolved in purified solvent and kept in a cool and dark place overnight. The resultant hydroperoxide content was determined iodometrically,[5] using a value of 25 000 M^{-1} for the molar extinction coefficient of I_3^- at 360 nm.[6]

Photolysis of the Sample after Singlet Oxygen Treatment

The automatic viscometer for the photolysis and viscosity measurement has been described recently.[7] The light source was a Bausch and Lomb super-pressure mercury arc connected with a high-intensity grating monochromator. The apparatus was also connected with a Unicam SP 1800 UV spectrophotometer and the absorbance of the sample during photolysis was measured. The sample was saturated with nitrogen during photolysis. The viscosity and absorbance were measured at 10- or 20-min intervals. The light intensities at different wavelengths were determined by ferrioxalate actinometry. With the viscosity and absorbance data, the number of chain scissions and the quantum yield was calculated by the procedure outlined by Amerik and Guillet.[8] The Mark–Houwink constants for the viscosity–molecular weight relationship in each solvent were determined, when necessary, by the rapid method described by Kilp and Guillet,[9] in which the random scission of polymer chains is used to produce samples of known polydispersity.

RESULTS AND DISCUSSION

Treatment of polyisoprene with singlet oxygen resulted in the formation of allylic hydroperoxide groups consistent with the scheme shown in equation (1). The amount of hydroperoxide was proportional to the duration of the treatment, as shown in Table 1. Furthermore, no degradation of molecular weight resulted from the treatment, which is again consistent with equation (1). Molecular weight changes occurred only after irradiation of the polymer with near-ultraviolet light. The quantum yield for degradation increased with hydroperoxide content (see Table 2). Polyisoprene treated with gaseous oxygen in the

Table 1 Hydroperoxide content in relation to treatment time

Singlet oxygen treatment (h)	Hydroperoxide content	
	mol-%	mol × 10^6
2	0·04	1·9
4	0·05	2·4
6	0·07	3·3
8	0·08	3·9
10	0·10	4·4
20	0·15	6·8

Table 2 Quantum yield (ϕ) in relation to hydroperoxide content, using 1,2-dichloroethane at 365 nm and 30 °C

ϕ	[—OOH](%)
0·0045	0·15
0·0039	0·10
0·0021	0·04

dark for 20 h in the absence of the microwave discharge showed the same rate of photolysis as the untreated control.

In 1,2-dichloroethane solution, the quantum yield for photolysis decreased with wavelength, as shown in Table 3. The quantum yields were similar in cyclohexane and 1,2-dichloroethane, but substantially reduced in benzene (Table 4).

Table 3 Effect of wavelength on quantum yield, using 1,2-dichloroethane at 30 °C

λ (nm)	ϕ_s
365	0·003
340	0·025
313	0·051
295	0·056
280	0·111

Table 4 Effect of solvent on ϕ_s for polyisoprene at 30 °C with hydroperoxide content 0·06%

λ (nm)	1,2-Dichloroethane	Cyclohexane	Benzene
280	0·111	0·024	0·006
313	0·051	0·031	0·004

In the latter case, most of the light is being absorbed by the benzene, and little of this excitation is apparently transferred to the peroxy groups.

The relatively low values of the quantum yields and the wavelength effect observed suggest that the photolysis of polymeric hydroperoxides is not as efficient as was previously believed. It seems likely that it proceeds from an unbound dissociative state but that the presence of the polymeric environment may enhance the recombination of the primary radical pairs effectively to reduce the quantum yield. This would be consistent both with the low quantum efficiencies and the wavelength effect, since radical separation would be expected to be greater when higher energy photons are absorbed by the peroxy group. This would lead to higher efficiencies at shorter wavelengths.

The low quantum yields observed in this study for singlet oxygen-treated polyisoprene suggest that singlet oxygen probably does not play an important intermediary role in the photo-oxidation of saturated polymers where the double bond content is restricted to chain ends only.

ACKNOWLEDGEMENT

The authors acknowledge the generous financial support of the National Research Council of Canada.

REFERENCES

1. A. M. Trozzolo and F. H. Winslow, *Macromolecules*, **1**, 98 (1968).
2. B. Rånby and J. F. Rabek, *Photodegradation, Photo-oxidation and Photostabilization of Polymers*, Wiley, New York, 1975, Ch. 5.
3. K. Furukawa, E. W. Oray and E. A. Ogryzlo, *Ann. N.Y. Acad. Sci.*, **171**, 175 (1970).
4. M. L. Kaplan and P. G. Kelleher, *Rubb. Chem. Technol.*, **45**, 423 (1972).
5. R. D. Mair and A. J. Gaupner, *Anal. Chem.*, **36**, 194 (1964).
6. A. D. Awtrey and R. E. Connick, *J. Amer. Chem. Soc.*, **73**, 1842 (1951).
7. T. Kilp, W. Panning, B. Houvenaghel-Defoort, and J. E. Guillet, *Rev. Sci. Instr.*, **47**, 1496 (1976).
8. Y. Amerik and J. E. Guillet, *Macromolecules*, **4**, 375 (1971).
9. T. Kilp and J. E. Guillet, *Macromolecules*, **10**, 90 (1977).

28

Improving the Self-adhesion of EPDM by means of a Singlet Oxygen Reaction

E. TH. M. WOLTERS, C. A. VAN GUNST, and H. J. G. PAULEN

Central Laboratory, DSM, Geleen, The Netherlands

INTRODUCTION

In rubber technology, a compound is considered to possess tack, or self-adhesion, when two specimens of the same rubber adhere to each other when brought into contact.[1] In most manufacturing processes (for instance, tyre production) the compound is required to possess this property prior to vulcanization.

Natural rubber has a tack level which is satisfactory for most purposes; this is primarily due to the fact that rubber molecules can diffuse across the contact area.[2] Another requirement for satisfactory tack is the crystallization under tension.[3]

EPDM, a synthetic rubber composed of ethylene, propylene and a third monomer, for example ethylidenenorbornene, possesses a high UV and ozone resistance in comparison with natural rubber. It could therefore become an attractive substitute for natural rubber (especially as tyre rubber) if it had a sufficiently high level of tack.

Crystallization under tension presents no problem for this rubber,[4] but its diffusion is appreciably less than that for natural rubber.[5] One of the means pursued to compensate for this low diffusion ability of EPDM involves the use of a tackifier, such as a condensate of *p-tert*-octylphenol and formaldehyde, but the tack level achieved in this way was still inadequate.

We have now found that EPDM with suitable physical constants and containing such a tackifier appears to develop supertack when exposed to visible light, and further research has shown that a singlet oxygen reaction is involved. Such a reaction requires, in addition to light and oxygen, the presence of a sensitizer and a substrate.

SENSITIZER

EPDM compounds contain large amounts of carbon black and extender oil, as well as minor amounts of tackifier and other chemicals necessary for the

Table 1 Effect of oil on tack

Oil	Clay-gel analysis			DSM tack (g per 5 mm)	Relative O_2 uptake	Transmission at 400 nm (%)
	Saturated hydrocarbon	Aromatic	Polar groups			
Paraffin oil	—	—	—	320	1	100
Sunpar 150 (paraffinic)	86	14·0	0·3	340	28·5	100
BP oil 620 (paraffinic)	76	23·5	2·5	720	61·5	95
Shellflex 68 (451) (paraffinic)	73	25·5	7·5	340	22	100
Sunthene 4240 (naphthenic)	53·5	45·0	2·0	1 480	33	100
Flexon 391 (naphthenic)	—	—	—	3 600	56·5	94
Sundex 790 (aromatic)	22·5	68·5	9·0	>15 000	100	65
Dutrex 55 (aromatic)	—	78	—	>15 000	100	65
Sundex 8125 (aromatic)	18	68	14	>15 000	100	52

vulcanization process. The sensitizer is present in the extender oil. In Table 1 different types of oil are listed in the first column in order of increasing aromatic content and number of polar groups. To test whether this aromaticity, which is due to condensed aromatic systems, causes oil to act as a $^1O_2$ sensitizer, we carried out the following experiment.

A solution of tetramethylethylene was irradiated and the oxygen uptake measured for each of the oil types functioning as sensitizer. There is a good correlation between the aromatic content and the oxygen consumption. The same correlation is found between the sensitizing capacity of an oil type and the tack level of EPDM compounded with this type; this is also seen in Table 1.

A quick method to determine the $^1O_2$-sensitizing capacity of an oil is to measure the transmission at 400 nm. This is illustrated in the fourth column of Table 1, from which it can be seen that the sensitizing capacity increases as the transmission decreases.

When tetraphenylporphyrin (TPP), a well known $^1O_2$ sensitizer, is used instead of an aromatic oil, an increase in tack is also achieved (Figure 1).

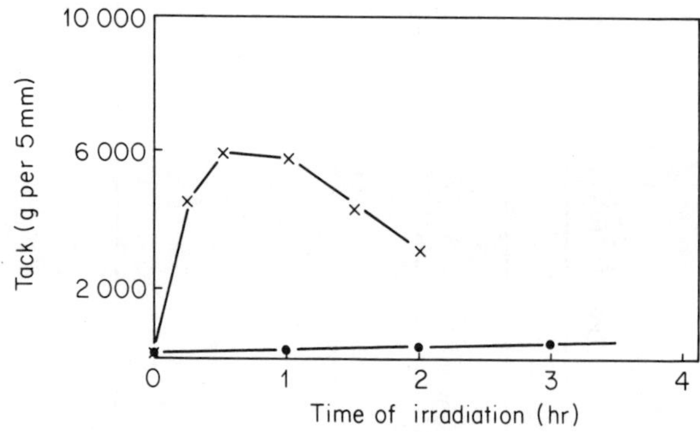

Figure 1 Effects of oils and sensitizer on tack after irradiation (medicinal paraffin instead of aromatic oil). ●, Without TPP; ×, with 4×10^{-4} mol of TPP

THIRD MONOMER (SUBSTRATE)

EPDM consists of ethylene, propylene and a third monomer, the last component being bound by one of its double bonds, while its second double bond is used in the vulcanization process. Not all of the third monomers used in EPDM are equally reactive towards $^1O_2$, and it can be deduced from their structures that ethylidenenorbornene (**1**), being the only one with its double bond alkyl-substituted three times, will be the most reactive.[6] This is confirmed by tack values obtained after irradiation of EPDM samples containing various third monomers. EPDM with dicyclopentadiene [DCPD, (**2**)], propenylnorbornene (**3**), or hexa-1,4-diene (**4**) was found to give much lower tack levels than the compound with ethylidenenorbornene [EN, (**1**)] (Table 2).

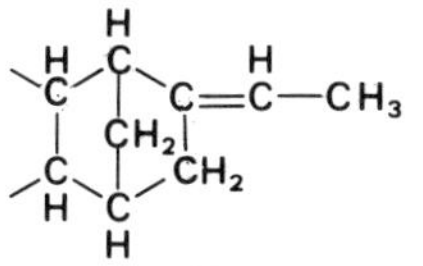

(1) 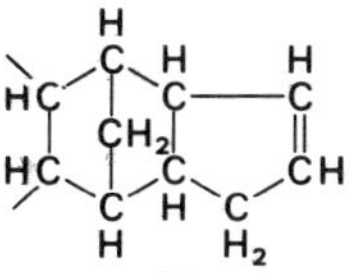 (2)

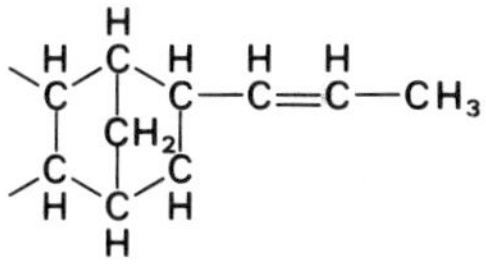

(3) 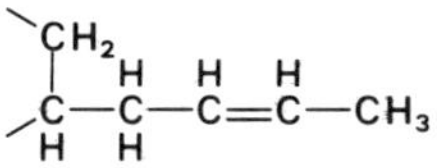 (4)

Table 2 Effect of third monomer on tack

Third monomer	DSM tack (g per 5 mm)	
	Dark	Irradiation for 4 hr
None	550	600
Hexa-1,4-diene	120	140
Dicyclopentadiene	40	80
Propenylnorbornene	900	1 400
Ethylenenorbornene	600	14 000

It should be noted that the E–P–DCPD terpolymer does not react with singlet oxygen, as is confirmed by the tack values. On subjecting a solution of this terpolymer to dye-sensitized photo-oxidation conditions, absolutely no oxygen consumption was observed, whereas an E–P–EN terpolymer rapidly consumed oxygen. This reaction can be written as

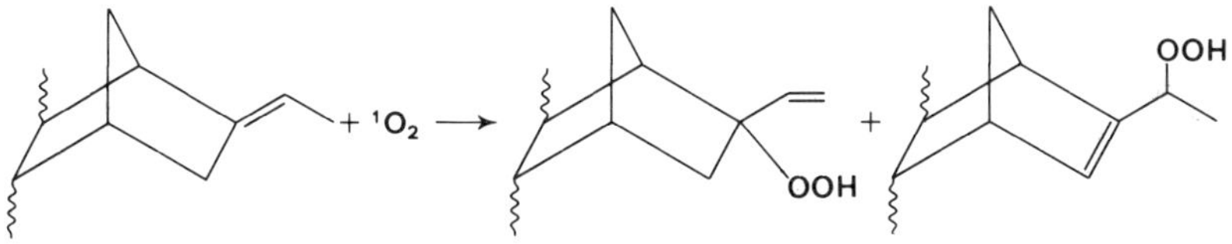

$$+ \, {}^1O_2 \longrightarrow \quad \text{OOH} \quad + \quad \text{OOH}$$

Two different alkylhydroperoxides are formed, the tertiary peroxide in amounts almost twice those of the secondary peroxide.[7]

It is interesting to compare the above results with those reported on the metal-catalysed autoxidations of various EPDM rubbers at 70 °C. Here it was found that an E–P–EN terpolymer was less reactive than an E–P–DCPD terpolymer.[8] This opposite order of reactivity illustrates the fundamental difference in mechanism of dye-sensitized photo-oxidation and metal-catalysed autoxidation.

QUENCHER

Addition of a $^1O_2$ quencher causes a reduction in reaction rate, and consequently the tack level is also lowered accordingly. By normal irradiation in air a tack level of 15 000 g per 5 mm is reached. Addition of 1% of A.O. 4010, the trade name of *N*-cyclohexyl-*N*′-phenyl-*p*-phenylenediamine, a known $^1O_2$ quencher,[9] reduces the tack to only 2500 g per 5 mm. Irradiation in nitrogen results, of course, in even lower values (Figure 2).

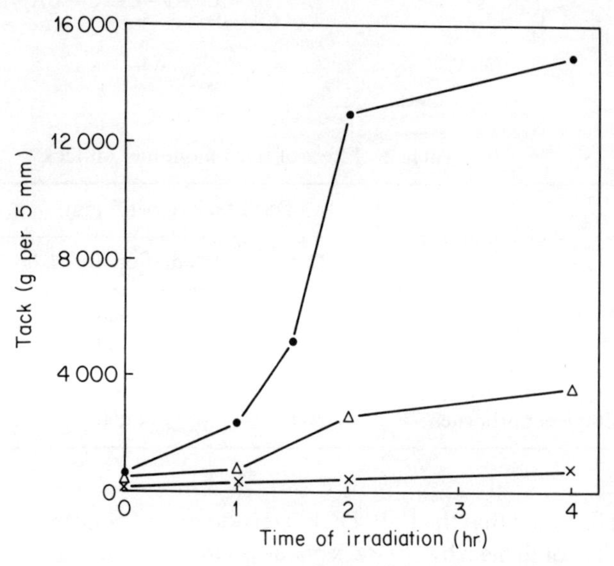

Figure 2 Interfering effects on the development of tack by irradiation (Keltan 712 TE). ●, In air; △, +1% of A.O. 4010; ×, in nitrogen

SELF-ADHESION

Having concluded that the high tack level obtained is caused by a $^1O_2$ reaction, there remains the question of why the products of such a reaction increase the self-adhesion so considerably. In our opinion, the formation of hydroperoxides leads to a crosslinked surface. This crosslinking, in combination with the increased polarity of the surface, may cause a second surface phase to be formed. This phase consists largely of tackifier, and is ultimately responsible for the high self-adhesion observed.

REFERENCES

1. W. C. Wake, in *Adhesion and Adhesives* (Ed. R. Houwink and G. Salomon), Elsevier, Amsterdam, 1965, p. 413.
2. S. S. Vojutskü, *J. Adhesion*, **3**, 69 (1971).

3. J. D. Skewis, *Rubb. Chem. Technol.*, **39**, 217 (1966).
4. U. Flisi, A. Valvassori, and G. Novajra, *Rubb. Chem. Technol.*, **44**, 1093 (1971).
5. R. P. Campion, *J. Adhesion*, **7**, 1 (1975).
6. C. S. Foote, *Accounts Chem. Res.*, **1**, 104 (1968).
7. E. F. J. Duynstee and M. E. A. H. Mevis, *Eur. Polym. J.*, **8**, 1375 (1972).
8. H. Schnecko and J. S. Walker, *Eur. Polym. J.*, **7**, 1047 (1971).
9. J. P. Dalle, *Photochem. Photobiol.*, **15**, 411 (1972).

29

Photosensitized Singlet Oxygen Oxidation of Polysulphides

H. S. LAVER and J. R. MACCALLUM

Department of Chemistry, University of St. Andrews, Fife, Scotland

INTRODUCTION

The reaction of singlet oxygen with dialkyl sulphides has been investigated in some detail.[1–3] The overall chemical change is as follows:

$$R_2S \xrightarrow{^1O_2} R_2SO \xrightarrow{^1O_2} R_2SO_2 \qquad (1)$$

The efficiency of conversion is solvent dependent. For example, the use of aprotic solvents and high temperatures results in very low yields of sulphoxide or sulphone in the photosensitized oxidation of diethyl sulphide. The mechanism of the reaction is thought to involve an unstable persulphoxide intermediate:

$$R_2S \xrightarrow{^1O_2} R_2\overset{+}{S}OO^- \xrightarrow{R_2S} 2R_2SO \qquad (2)$$

The occurrence in vulcanized rubber of sulphides and polysulphides, together with alkene functions, provides competitive routes for reactions involving singlet oxygen. This paper reports the results of a preliminary examination of the reaction of singlet oxygen with poly(propylene sulphide).

EXPERIMENTAL

Poly(propylene sulphide) was prepared by adding fresh sodium wire to re-distilled propylene sulphide and leaving the mixture for several days. The resulting polymer was dissolved in tetrahydrofuran and re-precipitated several times from methanol before drying under vacuum for 2 weeks.

Rhodamine B was used in the dye-sensitized photo-oxidation as it was found to be more compatible with the polysulphide than, for example, methylene blue. Re-distilled methylene dichloride (spectroscopic grade) was used as solvent. Irradiation and oxygen uptake experiments were carried out in the apparatus shown in Figure 1.

Singlet oxygen was also generated in a gas flow system using a microwave discharge. Silver and oxidized silver gauze were placed in the gas stream between

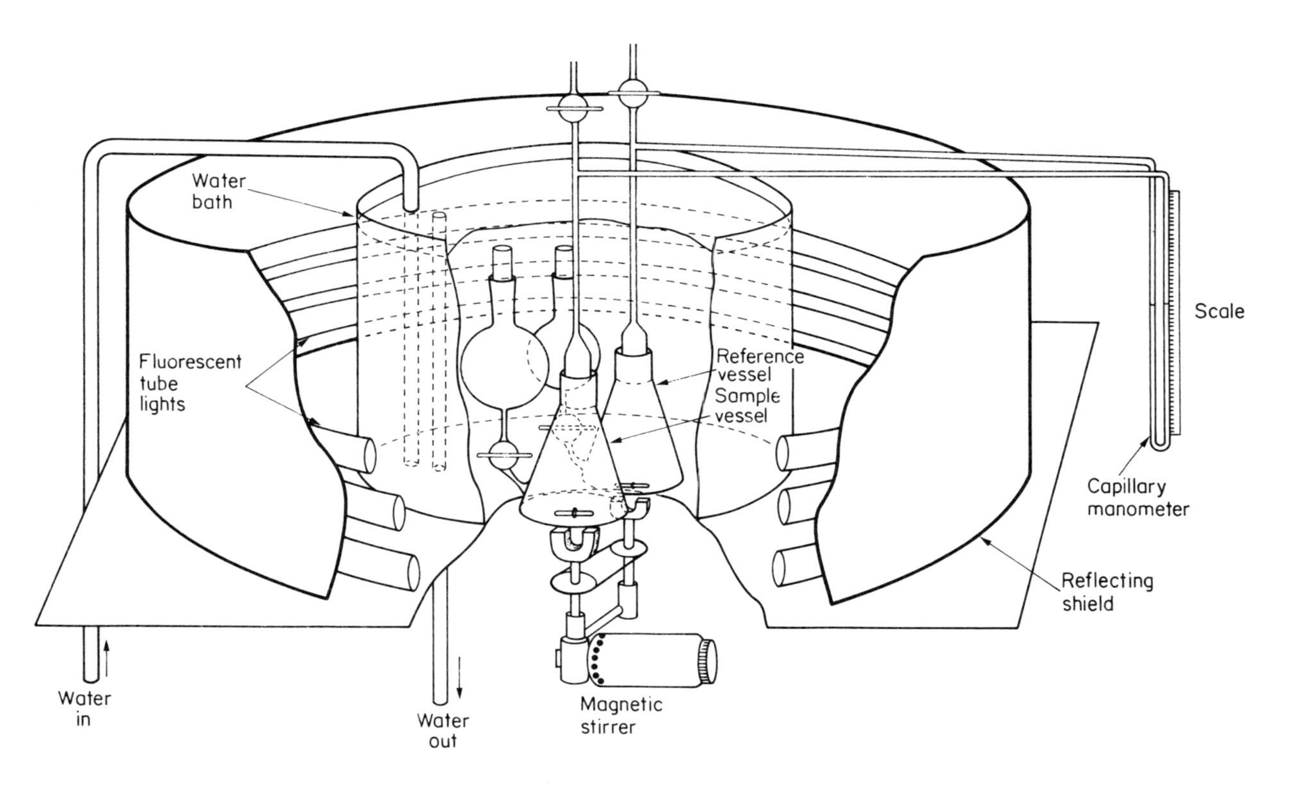

Figure 1 Apparatus used to study the photosensitized oxidation of poly(propylene sulphide).

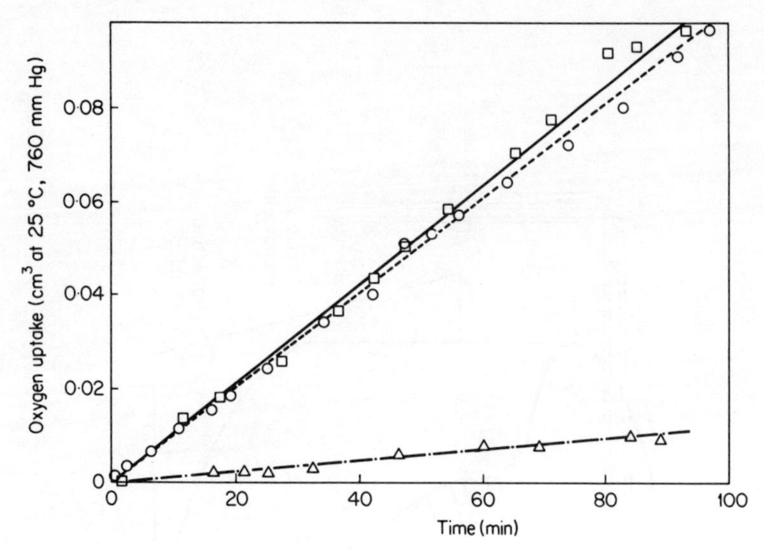

Figure 2 Oxygen uptake *versus* time of radiation. □, polymer, Rhodamine B, solvent, O_2 at 1 atm pressure; ○, polymer, Rhodamine B, solvent, O_2 at 1 atm pressure, *β*-carotene; △, polymer, solvent, O_2 at 1 atm pressure

the discharge and polymer sample. Neither oxygen atoms nor ozone could be detected in the oxygen arriving at the sample.[4]

Viscosity measurements were made at 30 °C using a viscometer which could be degassed and sealed.

RESULTS

Figure 2 shows the effects of various additives on the rate of oxygen consumption by a solution of polymer in CH_2Cl_2 at $25 \pm 0{\cdot}03$ °C, under conditions of constant intensity of irradiation. The total extent of oxidation was less than 0·5%. The same concentrations of dye ($4{\cdot}74 \times 10^{-5}$ M), polymer ($7{\cdot}71 \times 10^{-2}$ M), and *β*-carotene ($2{\cdot}98 \times 10^{-4}$ M) were used in all viscometry and oxygen uptake runs.

Figure 3 illustrates the change in viscosity during photolysis in an inert atmosphere. The infrared spectrum of the polymeric product of photo-oxidation was examined by casting films of the reaction mixture on potassium bromide discs.

Figure 4 represents the changes observed in the IR spectra. Films of poly(propylene sulphide) were cast on to KBr plates and subjected to the gas flow downstream from the microwave discharge in O_2. Singlet oxygen was shown to be present under these conditions.[4] After several hours' treatment no observable change was detected in the IR spectrum of the sample.

DISCUSSION

From the results obtained in the oxygen uptake experiments, we conclude that the polysulphide does not react readily with singlet O_2. The addition of an

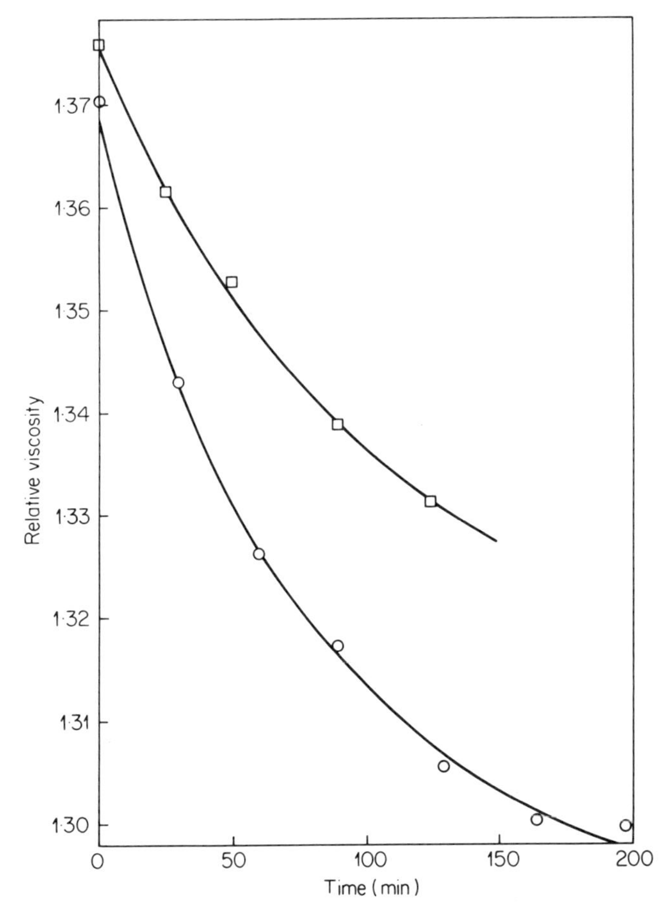

Figure 3 Change in relative viscosity of solutions of polymer with and without sensitizer. ○, Polymer, Rhodamine B, solvent, N_2 at 1 atm pressure; □, polymer, solvent, N_2 at 1 atm pressure

efficient $^1O_2$ quencher, β-carotene, has little effect on the rate of reaction.

It has been shown that Rhodamine B, compared with, for example, methylene blue, is a poor sensitizer for the production of $O_2\ ^1\Delta_g$.[5] However, the requirement of polymer and dyestuff to be co-soluble severely limited the choice of dyestuff. The following scheme can be considered as describing the course of a singlet oxygen reaction:

Rate constant

$$D \longrightarrow {}^1D^* \qquad k_a \tag{3}$$

$${}^1D^* \longrightarrow {}^3D^* \qquad k_i \tag{4}$$

$${}^3D^* + {}^3O_2 \longrightarrow D + {}^1O_2 \qquad k_{O_2} \tag{5}$$

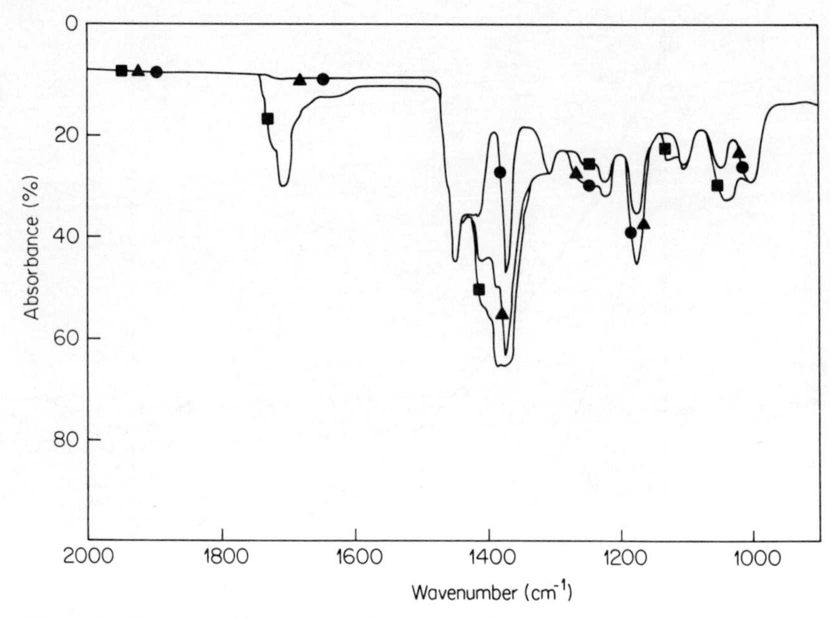

Figure 4 Changes taking place in IR spectra of poly(propylene sulphide) during radiation. ●, Polymer stored in dark; ▲, polymer after 40 h radiation; ■, polymer + Rhodamine B after 40 h radiation

$$^{1}O_2 + S \longrightarrow \text{Product} \qquad k_r \tag{6}$$

$$^{1}O_2 + Q \longrightarrow {}^{3}O_2 + Q \qquad k_Q \tag{7}$$

$$^{1}O_2 \longrightarrow {}^{3}O_2 \qquad k_d \tag{8}$$

where D = dyestuff, S = substrate, and Q = quencher.
Application of the steady state assumption to the concentration of $^{1}O_2$ allows equation (9) to be deduced for the rate of consumption of oxygen:

$$-\frac{d[^{3}O_2]}{dt} = \frac{\beta k_r[S][^{3}O_2]}{k_r[S] + k_Q[Q] + k_d} \tag{9}$$

where β incorporates the quantum yield for production of $^{3}O_2$.

Provided that $k_r[S] + k_Q[Q]$ is not very much greater than k_d then a low value for the yield of $^{1}O_2$ from the dyestuff can be compensated for by a low k_d. Thus for methanol, a good solvent for methylene blue, $k_d = 2 \times 10^5\ s^{-1}$,[6] whereas for methylene chloride $k_d = 1 \times 10^4\ s^{-1}$.[6] The most probable reaction is that the polymer in the presence of dyestuff undergoes an abstraction reaction with the ultimate production of carbonyl functions, as shown by the band appearing at about 1705 cm^{-1} in Figure 4. It can be seen that the polymer is photolysed in the absence of dye molecules, although the reaction is much slower than when a photosensitizer is present. We propose the following mechanism:

$$RB + h\nu \longrightarrow RB^1 \tag{10}$$

$$RB^1 \longrightarrow RB^3 \tag{11}$$

$$RB^3 + O_2 \longrightarrow RB + {}^1O_2 \tag{12}$$

$$RB^3 + PH \longrightarrow (RBPH) \tag{13}$$

$$(RBPH) \longrightarrow R\dot{B}H + P\cdot \tag{14}$$

$$P\cdot + O_2 \longrightarrow PO_2^{\cdot} \tag{15}$$

$$PO_2^{\cdot} + PH \longrightarrow PO_2H + P\cdot \tag{16}$$

$$PO_2H + h\nu \longrightarrow \text{carbonyl functions and chain scission} \tag{17}$$

where RB represents Rhodamine B, PH represents

$$\left(\begin{array}{c} CH_3 \\ | \\ -CH-S-CH_2- \end{array} \right)_n,$$

and P· represents either

$$\left(\begin{array}{c} CH_3 \\ | \\ -\dot{C}-S-CH_2- \end{array} \right)_n \quad \text{or} \quad \left(\begin{array}{c} CH_3 \\ | \\ -CH-S-\dot{C}H- \end{array} \right)_n.$$

The conditions we chose for this initial study can be classified as 'aprotic'—methylene chloride solution and bulk polymer. It is possible that the use of a protic solvent may enhance the reactivity of singlet oxygen towards the polymer, but attempts to do this have met with problems of solubility of the polymer. However, the polymer also differs from model dialkyl sulphides in that steric conditions will be somewhat more stringent in the macromolecule. Furthermore, the significance of the effect of the proximity of the sulphur atoms intramolecularly is unknown.

REFERENCES

1. G. O. Schenck and C. H. Krauch, *Chem. Ber.*, **96**, 517 (1963).
2. C. S. Foote, R. W. Denny, L. Weaver, Y. Chang, and J. W. Peters, *Ann. N.Y. Acad. Sci.*, **171**, 139 (1970).
3. C. S. Foote and J. W. Peters, *J. Amer. Chem. Soc.*, **93**, 3795 (1971).
4. C. T. Rankin, *PhD Thesis*, University of St. Andrews, 1972.
5. C. Balny, J. Canva, P. Douzou, and J. Bourdon, *Photochem. Photobiol.*, **10**, 375 (1969).
6. D. R. Adams and F. Wilkinson, *J. Chem. Soc., Faraday Trans. II*, **68**, 586 (1972).

30

Singlet Oxygen and Chemiluminescence in the Oxidation of Polymers Resembling Humic Acids

S. SLAWIŃSKA and T. MICHALSKA

Institute of Physics, Technical University, Szczecin, Poland

In the biosphere, certain dark-coloured macromolecular substances of biogenic and pyrogenic origin are found to occur in melanins, humus acids, coal, soot and smoke. These carbon-containing polymers with a high degree of polydispersity are present in human skin, hair and brain (substantia nigra), in microorganism and insect dyes such as melanins, in the soil and in the aquatic environment in the form of fulvic and humic acids (HA), as well as in the atmosphere in soot and smoke. They play both beneficial and detrimental roles in the environment and organisms. Some common properties are found in these polymers and it is possible to classify them as one group of carbonaceous polymers (CP):

(1) CP are built of aromatic sub-units surrounded by various side-chains, e.g. carbohydrates, phenolic acids, polypetides, and metals in the case of soil HA.[1] Interactions such as hydrogen bonding, Van der Waals forces, and ionic and covalent bonds present among the constituents of HA result in the formation of extremely complex supramolecular structures[2] (Figure 1).

(2) The structure of a molecular sieve with abundant spongy microspaces and ionic groups such as COOH, OH and NH_2 ensures the efficient sorption of low molecular weight compounds. Sorption of O_2, which can easily penetrate into the CP matrix and xenabiotics such as pesticides and carcinogens is of particular interest.

(3) CP exhibit a uniquely efficient absorption over the UV, visible and IR ranges of the spectrum (dark colour) resulting from the multiple or continuous system of energy levels. They also reveal a very stable singlet EPR signal, reflecting the existence of unpaired electrons, most probably of semiquinone nature. Conditions for the formation of electronic excited states involving the ($n \leftarrow \pi^*$) transitions of >C=O and >NH groups, the ($\pi \leftarrow \pi^*$) transitions of the aromatic network, and for the formation of excited singlet oxygen are particularly favourable. CP are distributed mainly in the surface layers of

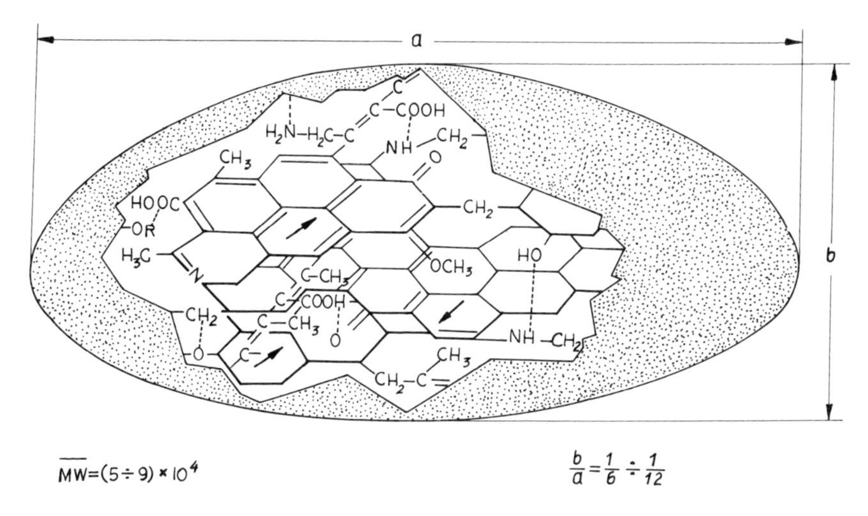

Figure 1 Polycyclic aromatic core of soil humic acids. The arrows symbolize unpaired electrons responsible for the EPR signal

organisms and soil, and therefore provide efficient absorption of solar radiation, which penetrates the biosphere, and conditions for possible reactions with O_2. However, the influence of light on the system CP–O_2 is still unknown. Our interest in this field was additionally stimulated by the possibility of an increase in UV radiation following the depletion of the ozone layer.

CP oxidized with O_2 or H_2O_2, or photo-oxidized by visible or UV light, exhibit a weak chemiluminescence (CL) in the visible region of the spectrum.[3–9] The kinetics and spectral distribution of CL accompanying the oxidative degradation of phenolic and quinoid groups in HA with alkaline H_2O_2 have already been investigated.[10]

This work deals with the kinetics and spectra of ultra-weak CL that accompanies the autoxidation and photo-oxidation of HA. These processes may be considered as a model for a degradative oxidation of HA occuring in the natural environment.[4] The chemiluminescent method used in this work may answer the question of whether or not singlet oxygen is involved in the degradative oxidation of HA.

CL spectra, presented for several HA in Figure 2, were measured with a set of calibrated cut-off filters (GOST 9411–66) and an EMI 9558BQ photomultiplier, cooled to −70 °C. The single photoelectron counting method and a flow system of solvents were used. Spectra were corrected for absorption. All CL spectra contain three emission bands with maxima at 490–500, 570, and 635 nm. The red emission CL band with maximum at 635 nm corresponds to the transition in the $^1O_2$ dimole:

$$2(^1\Delta_g) \longrightarrow 2(^3\Sigma_g^-) + h\nu(634\ \text{nm}) \qquad (1)$$

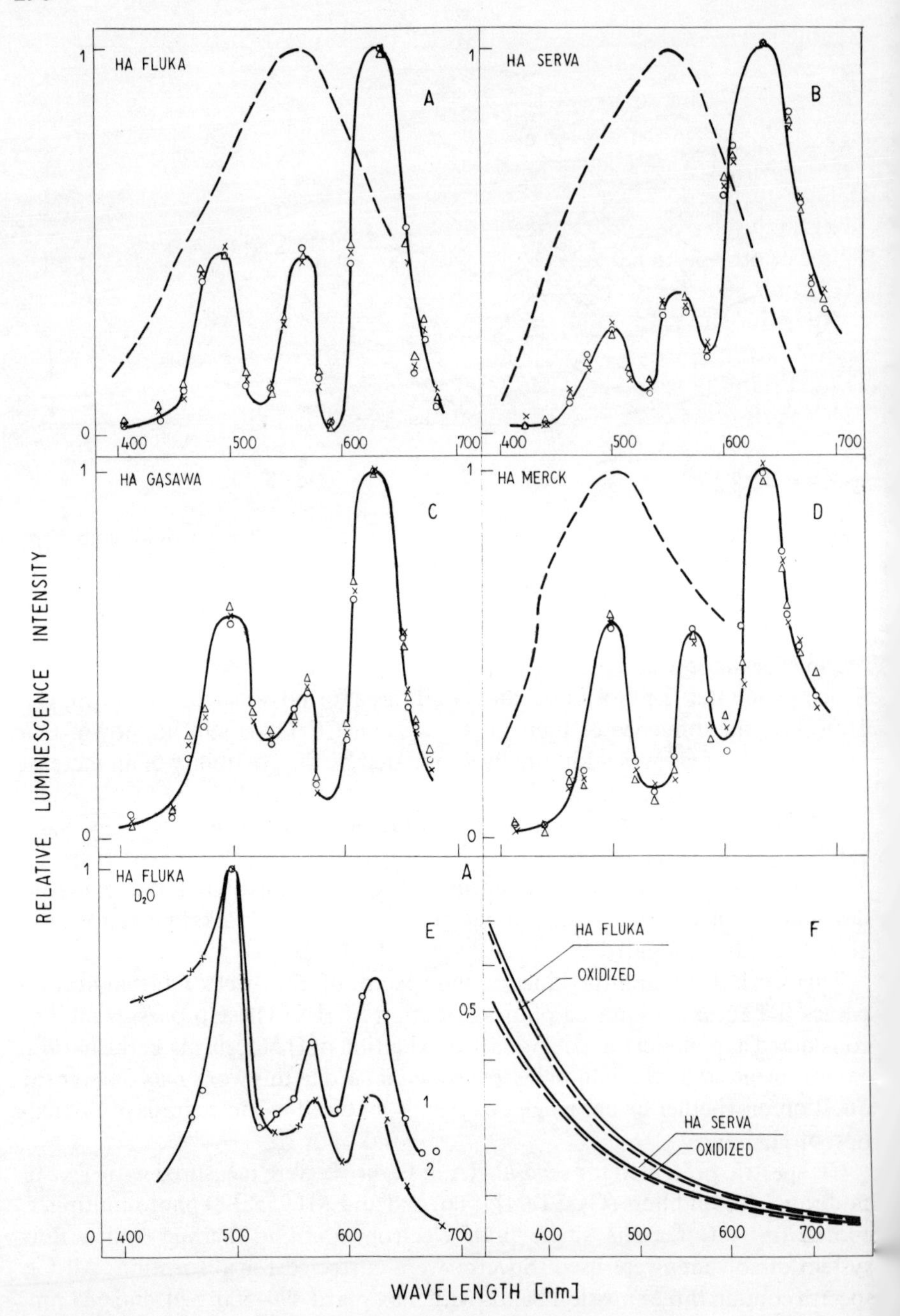

Figure 2 Chemiluminescence spectra of humic acids oxidized with O_2 (A, B, D) and H_2O_2 (C) and photo-oxidized (E) in 0·1 N NaOH. [HA] = 0·03% for A, B, C, and D and 0·01% for E in D_2O: 1, after 30 min; 2, after 4·5 h of irradiation. F, absorption spectra. Dashed lines in A, B, and D are fluorescence spectra. Temperature, 25 °C

Synthetic HA obtained by oxidative polymerization of hydroquinone, pyrocatechol and adrenochrome, as well as the simple model compound tetrahydroxy-*p*-benzoquinone, give CL maxima at the same positions.

The fluorescence spectra presented in Figure 2 are very broad and have the main maxima at 540–580 nm. The excitation beam ($\lambda = 366$ nm) was perpendicular to the direction of observation. Several emitters or isolated energy levels in the HA macromolecule contribute to the observed fluorescence spectra. Three of them, with maxima at 420, 500, and 580 nm, may be revealed by using excitation wavelengths of 260 and 480–500 nm.

Emission CL bands at 500 and 570 nm may arise from the superposition of transitions in singlet oxygen dimoles and radiative deactivation of electronic excited states in HA. The latter are generated either directly in an exothermic reaction of HA oxidative degradation, e.g. (a) via dioxetane rings:

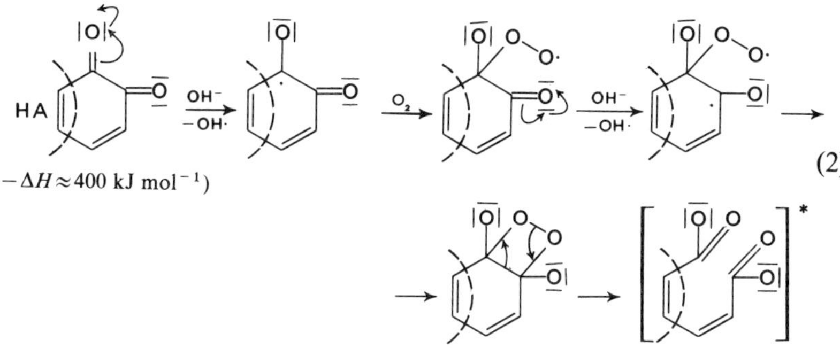

($-\Delta H \approx 400$ kJ mol^{-1}) (2)

(b) oxidative decarboxylation resulting from the nucleophilic attack by $HO_2^{\overline{\cdot}}$ radical ions:

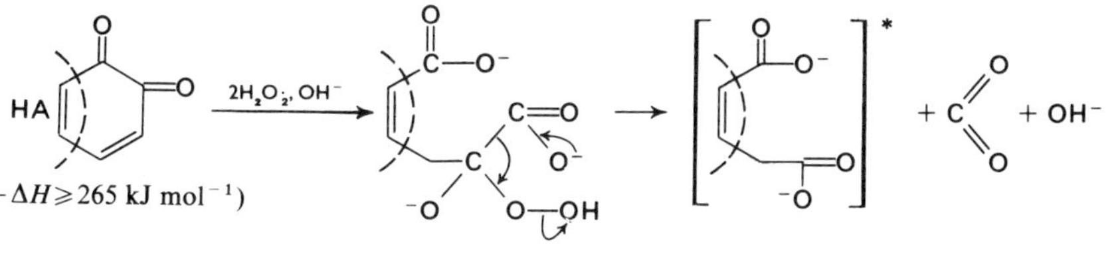

($-\Delta H \geqslant 265$ kJ mol^{-1}) (3)

or (c) indirectly as a result of energy transfer:

$$2O_2(^1\Delta_g\ ^1\Sigma_g^+) \text{ or } 2O_2(^1\Delta_g)_{v=1} + \mathrm{HA} \longrightarrow 2O_2(^3\Sigma_g^-) + \mathrm{HA}^* \quad (4)$$

$$\mathrm{HA}^* \longrightarrow \mathrm{HA} + h\nu_{F1}$$

$$\lambda_{F1} = 500 \text{ or } 570 \text{ nm.}$$

The last process is highly probable because of the broad fluorescence overlapping the emission spectrum of the $^1O_2$ dimole.

The absorption spectra of HA indicate a typical monotonic increase in optical density (O.D.) with decreasing wavelength (Figure 2F). The O.D. for all measured wavelengths decreases during the self-oxidation and photo-oxidation. Relative changes (ΔO.D./O.D.), already measurable after 15 min of self-oxidation, are greater for the long- than for the short-wavelength region one. The ratio $O.D._{400}/O.D._{600}$, which is a conventional measure of the aromaticity or degree of polymerization (DP) of HA, increases in almost all cases. This result indicates an oxidative degradation and the formation of products with lower molecular weights than those of the original HA.

In order to confirm the contribution of $^1O_2$ to the observed CL of HA, quenchers of $^1\Delta_g(O_2)$ and solvents in which the lifetime of $^1O_2(^1\Delta_g)$ is longer than that in water were used. Additionally, fluorescein, which is a strong fluorescer and may accept the excitation energy from $^1O_2$, was used. Kinetic curves of CL for HA oxidized in D_2O exhibit a higher CL intensity than the curves for HA oxidation in H_2O by a factor of about 2 (Figure 3B). Similar results were obtained for Serva and Fluka HA oxidized with H_2O_2 (Figure 3A). Addition of fluorescein to the reaction mixture results in increased CL (sensitized chemiluminescence) which is markedly higher in D_2O than in H_2O solution. This effect is observed for HA oxidized with both O_2 and H_2O_2 (Figures 3A and 3B).

These results indicate a strong isotope effect of D_2O. Since D_2O solutions contained about 0·5% of water, which strongly quenches the singlet oxygen, the observed increase in CL intensity was lower than that calculated from the ratio of the lifetimes of $^1O_2(^1\Delta_g)$ in D_2O (20 μs) and in H_2O (2 μs).[11]

Separate experiments with organic solvents showed that the CL intensity was the highest in 95% *n*-butanol, followed by 92% ethanol and 95% methanol, and was the lowest in H_2O. The results obtained correspond to the ($^1\Delta_g$) lifetime in these solvents. 5,5-Dimethylcyclohexa-1,3-diene, a quencher of $^1O_2(^1\Delta_g)$, markedly suppressed the CL intensity and its effect was more pronounced in ethanol than in H_2O; this result was expected on account of the absence of competitive quenching by H_2O.

Similar results were obtained when alkaline solutions of different HA saturated with O_2 or air were irradiated with a high-pressure mercury lamp ($\lambda > 300$ nm) for several hours. The relationship between the steady-state CL intensity (I_{max}) and irradiation time (τ_{ir}) shown in Figure 3C exhibits a typical shape of photo-oxidation rate without an induction period.[12] The CL which accompanies the photodegradation has similar emission spectra (Figure 2E), changes of $O.D._{400}/O.D._{600}$ values and the isotope effect of D_2O upon the rate of oxidative degradation. The decay of photoinduced CL, $I = f(t)$ (Figure 3C), follows complex kinetics with several steps: the first two steps conform to a first-order reaction with rate constants $k_1 = 0{\cdot}3\ s^{-1}$ and $k_2 = 0{\cdot}002\ s^{-1}$. Free-radical inhibitors such as di-*tert*-butyl-*p*-cresol, α-naphthol, and hydroquinone and quenchers of singlet oxygen such as tetramethylethylene, β-carotene (in the methanol–benzene soluble sub-fraction of HA), and cysteine decrease the CL intensity by 30–70%.

The results obtained strongly suggest that $^1O_2$ generation occurs in the

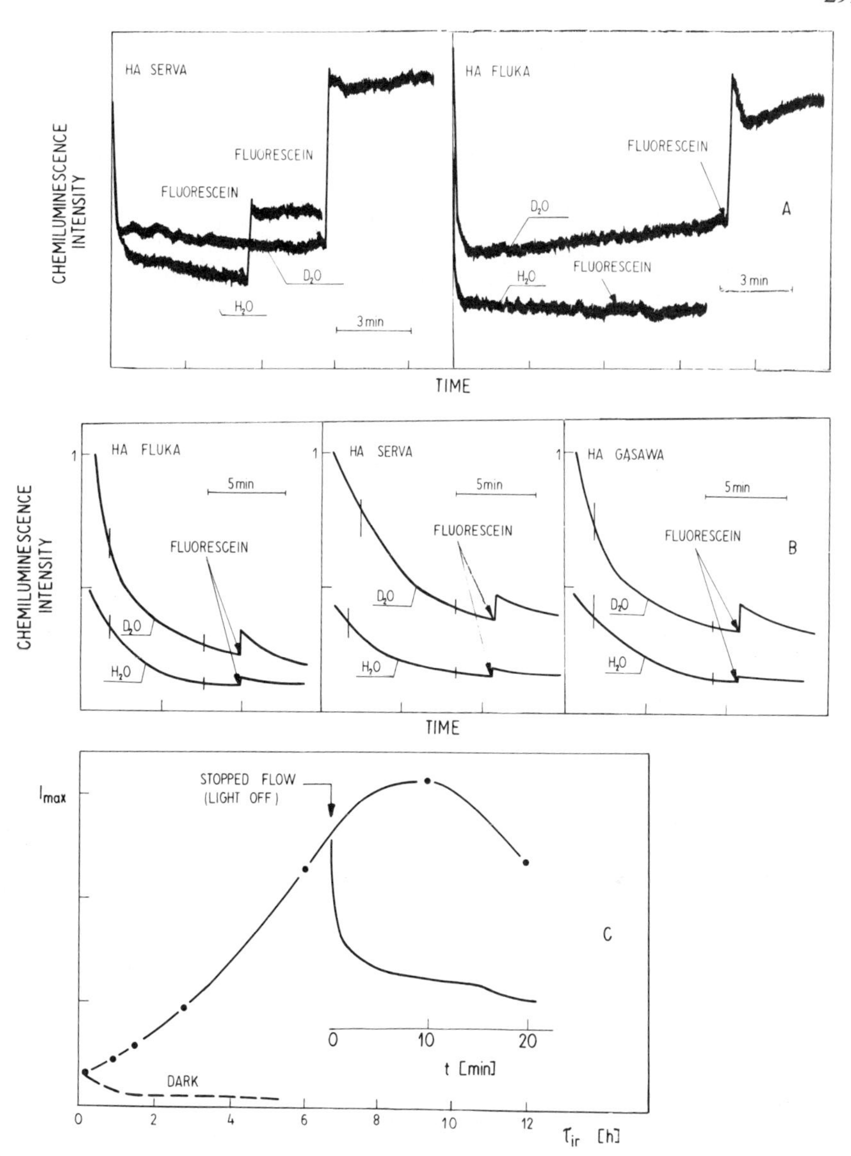

Figure 3 Kinetic curves of chemiluminescence of humic acids in H_2O and D_2O oxidized with H_2O_2 (A) and O_2 (B). C, maximum light intensity of photoinduced chemiluminescence (I_{max}) *versus* irradiation time (τ_{ir}) for the photo-oxidation of a Merck humic acid, and photochemiluminescence decay, $I = f(t)$

processes of HA oxidation with O_2 and/or H_2O_2. However, the precise mechanisms of $^1O_2$ formation are still obscure. Three reactions seem to be the most probable:

(1) The disproportionation of the $O_2^{\overline{\cdot}}$ radical ion, formed from semiquinone and molecular oxygen:[13]

$$SQ\cdot + O_2 \longrightarrow Q + O_2^{\overline{\cdot}}$$
$$2O_2^{\overline{\cdot}} \longrightarrow {}^1O_2^* + O_2^{2-} \quad (5)$$

(2) In the case of HA photo-oxidation, the mechanism of intramolecular energy transfer or formation of a charge-transfer complex (CTC), proposed by Rabek and Rånby[12,14] for photo-oxidative degradation of polystyrene might apply:

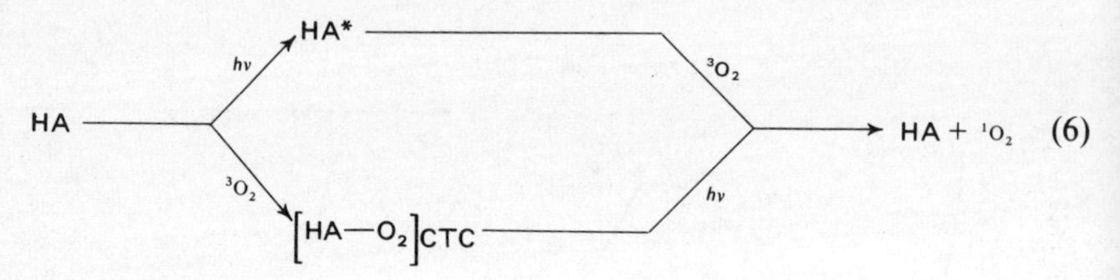

(3) A spin-allowed energy transfer from a chemically or photochemically excited $\rangle C{=}O$ group of HA oxidation products to the triplet state of oxygen:

$$^3(n,\pi^*) + {}^3O_2 \longrightarrow {}^1(n,\pi) + {}^1O_2^* \quad (7)$$

In our opinion, the generation of electronic excited states during HA oxidations, particularly of $^1O_2$, may have significant implications for certain natural systems that contain HA. Biologically important processes would be expected, such as a 'dark' photodynamic effect, hydroxylation, and degradation. Further studies in this unexplored field are desirable.

ACKNOWLEDGEMENTS

The authors express their gratitude to Dr. J. Sławiński for stimulating discussions and to Miss B. Bakiewicz for her technical assistance.

REFERENCES

1. N. M. Atherton, P. A. Cranwell, A. J. Floyd, and R. D. Haworth, *Tetrahedron*, **23**, 1653 (1967).
2. D. S. Orlov, Ya. M. Ammosova, and G. I. Glebova, *Geoderma*, **13**, 211 (1975).
3. D. Sławińska and J. Sławiński, *Nature, Lond.*, **213**, 902 (1967).
4. D. Sławińska, *Chemiluminescence in the Processes of Formation and Degradation of Humus Substances and its Analytical Application*, Prace Naukowe Politechniki Szczecińskiej, Instytut Fizyki, Szczecin, 1973, pp. 33, 37, 50.
5. H. H. Seliger, W. H. Bigley, and J. P. Hamman, *Science. N.Y.*, **185**, 253 (1974).
6. J. Stauff and H. Fuhr, *Angew. Chem.*, **87**, 132 (1975).
7. D. Sławińska, J. Sławiński, and T. Sarna, *Photochem. Photobiol.*, **21**, 393 (1975).
8. D. Sławińska and J. Sławiński, *Polish J. Soil Sci.*, **8**, 37, 49 (1975).

9. D. Sławińska and T. Michalska, in *Biochemistry in Environmental Protection*, (ed. Szczecińskie Towarzystwo Naukowe, Szczecin), in press.
10. D. Sławińska and J. Sławiński, *Rocz. Chem.*, **44**, 1955, 2415 (1970).
11. R. Nilsson, P. B. Merkel, and D. R. Kearns, *Photochem. Photobiol.*, **16**, 117 (1972).
12. B. Rånby and J. F. Rabek, *Photodegradation Photo-oxidation and Photostabilization of Polymers*, Wiley, London, New York, 1975, pp. 103–105, 273.
13. J. Stauff, *Photochem. Photobiol.*, **4**, 1199 (1965).
14. J. F. Rabek and B. Rånby, *Polym. Eng. Sci.*, **15**, 40 (1975).

31

Singlet Oxygen Oxidation of Lignin Structures

G. GELLERSTEDT, K. KRINGSTAD, and E. L. LINDFORS

Chemistry Department, Swedish Forest Products Research Laboratory, Box 5604, S-114 86 Stockholm, Sweden

INTRODUCTION

During the last decade there has been increasing interest in the properties and reactions of singlet oxygen and both its fundamental and the applied chemistry have been extensively reviewed. The formation and participation of singlet oxygen in reactions that occur in nature have also been considered.[1–5] Singlet oxygen can be formed in different ways, *viz*. by light-induced energy transfer to oxygen, by decomposition of peroxides or ozonides or by a microwave discharge.[2] In particular, the facile formation of singlet oxygen by light-induced energy transfer has led to the conclusion, supported by several experiments, that singlet oxygen participates in reactions leading to the degradation of various synthetic polymers.[6]

Native lignin and lignins modified by various pulping processes contain functional groups which are known to be reactive towards singlet oxygen. Thus, native lignin contains phenylpropane units in phenolic and non-phenolic units. Cinnamyl alcohol and cinnamaldehyde structures are present as end groups.[7] Under various pulping conditions, stilbene and 1,4-diphenylbutadiene structures may be formed as part of the lignin polymer.[8] These structures together with the corresponding singlet oxygen reactions are outlined below.

In recent years, great progress has been made in the field of lignin-retaining bleaching. Thus, mechanical pulp is today bleached on an industrial scale to a brightness level of about 80% SCAN units and applied as a substitute for bleached chemical pulp in certain paper grades. This development is highly desirable from at least two points of view: better utilization of the wood raw material and a considerable reduction of water and air pollution. The improved bleaching technique has not, however, eliminated the tendency to yellowing of such pulps (Figure 1) and this disadvantage considerably limits their field of application.[9] Therefore, a better understanding of the chemical reactions involved in the yellowing process is desirable in order to devise possible methods for the stabilization of such pulps against the influence of exposure to daylight

Singlet oxygen reaction

$^1O_2 + C{=}C{-}C(H) \rightarrow C{-}C{=}C(O{-}OH)$; $C{-}C{-}C$ (dioxetane, O–O) (1)

$^1O_2$ + diene $\rightarrow$ cyclic endoperoxide (2)

$^1O_2$ + $-CH$ (aryl) $\rightarrow$ $-C{-}O{-}OH$ (aryl) (3)

$^1O_2$ + R-phenol (OH) $\rightarrow$ R-phenoxy (O•) $+ HO_2\bullet$ (4)

Lignin structure

CH_2OH (CHO), CH=HC, OCH_3, O (native) **(1)** (formed)

CH=CH–CH=CH, OCH_3, CH_3O (formed) **(2)**

C–C–HC–OR_1, OCH_3, OH (OR_2) (native) **(3)**

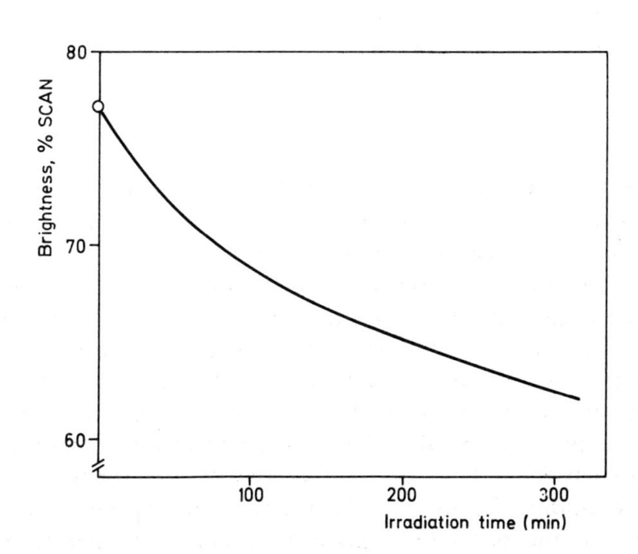

Figure 1 Light-induced yellowing of bleached mechanical pulp

RESULTS AND DISCUSSION

The light-induced yellowing of lignin-rich pulps can be divided into three separate steps, *viz.* the absorption of light by lignin, the formation of radicals,

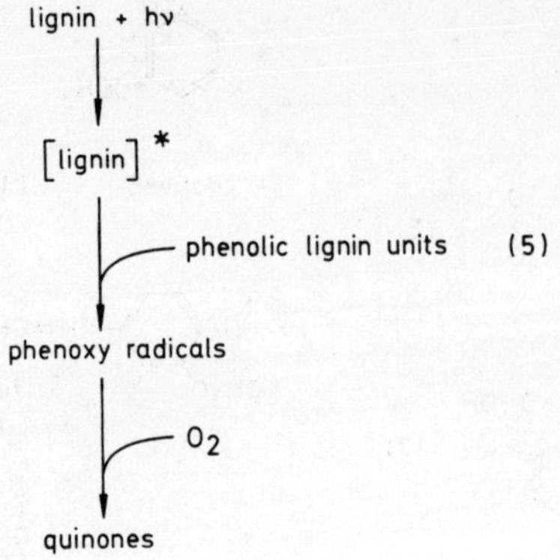

lignin + hv
↓
[lignin]*
↓ ← phenolic lignin units (5)
phenoxy radicals
↓ ← O_2
quinones

in particular from phenolic units in lignin, and reactions between the phenoxy radicals thus created and oxygen with the formation of quinoid structures. The first step, the absorption of light, is known to be restricted to a small number of structural groups in lignin.[10] The most important include end groups of the

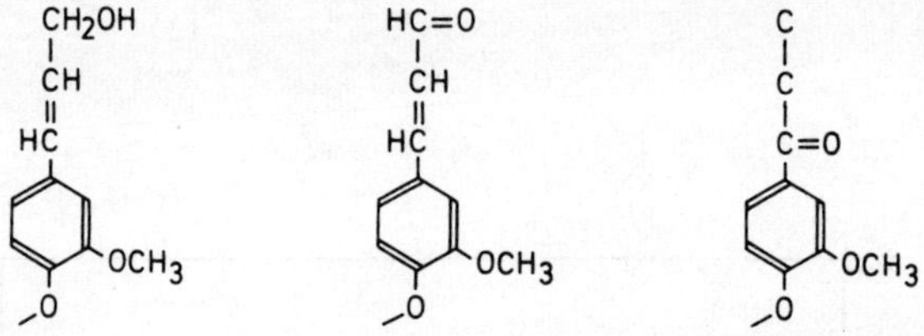

(4) (5) (6)

cinnamyl alcohol (**4**) and cinnamaldehyde (**5**) types and carbonyl groups conjugated with aromatic rings (**6**). These structures absorb light above 300 nm, which is the lower wavelength limit of daylight radiation.

It has been demonstrated using lignin model compounds that singlet oxygen is formed in these reactions. Thus, irradiation of coniferyl alcohol methyl ether with near-ultraviolet light in the presence of oxygen produces products which are consistent with a mechanism involving energy transfer from the excited model compound to oxygen, formation of singlet oxygen, addition of singlet oxygen across the double bond of the model compound and finally cleavage of the intermediately formed dioxetane structure with the formation of two aldehydes. Concomitantly the nucleophilic attack by the solvent on the dioxetane structure leads to the formation of glycerol structures and hydrogen peroxide. The presence of a singlet oxygen quencher such as β-carotene inhibits the reaction to a large extent.[11]

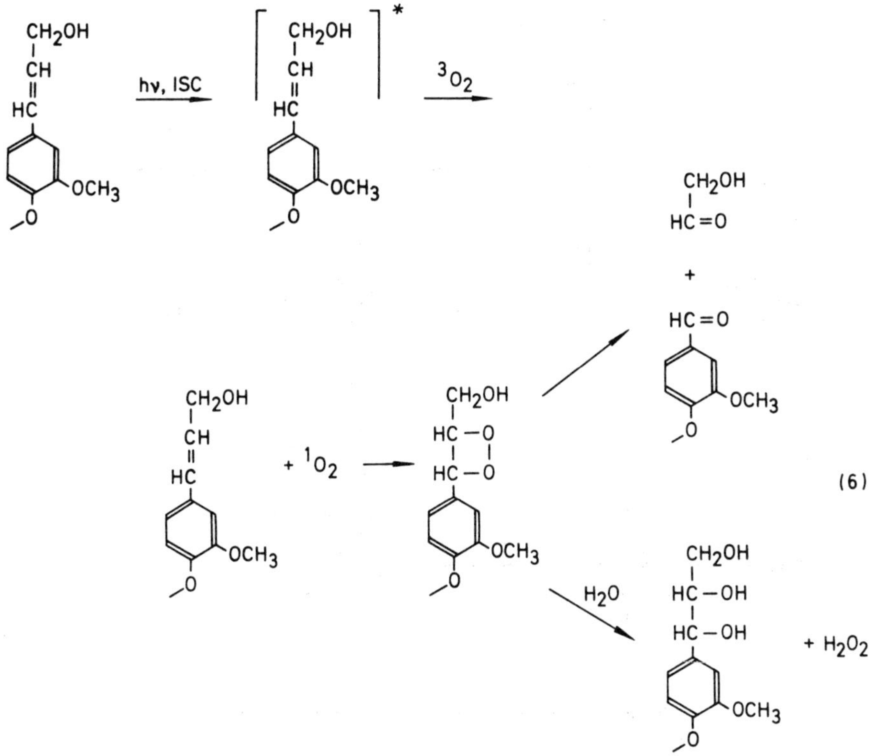

$$\text{(6)}$$

The second and third steps in the light-induced oxidation of lignin, *viz.* the formation of phenoxy radicals and quinones, respectively, have also been examined with regard to singlet oxygen participation. Thus, irradiation of a simple lignin model compound such as apocynol in the presence of oxygen and benzophenone acting as a sensitizer yields methoxybenzoquinone and acetaldehyde as a dominating colour-forming reaction.[12] The formation of phenoxy radicals in this reaction can, in principle, take place in two different ways. Excited benzophenone, in a direct reaction, may abstract hydrogen atoms from phenolic hydroxyl groups, thus creating phenoxy and benzhydrol radicals. In the presence of oxygen the latter are re-oxidized to benzophenone. Alternatively, excited benzophenone may transfer its energy to oxygen, leading to the formation of singlet oxygen. Singlet oxygen may then react with phenolic hydroxyl groups, forming phenoxy and hydroperoxy radicals.[13,14]

Results indicating that both alternatives apply to the light-induced oxidation of apocynol in solution were obtained by monitoring the amount of acetaldehyde formed when apocynol was irradiated in the presence or absence of singlet oxygen quenchers. Thus, the presence of β-carotene during the irradiation of a mixture of apocynol and benzophenone caused a significant decrease in the rate of formation of acetaldehyde compared with the non-inhibited reaction

$$>C{=}O \xrightarrow{h\nu,\ ISC} {}^{3}[>C{=}O]^{*}$$

$${}^{3}[>C{=}O]^{*} + \text{apocynol} \longrightarrow \text{phenoxyl radical} + \cdot\overset{|}{\underset{|}{C}}{-}OH \xrightarrow{+O_2;\ -HO_2\cdot} >C{=}O$$

$${}^{3}[>C{=}O]^{*} \xrightarrow{O_2} {}^{1}O_2$$

$$\text{apocynol} + {}^{1}O_2 \longrightarrow \text{phenoxyl radical} + HO_2\cdot \qquad (7)$$

$$\text{phenoxyl radical} \longleftrightarrow \text{quinonoid radical} \xrightarrow{+O_2;\ -HO\cdot} CH_3CHO + \text{methoxy-}p\text{-benzoquinone}$$

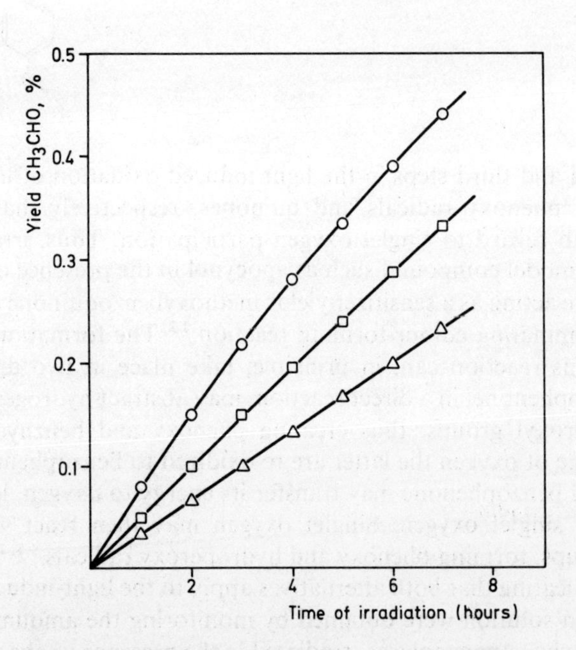

Figure 2 Yield of acetaldehyde after irradiation of apocynol in the presence of inhibitors at pH 6.[12] ○, Apocynol (0·2 mol l^{-1}); □, apocynol + β-carotene (0·005 mol l^{-1}); △, apocynol + BHT (0·005 mol l^{-1})

(Figure 2). A still more pronounced retarding effect was obtained, however, when apocynol was irradiated in the presence of BHT (2,6-di-*tert*-butyl-4-methylphenol). BHT can act as a singlet oxygen quencher although it is not as efficient as β-carotene. However, being a reactive phenol, BHT can also compete successfully with the apocynol present as a hydrogen donor to excited benzophenone. The latter reaction mechanism obviously is the most important in solution. It is, however, reasonable to assume that its importance must be less compared with the mechanism involving singlet oxygen in a solid matrix such as pulp fibres where various functional groups have a restricted mobility.

The light-induced yellowing of lignin can thus be summarized as follows. Light is absorbed by structures containing conjugated double bonds or carbonyl groups. In the excited state these groups may transfer energy to oxygen, giving rise to singlet oxygen. Subsequently, by the participation of singlet oxygen, conjugated double bonds are partly broken with the formation of new carbonyl structures. Excited carbonyl structures as well as singlet oxygen are able to create phenoxy radicals by hydrogen abstraction.

The decrease in brightness which takes place when lignin-rich pulps are exposed to daylight radiation is associated with the formation of quinoid

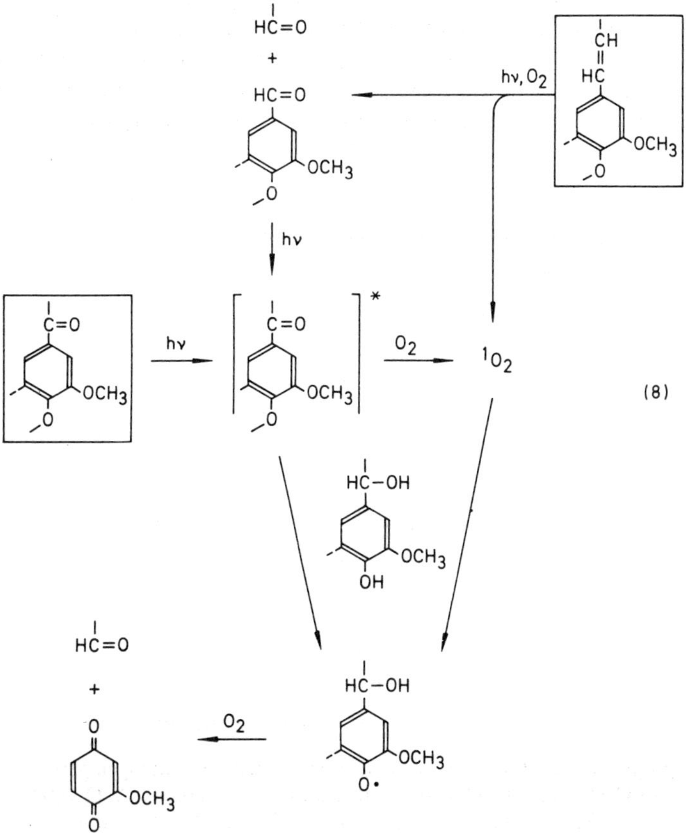

(8)

structures formed in subsequent reactions between phenoxy radicals and oxygen.[15] The primary oxidation product from softwood lignin structures is methoxybenzoquinone. This quinone is, however, extremely unstable at pH values of about 6 and above, and hydroxymethoxybenzoquinone as well as

$$\text{methoxybenzoquinone } (OCH_3) \xrightarrow[\text{pH} \geq 6]{1)\ +OH^{\ominus}\quad 2)\ \text{ox.}} \text{hydroxymethoxybenzoquinone anion } (O^{\ominus}, OCH_3) \text{ (unstable)} \qquad (9)$$

other deeply coloured products are rapidly formed in autoxidative processes.[12] The pH dependence observed with model compounds is in accordance with the behaviour of mechanical pulp. It has thus been observed that the light-induced oxidation of mechanical pulp under different pH conditions leads to different brightness losses with the maximum stability at about pH 6.

The oxidation of lignin by oxygen on exposure to light is an undesirable process leading to a deterioration of pulp and paper characteristics during storage. In other processes of interest to the pulp industry the formation and participation of oxygen in lignin reactions may also lead to a deterioration of quality. Thus, the lignin-retaining bleaching of mechanical pulps with hydrogen peroxide in alkaline solution does not result in complete removal of chromophoric structures.[16] The possibility of singlet oxygen being formed by the de-

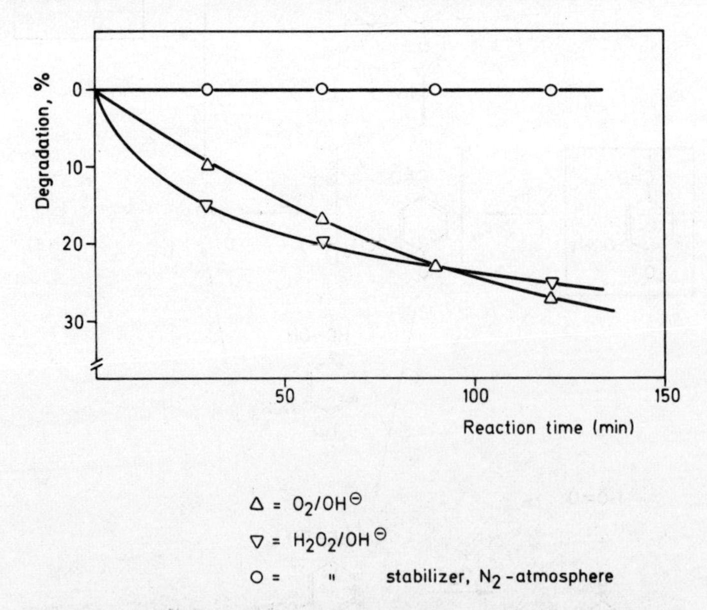

Figure 3 Oxidation of *p*-creosol with oxygen and hydrogen peroxide at pH 11·0 and 40 °C. Rate of degradation of *p*-creosol Δ, O_2/OH^-; ∇, H_2O_2/OH^-; ○, H_2O_2/OH^- stabilizer, N_2 atmosphere

composition of alkaline solutions of hydrogen peroxide and the recent observation that disproportionation of hydrogen peroxide in alkaline solution leads to the formation of singlet oxygen[17] suggest that the bleaching of pulp with hydrogen peroxide may involve singlet oxygen reactions. Being strongly electrophilic, singlet oxygen should readily attack ionized phenolic groups in lignin in reactions similar to those encountered in the alkaline autoxidation of phenols.[18]

The rate of disappearance of the phenol during the oxidation of *p*-creosol with alkaline hydrogen peroxide supports this view. Thus, the oxidation of *p*-creosol with hydrogen peroxide and oxygen separately (Figure 3) reveals that, provided that a sufficient amount of hydrogen peroxide is present as a source of singlet oxygen, this oxidation proceeds at a faster rate than the ground-state oxygen oxidation. Hydrogen peroxide does not react with *p*-creosol, as demonstrated by

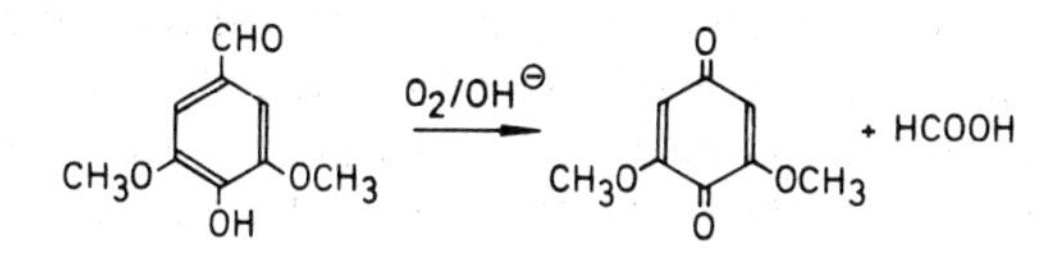

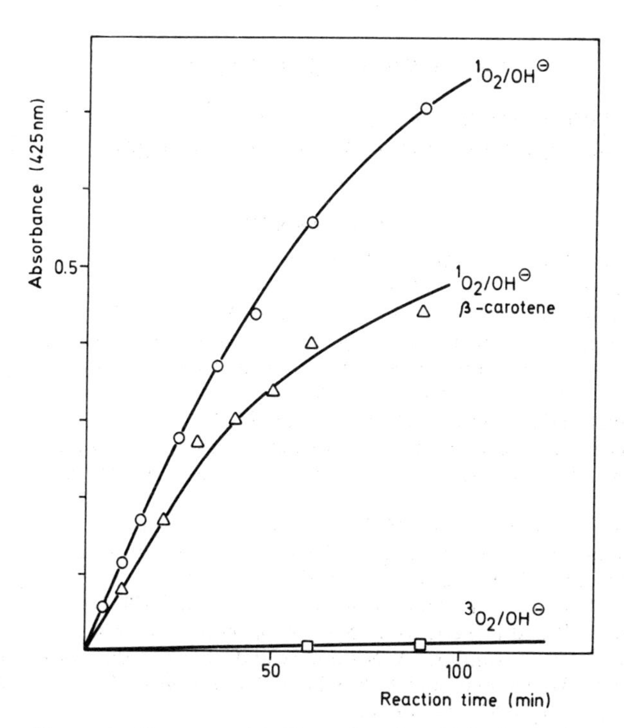

Figure 4 Oxygen oxidation of syringaldehyde under alkaline conditions. Rate of formation of 2,6-dimethoxybenzoquinone measured as change in absorbance at 425 nm

the complete lack of reactivity when *p*-cresol is treated with alkaline hydrogen peroxide in the presence of a hydrogen peroxide stabilizer and with simultaneous constant saturation of the reaction solution with nitrogen.

These results indicate that singlet oxygen under alkaline conditions is a more powerful oxidant of phenols than is ordinary oxygen. In order to test further this observation, some simple phenols related to lignin were oxidized under alkaline conditions not only with singlet oxygen but also with triplet oxygen. Singlet oxygen was generated by visible-light irradiation in the presence of Rose Bengal acting as a sensitizer. The oxidation of syringaldehyde was followed by monitoring the absorbance at 425 nm caused by 2,6-dimethoxybenzoquinone formed during the reaction (Figure 4). Obviously the great difference in reactivity observed in these experiments must be attributed in part at least to singlet oxygen oxidation since the singlet oxygen quencher β-carotene is able to inhibit the reaction to some extent.

On the basis of these results it may be concluded that singlet oxygen is an important oxidant in reactions of interest to the pulp and paper industry. In future studies, attention should therefore be paid to this interesting oxidant. A better understanding of the properties of singlet oxygen may be utilized to the advantage of the wood processing industry.

ACKNOWLEDGEMENT

Part of this work was supported by a grant from Cellulosaindustrins stiftelse för teknisk och skoglig forskning samt utbildning, which is gratefully acknowledged.

REFERENCES

1. C. S. Foote, *Pure Appl. Chem.*, **27**, 635 (1971).
2. D. R. Kearns, *Chem. Rev.*, **71**, 395 (1971).
3. A. Zweig and W. A. Henderson, Jr., *J. Polym. Sci. A1*, **13**, 717, 993 (1975).
4. B. Rånby and J. F. Rabek, *Photodegradation, Photo-oxidation and Photostabilization of Polymers*, Wiley, London, New York, Sydney and Toronto, 1975, p. 254.
5. J. Bland, *J. Chem. Educ.*, **53**, 274 (1976).
6. J. F. Rabek and B. Rånby, *Polym. Eng. Sci.*, **15**, 40 (1975).
7. K. V. Sarkanen and C. H. Ludwig, *Lignins, Occurrence, Formation, Structure and Reactions*, Wiley-Interscience, New York, London, Sydney and Toronto, 1971, p. 195.
8. J. Gierer, *Sven. Papperstidn.*, **73**, 571 (1970).
9. I. Fineman and S. Pettersson, *Sven. Papperstidn.*, **78**, 219 (1975).
10. S. Y. Lin and K. P. Kringstad, *Tappi*, **53**, 658 (1970).
11. G. Gellerstedt and E. Pettersson, *Acta Chem. Scand.*, **B29**, 1005 (1975).
12. G. Gellerstedt and E. Pettersson, *Sven. Papperstidn.*, in press.
13. T. Matsuura, N. Yoshimura, A. Nishinaga, and I. Saito, *Tetrahedron*, **28**, 4933 (1972).
14. C. S. Foote, M. Thomas, and T. Y. Ching, *J. Photochem.*, **5**, 172 (1976).
15. G. J. Leary, *Tappi.*, **51**, 257 (1968).
16. C. W. Bailey and C. W. Dence, *Tappi*, **52**, 491 (1969).
17. L. L. Smith and M. J. Kulig, *J. Amer. Chem. Soc.*, **98**, 1027 (1976).
18. F. Imsgard, *Autoxidation in Alkaline Media of Phenolic Compounds Related to Lignin, PhD Thesis*, Royal Institute of Technology, Stockholm, 1976.

32

The Use of Polymer-based Singlet Oxygen Sensitizers in the Study of Light-induced Yellowing of Lignin

G. BRUNOW, I. FORSSKÅHL, A. C. GRÖNLUND, G. LINDSTRÖM, and K. NYBERG

Department of Chemistry, University of Helsinki, Et. Hesperiank 4, 00100 Helsinki 10, Finland

A limitation to the use of high-yield pulps is imposed by the tendency towards photochemical yellowing both before and after the bleaching process. Earlier investigations[1–5] have shown that the lignin component in high-yield pulps is responsible to a great extent for the colour formation and that the major reaction is a photoinduced oxidation of lignin. Particularly aromatic α-carbonyl groups,[3–6] but also ring-conjugated carbon–carbon double bonds[7] and biphenyl structures,[3] all absorbing in the UV part of the daylight spectrum, play an active role in the yellowing process. According to present knowledge, such groups act as sensitizers, causing the formation of singlet oxygen. The singlet oxygen reacts with phenols, generating phenoxy radicals which further react to form coloured compounds[8] [reactions (1) and (2)].

$$>C{=}O \xrightarrow{h\nu} >C{=}O^{*} \xrightarrow{^{3}O_2} >C{=}O + {}^{1}O_2 \qquad (1)$$

$$\text{lignin–C}_6\text{H}_3(\text{OCH}_3)\text{OH} \xrightarrow{^{1}O_2} \text{lignin–C}_6\text{H}_3(\text{OCH}_3)\text{O}\cdot \xrightarrow{^{3}O_2} \text{Yellow products} \qquad (2)$$

The participation of singlet oxygen in the light-induced oxidation of phenols has been observed in the dye-sensitized oxidation of certain phenols[9,10] and the presence of singlet oxygen as an active reagent in the yellowing reaction has been suggested in reactions with model lignin compounds.[7,8]

Recently it was shown[7] that conjugated double bonds in model lignin compounds are degraded by UV light, giving rise to new ring-conjugated carbonyl structures, which further can induce photodegradation and yellowing.

The present investigation concerning polymer-based α-carbonyl sensitizers was undertaken in order to study the formation of singlet oxygen via α-carbonyl sensitization and the role of singlet oxygen in the yellowing of lignin. Acetoguaiacone was chosen as the aromatic α-carbonyl compound and the sodium salt of the phenol was attached to a chloromethylated polymer by means of an ether bond [reaction (3)].

$$\text{Ⓟ-}C_6H_4\text{-}CH_2Cl + NaO\text{-}C_6H_3(OCH_3)\text{-}C(=O)CH_3 \xrightarrow[\text{2 hrs, 110°C}]{\text{DMF}} \text{Ⓟ-}C_6H_4\text{-}CH_2O\text{-}C_6H_3(OCH_3)\text{-}C(=O)\text{-}CH_3 + NaCl$$

Ⓟ-AG

In order to optimize the yield of the polymeric benzyl ether, a solvent had to be used which swelled the polymer adequately and which did not contain hydroxyl or ethoxyl ions. The best procedure for the preparation was a 2 h reflux of equivalent amounts of the dry sodium salt of acetoguaiacone and of the chloromethylated polymer in dimethylformamide. The resulting polymer was filtered off and was repeatedly washed with several solvents and dried, and was then ready for use. The efficiency of the conversion was checked by means of the carbonyl absorption in the infrared spectrum of the resulting polymeric acetoguaiacone benzyl ether (abbreviated to Ⓟ-AG), and by measuring the amount of released chloride in the reaction mixture by potentiometric titration with silver nitrate. From the potentiometric method an 80% conversion has been calculated; the results are in good agreement with those from gravimetric analysis of the released chloride determined as silver chloride from the same reaction mixture. It was shown in a separate run that the chloromethylated resin did not release chloride in the absence of phenolate ions.

We have used the polymer-bound acetoguaiacone sensitizer in the study of the mechanism of light-induced yellowing of high-yield pulps with the aid of model compounds. Some polymer-based dye sensitizers were prepared according to the literature[11–13] in order to compare their efficiency with the synthesized Ⓟ-AG. The phenolic model compound was irradiated in air with UV light at 350 nm in the presence of the sensitizer and the rate of the disappearance of the phenolic compound was measured by gas chromatography (Figure 1).

From the results obtained, the conclusion can be drawn that the α-carbonyl

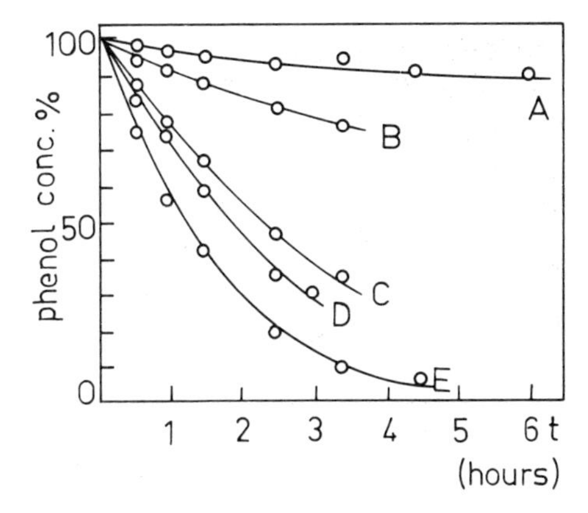

Figure 1 Rate of degradation of 2,6-dimethoxyphenol in the presence of different sensitizers. Irradiation in air. Sensitizers: A = Ⓟ-Rose Bengal, 3×10^{-3} M;[11,12] B = Rose Bengal, 5×10^{-3} M in solution; C = Rose Bengal on Amberlite IRA-410, 10×10^{-3} M;[13] D = methylene blue on Dowex 50W, 10×10^{-3} M;[13] E = Ⓟ-AG, 8×10^{-3} M (this work). The calculation of the concentrations of the dye sensitizers is based on the given capacity of the resins, assuming a reaction with all the functional groups, except in the case of A where it has been reported[12] that only one in five chloromethyl residues is reactive, and E where an 80% conversion has been measured

sensitizer is at least as efficient a singlet oxygen sensitizer as are the polymer-bound dyes.

The participation of singlet oxygen in the photodegradation of model lignin compounds in the presence of aromatic α-carbonyl groups was verified by means of quenching experiments with β-carotene.[14] In Figure 2 the retarding effect is shown using different amounts of quencher in the reaction of *tert*-butylguaiacylcarbinol sensitized by non-bound *p*-methoxypropiophenone. The same effect was observed in the presence of the polymer-bound acetoguaiacone.

It has been shown[15] that polystyrene sensitizes the formation of singlet oxygen. We could also demonstrate a catalysing effect of the chloromethylated resin by itself. The reaction was also quenched by β-carotene.

The lifetime of singlet oxygen is dependent on the solvent to a large extent.[16] The solvent used in the experiments was 1,2-dimethoxyethane, in which the lifetime of singlet oxygen has not been reported, but it can be expected to be of the same order as the value reported for dioxane (32×10^{-6} s).[17] A comparison between the rates of disappearance of *tert*-butylguaiacylcarbinol with Ⓟ-AG as sensitizer in 1,2-dimethoxyethane and in carbon tetrachloride (Figure 3) was

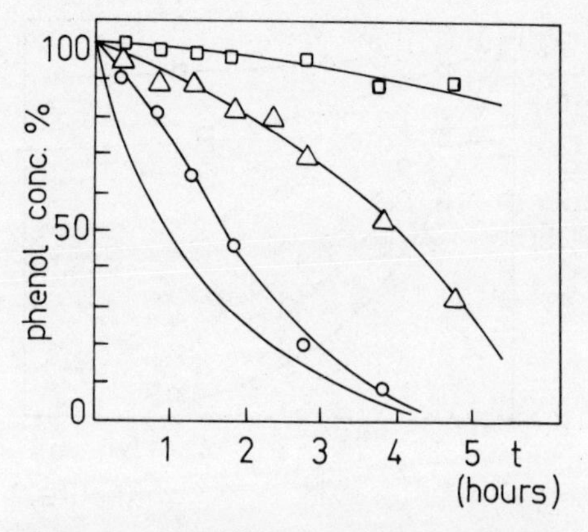

Figure 2 Quenching of the *p*-methoxypropiophenone-sensitized reaction of *tert*-butylguaiacylcarbinol in air. □, $2{\cdot}5 \times 10^{-3}$ M *β*-carotene; △, $0{\cdot}5 \times 10^{-3}$ M *β*-carotene; ○, $0{\cdot}25 \times 10^{-3}$ M *β*-carotene; ——, no quencher

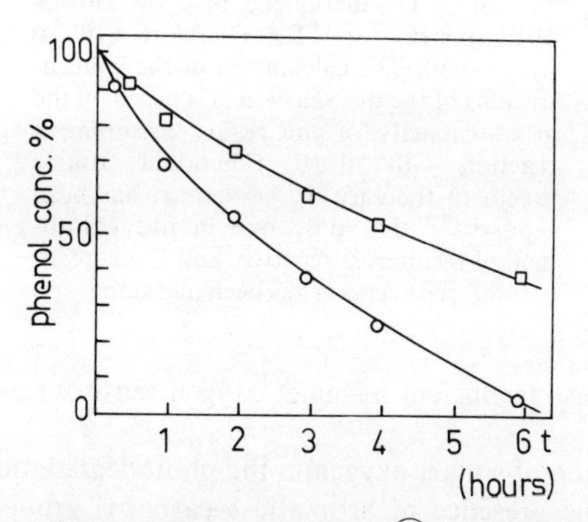

Figure 3 Solvent effect in the (P)-AG-sensitized degradation of *tert*-butylguaiacylcarbinol. □, Carbon tetrachloride; ○, 1,2-dimethoxyethane

carried out. The rate of degradation is higher in dimethoxyethane than in carbon tetrachloride although the lifetime of singlet oxygen is thought to be much longer in the latter solvent, for which the reported value is $\tau = 700 \times 10^{-6}$ s.[16]

The reactions were performed in both the solid and liquid phases in the presence of model lignin compounds. The rate of reaction is much lower in the solid phase than in solution. In the solid phase the way in which the phenol and the sensitizer are mixed influences the reaction rate considerably and the highest reaction rates were achieved when the phenol was first dissolved in a solvent and then evaporated on the sensitizer. This procedure minimizes the distance between the sensitizer and the substrate, causing a greater overall reaction rate, in conformity with the singlet oxygen mechanism. Chromatographic analysis shows that a similar mixture of products is obtained in both the solid and liquid phases. The structures of the products are under investigation.

REFERENCES

1. G. J. Leary, *Tappi*, **51**, 257 (1968).
2. K. Kringstad, *Tappi*, **52**, 1070 (1969).
3. S. Y. Lin and K. P. Kringstad, *Tappi*, **53**, 658 (1970).
4. K. P. Kringstad and S. Y. Lin, *Tappi*, **53**, 2296 (1970).
5. S. Y. Lin and K. P. Kringstad, *Norsk Skogsind.*, **25**, 252 (1971).
6. G. Brunow and B. Eriksson, *Acta Chem. Scand.*, **25**, 2779 (1971).
7. G. Gellerstedt and E. L. Pettersson, *Acta Chem. Scand.*, **B29**, 1005 (1975).
8. G. Brunow and M. Sivonen, *Paperi Puu*, **57**, 215 (1975).
9. T. Matsuura, N. Yoshimura, A. Nishinaga, and I. Saito, *Tetrahedron*, **28**, 4933 (1972).
10. I. Saito, N. Yoshimura, T. Arai, K. Omura, A. Nishinaga, and T. Matsuura, *Tetrahedron*, **28**, 5131 (1972).
11. E. C. Blossey, D. C. Neckers, A. L. Thayer, and A. P. Schaap, *J. Amer. Chem. Soc.*, **95**, 5820 (1973).
12. A. P. Schaap, A. L. Thayer, E. C. Blossey, and D. C. Neckers, *J. Amer. Chem. Soc.*, **97**, 3741 (1975).
13. J. R. Williams, G. Orton, and L. R. Unger, *Tetrahedron Lett.*, **1973**, 4603.
14. C. S. Foote and R. W. Denny, *J. Amer. Chem. Soc.*, **90**, 6233 (1968).
15. B. Rånby, *Kemia-Kemi*, **1**, 477 (1974).
16. P. B. Merkel and D. R. Kearns, *J. Amer. Chem. Soc.*, **94**, 7244 (1972).
17. R. H. Young, D. Brewer, and R. A. Keller, *J. Amer. Chem. Soc.*, **95**, 375 (1973).

33

Deactivation of Singlet Oxygen by Polyolefin Stabilizers

H. FURUE and K. E. RUSSELL

Department of Chemistry, Queen's University, Kingston, Ontario, Canada

The polyolefins polyethylene and polypropylene can be stabilized towards photodegradation by the incorporation of metal complexes such as bis(di-*n*-butyldithiocarbamato)nickel(II). A possible mode of action of the stabilizers involves retardation of the formation of hydroperoxide which can occur when singlet oxygen reacts with olefin groups within the polymer. In an earlier study,[1] the rate constants for the quenching of singlet oxygen by certain metal chelates in carbon disulphide solution were determined. Singlet oxygen has a relatively long lifetime in carbon disulphide and it was produced by flash photolysis of oxygenated carbon disulphide solutions containing methylene blue as sensitizer. The rate constants for quenching of singlet oxygen were obtained from the analysis of the disappearance of the acceptor 1,3-diphenylisobenzofuran (DPBF). Singlet oxygen has an even longer lifetime in carbon tetrachloride[2] and the study has been extended to the determination of rate constants for the quenching of singlet oxygen by a wide range of metal complexes in carbon tetrachloride.

The deactivation of singlet oxygen by the metal complex may occur in a simple collisional process or it may, in principle, involve a charge-transfer interaction between the singlet oxygen and the complex. In order to learn more about deactivation by metal complexes, we have made a parallel investigation using triplet pentacene, a species which is roughly isoenergetic with singlet oxygen.

RESULTS AND DISCUSSION

The solvent chosen for the present singlet oxygen study was carbon tetrachloride containing 2 vol.-% of methanol because of the very low solubility of methylene blue in pure carbon tetrachloride. The lifetime of singlet oxygen in this mixed solvent is 310 ± 50 μs and the rate constant for reaction with DPBF at 25 °C is $(5{\cdot}3 \pm 1{\cdot}7) \times 10^{8}$ M^{-1} s^{-1}. Rate constants, k_Q, for the quenching of singlet oxygen by a range of polyolefin stabilizers and related compounds are given in Table 1. The range of quencher concentrations varied from 5 to 10×10^{-7} M for

Table 1 Rate constants for the quenching of singlet oxygen by polyolefin stabilizers

Quencher	Solvent	$10^{-8}\,k_Q(\mathrm{M}^{-1}\,\mathrm{s}^{-1})$
2-Hydroxy-4-dodecycloxybenzophenone	CCl_4	<0·002
Irganox 1076	CCl_4	0·0046
Bis(di-*n*-butyldithiocarbamato)zinc(II)	CS_2	0·2 ± 0·1
2,2′-Thiobis(4-*tert*-octylphenolato)(*n*-butylamine)nickel(II)	CS_2	6·0 ± 3·2
Bis(2,4-pentanedionato)diaquonickel(II)	CCl_4	0·6 ± 0·06
Bis[2,2′-thiobis(4-*tert*-octylphenolato]nickel(II)	CS_2	5·0 ± 2·5
Bis(diethyldithiocarbamato)nickel(II)	CCl_4	66 ± 6
Bis(di-*n*-butyldithiocarbamato)nickel(II)	CCl_4	41 ± 6
Bis(di-*n*-butyldithiocarbamato)nickel(II)	CS_2	40 ± 15
Bis(diisopropyldithiophosphato)nickel(II)	CCl_4	76 ± 10
Bis(di-*n*-butyldithiocarbamato)copper(II)	CCl_4	<5
Tris(di-*n*-butyldithiocarbamato)cobalt(III)	CCl_4	12 ± 2
Tris(2,2′-bipyridine)ruthenium(II)chloride	CCl_4	<10

bis(di-*n*-butyldithiocarbamato)nickel(II) to as high as $1{\cdot}0 \times 10^{-2}$ M for 2-hydroxy-4-dodecycloxybenzophenone. In some cases, e.g. bis(di-*n*-butyldithiocarbamato)copper(II), only upper limits for k_Q are quoted, no quenching being observed at the highest quencher concentration which could be used without significant absorption of the monitoring light.

The values of k_Q for quenching of singlet oxygen by the polyolefin stabilizers range from $<2 \times 10^5$ to $7{\cdot}6 \times 10^9\ \mathrm{M}^{-1}\,\mathrm{s}^{-1}$. The highest values of k_Q in this study were obtained with the nickel dialkyldithiocarbamates and diisopropyldithiophosphate. Change of metal in the complex causes a large variation in quenching rate. Bis(di-*n*-butyldithiocarbamato)zinc(II) has a k_Q value of $2 \times 10^7\ \mathrm{M}^{-1}\,\mathrm{s}^{-1}$, the corresponding d^9 and d^8 complexes of copper and nickel have k_Q values of $<5 \times 10^8\ \mathrm{M}^{-1}\,\mathrm{s}^{-1}$ and $4{\cdot}0 \times 10^9\ \mathrm{M}^{-1}\,\mathrm{s}^{-1}$, respectively, and d^6 tris(di-*n*-butyldithiocarbamato)cobalt(III) has a k_Q value of $1{\cdot}2 \times 10^9\ \mathrm{M}^{-1}\,\mathrm{s}^{-1}$.

The lowest singlet–singlet absorptions in the diamagnetic square-planar d^8 nickel complexes are due to metal d–d transitions. Rough estimates of the triplet levels yield values in the 7000–8000 cm^{-1} range so that spin-allowed electronic energy transfer may occur fairly readily in these cases, yielding high values of k_Q. A similar argument applies to the cobalt complex. On the other hand, no d–d transition is possible for the zinc complex and the ligand π–π^* transition involves a much higher energy. There is thus a low probability for energy transfer to the quencher via equation (1) and the quenching rate constant is low.

$$ {}^1\Delta O_2 + {}^1Q \longrightarrow {}^3\Sigma O_2 + {}^3Q \qquad (1) $$

Tris(2,2′-bipyridine)ruthenium(II) chloride is a diamagnetic octahedral d^6 complex which phosphoresces with a maximum at $15{\cdot}7 \times 10^3\ \mathrm{cm}^{-1}$.[3] This triplet energy is much higher than the singlet oxygen level and the upper limit of $10 \times 10^8\ \mathrm{M}^{-1}\,\mathrm{s}^{-1}$ for the k_Q of this complex is thus not surprising. There is no obvious correlation between k_Q and paramagnetism of the complexes, nor is there evidence for an enhanced inter-system crossing mechanism.

The decay of triplet pentacene was studied in benzene solution at room temperature; the monitoring wavelength was 492 nm. The solubility of pentacene in benzene at room temperature is very low (ca. 10^{-5} M) and, in the absence of quenchers, the triplet pentacene decays by a first-order process with a rate constant, k_d, of $(1{\cdot}59 \pm 0{\cdot}06) \times 10^4$ s^{-1}. Quenching rate constants, k_Q, were determined by adding appropriate concentrations of quencher to the benzene solution of the pentacene and observing the increased rate of disappearance of the triplet pentacene. The rate constants for a number of metal complexes are given in Table 2. The palladium and platinum complexes absorb strongly at the monitoring wavelength and the values of k_Q are only upper limits.

As with the quenching of singlet oxygen, the rate constants for quenching triplet pentacene vary considerably with the metal in the complex. There appears to be no correlation between k_Q and paramagnetism of the complex; the paramagnetic bis(diisopropyldithiophosphato)bis(pyridine)nickel(II) has a rate constant of $3{\cdot}6 \times 10^8$ M^{-1} s^{-1} whereas k_Q values for known diamagnetic complexes range from $1{\cdot}8 \times 10^5$ to $4{\cdot}7 \times 10^9$ M^{-1} s^{-1}. The maximum k_Q of $4{\cdot}7 \times 10^9$ M^{-1} s^{-1}, observed for bis(dithiobenzil)nickel(II), is close to the maximum quenching rate constant ($6 \pm 1 \times 10^9$ M^{-1} s^{-1}) observed in quenching of triplet states in benzene solution at room temperature.[4,5] The absorption at 866 nm in the spectrum of bis(dithiobenzil)nickel(II) has been assigned to a ligand π–π^* transition, and the corresponding triplet energy could well be lower than that of triplet pentacene. The relatively high quenching efficiency of this complex is thus explained by energy transfer from the triplet pentacene to an acceptor of lower energy, under conditions of low steric hindrance. The next highest k_Q values are observed for the nickel dialkyldithiocarbamates, nickel diisopropyldithiophosphate and cobalt di-*n*-butyldithiocarbamate which have triplet energy levels estimated to be close to that of triplet pentacene. d–d transitions are involved and, since the orbitals are largely localized on the metal, quenching may be subject to steric hindrance by the ligand.[6] Bis(di-*n*-butyl-

Table 2 Rate constants for the quenching of triplet pentacene by metal chelates

Quencher	$10^{-8}\ k_Q$(M^{-1} s^{-1})
Bis(dithiobenzil)nickel(II)	47 ± 7
Bis(diisopropyldithiophosphato)nickel(II)	$7{\cdot}8 \pm 3{\cdot}4$
Bis(diisopropyldithiophosphato)bis(pyridine)nickel(II)	$3{\cdot}6 \pm 0{\cdot}6$
Bis(diethyldithiocarbamato)nickel(II)	$7{\cdot}0 \pm 0{\cdot}8$
Tris(di-*n*-butyldithiocarbamato)cobalt(III)	$3{\cdot}3 \pm 0{\cdot}7$
2,2′-Thiobis(4-*tert*-octylphenolato)(*n*-butylamine)nickel(II)	$0{\cdot}23 \pm 0{\cdot}15$
Bis(2,4-pentanedionato)diaquonickel(II)	$0{\cdot}54 \pm 0{\cdot}06$
Bis(di-*n*-butyldithiocarbamato)zinc(II)	$0{\cdot}0018 \pm 0{\cdot}0015$
Bis(diethyldithiophosphato)platinum(II)	$<0{\cdot}02$
Bis(diethyldithiophosphato)palladium(II)	$<0{\cdot}2$

dithiocarbamato)zinc(II) has a very low k_Q and this is explained by the lack of a low-lying triplet level in this complex which can take part in an efficient energy-transfer process. The lowest triplet level of the platinum complex is estimated to be roughly 15 000 cm^{-1} above the ground state and the low k_Q for this complex is readily explained.

There is a parallel between the ease of quenching of singlet oxygen by a metal complex and its ability to quench triplet pentacene. This is particularly apparent if the observed k_Q values are expressed as fractions of the maximum rate constants observed for quenching of singlet oxygen and triplet pentacene. For example, bis(dithiobenzil)nickel(II) has a k_Q for singlet oxygen quenching[7] of about $1{\cdot}5 \times 10^{10}$ M^{-1} s^{-1}, which is 75% of the highest k_Q value of 2×10^{10} M^{-1} s^{-1} observed for quenching of singlet oxygen by β-carotene.[2] The corresponding percentage for quenching of triplet pentacene by the benzil complex is also close to 75%. The corresponding percentages for quenching by nickel diethyldithiocarbamate are 33% and 12%. These results suggest that the same general mechanism applies to the quenching of singlet oxygen and triplet pentacene, that of spin-allowed energy transfer. No evidence has been obtained in this study for complex formation between the donor and the acceptor metal complex.

ACKNOWLEDGEMENT

This research was supported by an operating grant from the National Research Council of Canada.

REFERENCES

1. J. Flood, K. E. Russell, and J. K. S. Wan, *Macromolecules*, **6**, 669 (1973).
2. P. B. Merkel and D. R. Kearns, *J. Amer. Chem. Soc.*, **94**, 7244 (1972).
3. M. S. Wrighton, L. Pdungsap, and D. L. Morse, *J. Phys. Chem.*, **79**, 66 (1975).
4. P. J. Wagner and I. Kochevar, *J. Amer. Chem. Soc.*, **90**, 2232 (1968).
5. A. Farmilo and F. Wilkinson, *Chem. Phys. Lett.*, **34**, 575 (1975).
6. A. Adamczyk and F. Wilkinson, *J. Chem. Soc. Faraday Trans. II*, **68**, 2031 (1972).
7. A. Zweig and W. A. Henderson, *J. Polym. Sci. A1*, **13**, 717 (1975).

34

*Stabilization of Polymers Against Singlet Oxygen**

D. M. WILES

Division of Chemistry, National Research Council, Ottawa, Canada

INTRODUCTION

Although the formation and properties of singlet molecular oxygen ($^1O_2$, $^1\Delta_g$) have been under investigation for many years, there has been particular interest recently on the part of chemists who are concerned with the deleterious effects of $^1O_2$ on specific types of molecules. The reactivity of the $^1\Delta_g$ state, over ground-state triplet oxygen, derives from the ca. 22 kcal mol^{-1} of excess energy which it carries; $^1O_2$ therefore reacts rapidly with C—C unsaturation, amines, sulphides, polynuclear aromatics and many other functional groups.[1–3] One of the more important commercial implications of $^1O_2$ reactivity is in the area of the oxidative degradation of macromolecules. With the availability of several laboratory methods for the preparation of singlet oxygen,[4,5] its effects on plastics,[6–10] rubbers[11–13] and fibres[14,15] have been investigated in some detail. There is no doubt that $^1O_2$ can contribute to the photo-oxidative degradation of numerous polymers although the extent to which this actually occurs has not been established unambiguously for most systems.

SINGLET OXYGEN REACTIONS

Many commercially important polymers contain carbon–carbon unsaturation and it is well known that $^1O_2$ reacts with olefins (in the liquid and gaseous phases) to give hydroperoxides. Moreover, gaseous $^1O_2$ has been shown to produce surface OOH groups in films of polydienes[13] and polyolefins,[7,9] for example. Rabek and Rånby[16] have proposed that the abstraction of tertiary hydrogen atoms by $^1O_2$ leads to the formation of hydroperoxy groups in the photo-oxidation of polystyrene. Finally, since singlet oxygen has been observed to abstract hydrogen from substituted phenols, it is assumed[10] that this can be expected to happen in polymers which contain OH groups.

The formation of hydroperoxides on polymer chains critically alters their stability. The thermal instability of these groups means that they are able to

* Issued as NRCC No. 15534.

initiate the autoxidation of the polymers.[17] The photoinstability of the OOH moiety means that their facile photocleavage to alkoxy radicals, followed by the β-scission of these macro-alkoxy radicals, will result in loss of useful properties because of polymer backbone cleavage.

In principle, there are three concepts upon which to base the stabilization of macromolecules against the degradation which can result from $^1O_2$ attack: (1) prevention of the formation of $^1O_2$ on and in solid polymers; (2) deactivation of $^1O_2$, normally by quenching, before it has a chance to react within the polymer system; and (3) removal, normally by decomposition or scavenging, of the products from the $^1O_2$–polymer reactions.

PREVENTION OF $^1O_2$ FORMATION

In considering this concept, it must be remembered that the atmospheric concentration of $^1O_2$ is normally low in comparison with O_3, for example. This is not surprising in view of the fact that the former reacts much more rapidly with olefinic unsaturation than does the latter. It does mean, however, that it is necessary to consider mechanisms by which $^1O_2$ may be generated in a polymer, especially in the context of polymer stabilization.

There are at least two significant routes by which $^1O_2$ can be generated thermally. The first of these, the formation of O_3 complexes followed by their decomposition to give $^1O_2$, has been discussed by Murray and Kaplan.[18] Thus, in situations of significant primary O_3 attack on polymer systems, secondary oxidation owing to $^1O_2$ should also be expected. Singlet oxygen is produced in the self-termination of primary and secondary peroxy radicals[19] during the autoxidation of hydrocarbons. In the solid phase, where kinetic chain lengths are short, radical self-termination is likely to be common and $^1O_2$ could be contributing substantially to the observed oxidation.

There appears to be an important secondary process by which singlet oxygen is likely to be produced photochemically in polymers exposed to sunlight or other, comparable, light sources. This process involves the transfer of energy from the triplet state of a chromophoric impurity (sensitizer) to ground-state oxygen to produce $^1O_2$. Trozzolo and Winslow[6] first suggested that carbonyl groups, formed by thermal oxidation during processing, were likely to be key sensitizing impurities in the formation of $^1O_2$ during polyethylene irradiation. There are, however, numerous other additives or impurities present in all commercial polymer systems which could also act as sensitizers. It has been shown, for example, that polynuclear aromatic (PNA) compounds are present in the atmosphere and are absorbed in polyolefin films at up to 100 p.p.m. levels.[9] Many PNA compounds absorb near UV wavelengths and generate $^1O_2$ in the quenching of PNA triplets by $^3O_2$.

According to a simplistic concept, therefore, protection of polymers against $^1O_2$ reactions can be achieved by reducing complex formation with O_3, by reducing the formation (and subsequent reaction) of peroxy radicals, and by eliminating UV-sensitive impurities or additives in polymers. While it is possible to proceed further along these lines in the laboratory, this is not a novel practical approach towards enhanced polymer stability. Apart from singlet oxygen attack on polymers, it is desirable to adhere to the above concept in order to prolong the useful life of polymers anyway. Presumably, considerations of cost and other practicalities limit the purity of commercial polymers and their preservation during processing and storage.

QUENCHING OF SINGLET OXYGEN

For the reasons outlined above, it seems certain that $^1O_2$ cannot be prevented entirely from forming in polymers in which the thermal and/or photochemical routes described previously can proceed. Notwithstanding the high reactivity of $^1O_2$, however, compounds which will deactivate it back to $^3O_2$ in the liquid phase have been known for some time.[20] The deactivators (quenchers) include amines, sulphides and phenols, all of which are less effective than β-carotene, the most efficient quencher of $^1O_2$. It should be noted, however, that β-carotene is not a practical UV stabilizer for polymers owing, among other things, to its inherent photo- and thermal instability. Both aromatic[21] and aliphatic amines are efficient quenchers and a comparison of their efficiencies in various solvents has been published.[5] More recently, it has been reported from several laboratories[22–26] that nickel chelates which provide photostability to polyolefins are also efficient quenchers of singlet oxygen in solution. The data of Farmilo and Wilkinson[26] show that the quenching rate constant (k_q) decreases rapidly for a series of chelates with increasing triplet energy. Thus, although the precise mechanism by which $^1O_2$ is quenched by chelates is not known, energy transfer to a low-lying triplet level of some chelates is certainly possible.

Data are available[8,21,22,26–28] to show that efficient quenching of $^1O_2$ occurs in the liquid phase with certain aromatic amines, Ni(II), Co(II), and Fe(III) dialkyl dithiocarbamates, Ni(II) diisopropyldithiophosphate, Ni(II) Schiff-base complexes (aldimines or ketoximes), and of course β-carotene. What is needed, however, is unambiguous evidence that some or all of these compounds quench singlet oxygen in the solid phase. We have obtained some indication[8,27] that this is the case by observing the retardation of the $^1O_2$-induced bleaching of rubrene when the latter is in the form of a solid, co-deposited with a Ni(II) chelate. The same effect is obtained when the nickel chelate is deposited 'upstream' from the rubrene in a flow system involving $^1O_2$ generation by a microwave discharge.

Direct evidence of $^1O_2$ quenching in solid polymers is more difficult to obtain, although the data in Table 1 provide some indication of this. In some cases, additives which prolong the period from exposure to brittle failure of polypropylene films also quench singlet oxygen efficiently in solution. Thus, some of these same additives may be providing UV stability in part by quenching $^1O_2$. On the other hand, there are some compounds in Table 1 which either quench or stabilize, but not both. This can probably be accounted for on the basis that transition metal chelates can also operate as peroxide decomposers, radical

Table 1 Comparison of polpropylene film UV stabilization (hours) and $^1O_2$ ($^1\Delta_g$) quenching rate constants (k_q)

Additive	PP lifetime[a] (h)	k_q ($M^{-1} s^{-1} \times 10^{-8}$)[b]
None	110	—
β-Carotene	—	85[c]
Ni(II) diisopropyldithiophosphate	430	54[c]
Fe(III) diisopropyldithiocarbamate	—	38[c]
Ni(II) diisopropyldithiocarbamate	1310	34[c]
Ni(II) bis[2-hydroxy-5-methoxyphenyl-*N*-(*n*-butyl)aldimine]	380	24[d]
Co(II) diisopropyldithiocarbamate	—	19[c]
Ni(II) bis[2,2′-thiobis-(4-*tert*-octyl)phenolate]	250	1·3[d]
Ni(II) acetylacetonate	250	0·75[c]
Zn(II) diisopropyldithiocarbamate	186	<0·1[c]
2-Hydroxy-4-dodecyloxybenzophenone	350	<0·01[d]

[a] Time to brittle failue of 30-μm films containing 0·1 w-% of additive; Atlas carbon-arc Fade-Ometer irradiation.
[b] Additive (quencher) concentration 1×10^{-5} to 10^{-3} mol l^{-1}.
[c] In methylene chloride solution.
[d] In isooctane solution.

scavengers, excited chromophore quenchers, etc. Since the photo-oxidative degradation of polypropylene is complex,[29] an additive which stabilizes by more than one mechanism (as do many nickel chelates) would not necessarily impart a degree of stability to solid film which relates directly to its efficiency as a quencher of $^1O_2$.

Finally, there is another type of experiment[5,9] that indicates singlet oxygen quenching by Ni(II) dialkyldithiocarbamates in UV-irradiated polypropylene films. By using infrared spectroscopy to monitor the rate of build-up of photo-oxidation products during irradiation, it is convenient to determine the effects of particular additives. In Figure 1, data showing the rate of formation of polypropylene hydroperoxide (PPOOH) groups show such effects; analogous data are obtained by monitoring the rate of build-up of carbonyl groups. The inclusion of 0·004 M anthracene in a 22 μm commercial polypropylene film greatly enhances the rate of polymer photo-oxidation compared with the same film containing no PNA. This result is presumably due primarily to the anthracene acting as a photosensitizer, transferring electronic excitation energy to ground-state oxygen and causing the formation of $^1O_2$ within the film. Polypropylene film which contains 0·004 M anthracene and 0·002 M Ni(II) di-*n*-butyldithiocarbamate, however, photo-oxidizes much more slowly than the anthracene-doped film. The nickel chelate is a very efficient $^1O_2$ quencher (Table 1) and may have imparted greater UV stability to polypropylene than is observed in the additive-free control film because of this. Nevertheless, this nickel chelate is an efficient decomposer of polypropylene hydroperoxide[30] and it might also quench anthracene triplets. Thus, this type of experiment cannot

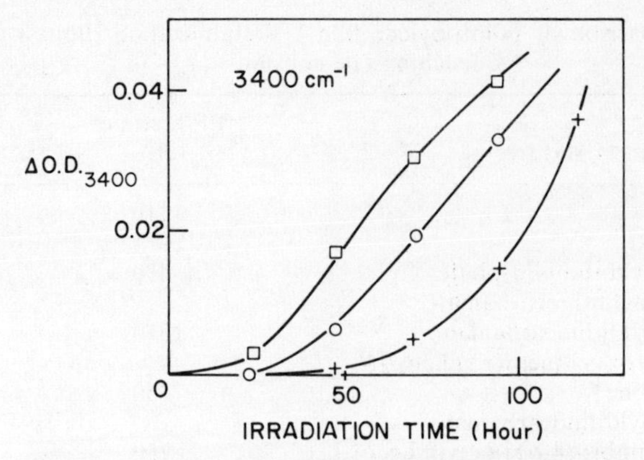

Figure 1 Effects of additives on the photodegradation of polypropylene films, 22 μm thick; xenon arc irradiation ($\lambda > 290$ nm); transmission IR spectra. ○, Polypropylene alone; □, polypropylene +0·004 M anthracene. +; polypropylene +0·004 M anthracene +0·002 M nickel dibutyldithiocarbamate

provide any indication of what proportion of the photo-oxidation of polypropylene derives from singlet oxygen attack. It is likely to be of greater significance, however, in the very early stages of exposure, before the formation and photolysis of hydroperoxide groups dominates the photodegradation. Further research appears to be necessary before the importance of $^1O_2$ in the photo-oxidation of other polymers can be established unequivocally.

SCAVENGING OF REACTION PRODUCTS

It may be assumed that during exposure to UV light, some $^1O_2$ will be formed in certain polymers, and some will react owing to the practical difficulty of completely preventing it. It is inevitable, therefore, that effective stabilization of these polymers will require the use of additives which cope with $^1O_2$–polymer reaction products. As discussed above, these products include hydroperoxide groups and the radicals (e.g. alkoxy, alkyl, and peroxy) which derive from them. Since these reactive entities are the important species in polyolefin photodegradation (possibly also for other polymers such as polystyrene) an appropriate additive 'package' should cope with them.

It has been shown,[31] for example, that some of the chelates which quench $^1O_2$ are also efficient decomposers of hydroperoxides. The vacuum photolysis of PPOOH (in polypropylene film) yields water as the major volatile product when wavelengths greater than 320 nm are used. The results of photogravimetric experiments in the presence and absence of specific additives illustrate (Figure 2) that some nickel chelates can reduce the rate of formation of water, presumably by decomposing PPOOH groups before they can absorb UV light.

It is not possible to quench electronically excited hydroperoxide groups (the first excited state is believed to be dissociative) so it is necessary to consider

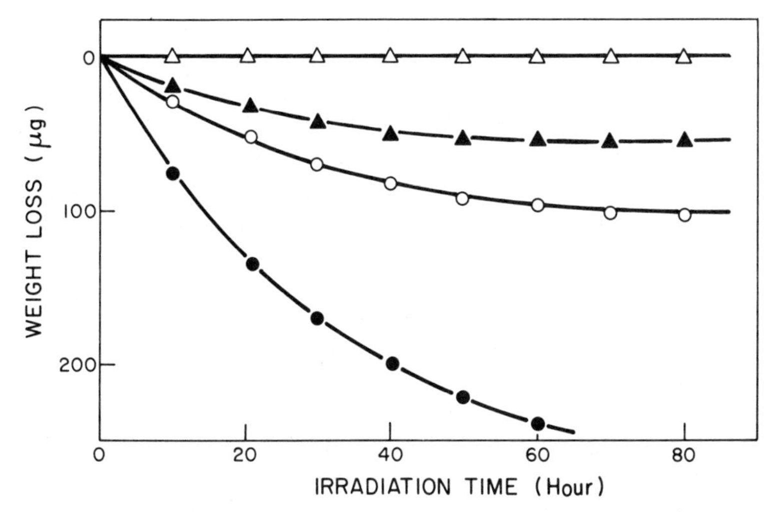

Figure 2 Effects of additives on the vacuum photolysis of polypropylene hydroperoxide in 22 μm (25 mg) films; photogravimetric technique; irradiation at 375 ± 35 nm. ●, PPOOH initially at 0·3 M; no additives. ▲, PPOOH initially at 0·07 M; no additives. ○, 0·3 M PPOOH, 0·1 wt.-% Ni(II) 2,2′-thiobis-(4-*tert*-octylphenolate)-*n*-butylamine; △, 0·07 M PPOOH, 0·1 wt.-% Ni(II) di(*O*-butyl-3,5-di-*tert*-butyl-4-hydroxybenzylphosphonate)

scavenging the photolysis products of OOH groups also. There is experimental evidence[31] that some polyolefin photostabilizers (e.g. some nickel chelates) may in fact scavenge OH, peroxy, and perhaps alkoxy radicals in solid polymers. This evidence was obtained by monitoring the build-up of photo-oxidation products (OOH and C=O) by infrared spectroscopy in polypropylene films irradiated in the presence of certain additives. Figure 3 shows that a nickel

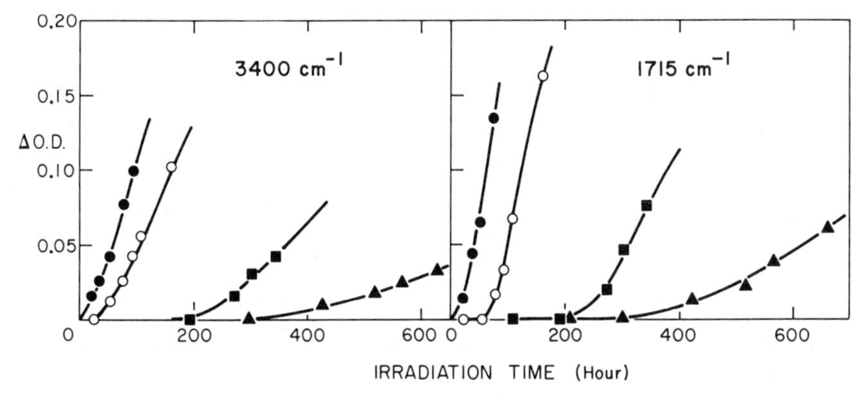

Figure 3 Effects of additives on the PPOOH-sensitized photo-oxidation of polypropylene films, 22 μm thick; xenon arc irradiation ($\lambda > 290$ nm); transmission IR spectra; PPOOH initially 0·03 M, carbonyl initially ca. 0·003 M. ●, No additive; ○, hydroperoxide-free film; ■, 0·11 wt.-% 2-hydroxy-4-dodecyloxybenzophenone; ▲, 1·0 wt.-% Ni(II) 2,2′-thiobis-(4-*tert*-octylphenolate)-*n*-butylamine

chelate and a hydroxybenzophenone significantly reduce the rates of formation of hydroperoxy (3400 cm^{-1}) and ketone carbonyl (1715 cm^{-1}) groups. It is concluded that these (and other additives) improve film photostability, at least in part, because they scavenge radical intermediates. Efficient photostabilization of polyolefins requires such scavenging anyway, whether or not these radicals derive from $^1O_2$–polymer reactions. The importance of free-radical trapping in polymer photostabilization has indeed been recognized for some time.[32,33]

CONCLUSIONS

1. Singlet molecular oxygen ($^1\Delta_g$) can be formed in solid polymers, thermally and/or photochemically.
2. $^1O_2$ does react with many polymers, primarily with C—C unsaturation to produce hydroperoxide groups, but possibly also in other ways.
3. Additives which will quench singlet oxygen are known.
4. In the case of polyolefins, singlet oxygen attack could be important only in the very early stages of UV exposure.
5. Photostabilizing additives (or combinations of them) can be selected for polyolefins which will decompose peroxides, scavenge radicals, and quench $^1O_2$ if this is needed. No separate additive should be required to protect against $^1O_2$ alone.
6. While point (5) may apply in principle to other types of polymers, published information about their photochemistry is not yet adequate to be sure that point (4) applies.

ACKNOWLEDGEMENTS

Information from many of the author's colleagues has been most helpful in the preparation of this manuscript.

REFERENCES

1. M. L. Kaplan, *Chem. Technol.*, **1971**, 621.
2. D. R. Kearns, *Chem. Rev.*, **71**, 395 (1971).
3. C. S. Foote, *Accounts Chem. Res.*, **1**, 104 (1968).
4. B. Rånby and J. F. Rabek, *Photodegradation, Photo-oxidation and Photostabilization of Polymers*, Wiley, New York, 1975, Ch. 5.
5. D. J. Carlsson and D. M. Wiles, *Rubb. Chem. Technol.*, **47**, 991 (1974).
6. A. M. Trozzolo and F. H. Winslow, *Macromolecules*, **1**, 98 (1968).
7. M. L. Kaplan and P. G. Kelleher, *J. Polym. Sci. B*, **9**, 565 (1971).
8. D. J. Carlsson, T. Suprunchuk, and D. M. Wiles, *J. Polym. Sci. B*, **11**, 61 (1973).
9. D. J. Carlsson and D. M. Wiles, *J. Polym. Sci. B*, **11**, 759 (1973).
10. J. F. Rabek and B. Rånby, *Polym. Eng. Sci.*, **15**, 40 (1975).
11. M. L. Kaplan and P. G. Kelleher, *Rubb. Chem. Technol.*, **45**, 423 (1972).
12. T. Mill, K. C. Irwin, and F. R. Mayo, *Rubb. Chem. Technol.*, **41**, 296 (1968).
13. M. L. Kaplan and P. G. Kelleher, *J. Polym. Sci. A1*, **8**, 3163 (1970).
14. G. S. Egerton and A. G. Morgan, *J. Soc. Dyers Colourists*, **1971**, 268.

15. J. Griffiths and C. Hawkins, *J. Soc. Dyers Colourists*, **1973**, 173.
16. J. F. Rabek and B. Rånby, *J. Polym. Sci. A1*, **12**, 273 (1974).
17. J. C. W. Chien and H. Jabloner, *J. Polym. Sci. A1*, **6**, 393 (1968).
18. R. W. Murray and M. L. Kaplan, *J. Amer. Chem. Soc.*, **90**, 537 (1968).
19. J. A. Howard and K. U. Ingold, *J. Amer. Chem. Soc.*, **90**, 1056 (1968).
20. C. S. Foote, R. W. Denny, L. Weaver, Y. Chang, and J. Peters, *Ann. N.Y. Acad. Sci.*, **171**, 139 (1970).
21. J. P. Dalle, R. Magous, and M. Masseron-Canet, *Photochem. Photobiol.*, **15**, 411 (1972).
22. D. J. Carlsson, G. D. Mendenhall, T. Suprunchuk, and D. M. Wiles, *J. Amer. Chem. Soc.*, **9**, 8960 (1972).
23. J. P. Guillory and C. F. Cook, *J. Polym. Sci. A1*. **11**, 1927 (1973).
24. B. Felder and R. Schumaker, *Angew. Makromol. Chem.*, **31**, 35 (1973).
25. J. Flood, K. E. Russell, and J. K. S. Wan, *Macromolecules*, **6**, 669 (1973).
26. A. Farmilo and F. Wilkinson, *Photochem. Photobiol.*, **18**, 447 (1973).
27. D. J. Carlsson, T. Suprunchuk, and D. M. Wiles, *Can. J. Chem.*, **52**, 3728 (1974).
28. R. H. Young, R. L. Martin, D. Feriozi, D. Brewer, and R. Kayser, *Photochem. Photobiol.*, **17**, 233 (1973).
29. D. J. Carlsson and D. M. Wiles, *J. Macromol. Sci. Rev. Macromol. Chem.*, **C14**, 65 (1976).
30. J. C. W. Chien and C. R. Boss, *J. Polym. Sci. A1*, **10**, 1579 (1972).
31. D. J. Carlsson and D. M. Wiles, *J. Polym. Sci., Polym. Chem. Ed.*, **12**, 2217 (1974).
32. J. H. Chaudet and J. W. Tamblyn, *SPE Trans.*, **1**, 57 (April 1961).
33. O. Cicchetti, *Adv. Polym. Sci.*, **7**, 70 (1970).

Index

Acetoguaiacone sensitizer, 312
Acetophenone, 53
N-Acetylglucosamine, 197
Acridine orange, 195
Acridine yellow, 195
Acriflavine, 195
Adamantylideneadamantane, 121
Allylic hydroperoxide, 164
Amino acids, 86
Anthracene, 31, 51, 55, 116, 145, 148, 150, 244, 266, 272
Anthracene endoperoxide, 119
Anthracene-9,10-epiperoxide, 150
Anthraquinone, 151
Anthraquinone dyes, 221
Anthrone, 153
Antioxidants, 91
Apocynol, 307
Autocatalytic stage, 212
Autooxidation, 211
Azide anion, 130
Azines, 85
Azomethine dyes, 85

Bayer–Villiger reaction, 183, 193
Benzophenone, 136
Bilirubin, 42, 45, 56, 59, 88
Biliverdin, 59, 88
7,7′-Binorbolidene, 121
Boltzmann distribution, 21
Bredt rule, 129
cis and *trans*-2-Butene, 114
tert-Butyl hydroperoxide, 232

Carbonyl compounds in photodegradation of polymers, 231
Δ^3-Carene, 114
β-Carotene, 28, 31, 32, 48, 49, 50, 53, 56, 70, 73, 83, 89, 114, 115, 118, 144, 154, 191, 217, 304, 307, 313, 322 323
Chargè-transfer interactions (CT complex), 25, 44, 48, 79, 94, 103, 129, 141, 246, 255
Chemiluminescence, 294
Chloranil, 246
2-Chloro-3,3-dimethyl-1-butene, 159
Chlorophyll, 1, 70, 136, 234
CIDNP, 180
Cinnamaldehyde, 304
Cinnamyl alcohol, 304
Clebsch–Gordon coefficients, 20
CNDO method, 225
Collision complex, 255
Complex states, 58
Cooperative transitions, 8
Corona discharge, 12
p-Cresol, 309
Criegee-type cleavage, 162
Cyclohexane, 234
N-Cyclohexyl-*N*″-phenyl-*p*-phenylenediamine, 286

DABCO (1,4-diazabicyclo-2,2,2-octane), 43, 78, 81, 82, 83, 89, 129, 140, 143
Degree of degradation, 248
Diamyl ketone, 232
Di-^{D}Anthracene, 154
trans-2,3-Dichloro-2-butene, 159
trans-1,2-Dichloroethylene, 159
Dichotomy, 142, 177
Dicyanoanthracene, 144
Dicyclopentadiene, 284
Diels–Alder adducts, 155, 183
N,*N*-Diethylfurfurylamine, 81
Dimethylaniline, 137
9,10-Dimethylanthracene, 59, 204
9,10-Dimethyl-1,2-benzanthracene, 57
2,3-Dimethyl-2-butene, 93, 114
Dimethylcyclohexane, 234
4,8-Dimethyl-4,8-dodecadiene, 164
1,1-Dimethylethylene, 232
2,5-Dimethylfuran, 114, 204
Dimethyl ketene, 175
Dionyl ketone, 235
p-Dioxene, 142
1,2-Dioxetane, 9, 121, 125, 182
1,3-Diphenylisobenzofuran, 29, 38, 58, 79, 103, 204, 316
9,10-Disubstitued anthracene, 147, 148

Dodecyltrimethylammonium chloride, 203
Dye-sensitized photo-oxygenation, 28

EHT method, 225
π-Electron selfconsistent theory, 225
Electron transfer, 59
Endoperoxides, 9, 120, 275
Energy fission, 57
Energy pooling, 8
Eosine, 136, 207, 221
Ergosterol, 116
Ethylidenenorbornene, 220, 282

Flavins, 136
Fluorescein, 267, 272
Fluorescence quenching, 44
Franck–Condon rules, 48, 53, 245
Free electron theory, 225
Freon-113, 8, 73, 103
Fulvic acids, 294

Guanosine, 139
Guanosine-5-monophosphate, 139

Haldane soup, 4
Hammett relationship, 44, 79, 124
Hematoporphyrin, 136
Hexa-1,4-diene, 284
Histidine, 86
HMO method, 225
Hock cleavage, 126
Humic acids, 294
Hunds' rule, 27
Hydroperoxides, 211, 216, 275, 280
Hydroxymethoxybenzoquinone, 308

Indoles, 186
INDO-method, 225

Kearns' correlation diagram, 30, 178
Ketenes, 174
α-Ketocarboxylic acids, 192

LCAO-MO configuration of oxygen, 6
Leucomalachite green, 103
Limonene, 114
Long-range interactions, 19
Lysozyme, 195

Malachite green, 1
Metal chelates, 33, 316
Methionine, 86, 90
Methoxybenzoquinone, 308
2-Methyl-2-butene, 114, 124
Methylcyclohexane, 234
Methylene blue, 29, 86, 136, 195, 204, 267, 272, 291
N-Methylindole, 49, 50, 52, 187
4-Methyl-4-octene, 164
2-Methyl-2-pentene, 124, 125
3-Methyl-2-pentene, 124
Micellar phase, 203
MINDO-method, 9, 225
Moloxide mechanism, 5

Naphthacene, 50, 55
Nickel chelates, 33, 237, 322
Nicotine, 78, 127
Nitrones, 83
Nitroxide, 82, 83
Norrish type I and type II reactions, 232
NO to NO_2 conversion, 2

Oct-1-ene, 234
Oxygen
  active form, 2
  diffusion rate constants in solvents, 37
  discovery, 1
  ground triplet state, 5
  higher energy states, 1
  potential energy curves, 7
Oxygenation reaction type I and type II, 136
Ozonolysis, 162

Pauli's principle, 27
PCILO-method, 225
Perepoxide, 182
Peroxides, trans-annular, 147
Peroxylactones, 174
Phenols, 90, 143
Phenoxy radicals, 305
2-Phenylnorbornene, 183
Phlogiston, 1
Photochemical smog, 2
Photodynamic inactivation, 195
Photo-oxidation of polyethylene, 231
Photo-peroxidation parameters, 57
α- and β-Pinene, 114
Piperidines, 82
Polyenes, 218
Polyisoprene, model compounds, 164
Polymer-based sensitizers, 312
Polymeric peroxide, 174
Polymer matrix, 213
Polymer peroxy radicals, 212
Polymethene pyrylium dye, 103
Poly(propylene sulphide), 288

Porphyrine, 1
Potential energy curves, 10
Propenylnorbornene, 284
Proteins, 86

Quadrupole–quadrupole interactions, 22
Quenchers
  effectiveness of, 39
  ionizing energy of, 18

Rhodamine, 289, 293
Rose bengale, 57, 58, 67, 85, 93, 139, 165, 186, 189, 220, 234, 267, 272

Scavenging, 324
Short-range interactions, 19
Singlet oxygen
  dimol emission, 8
  electronic configuration, 62, 135, 225
  energy pooling reactions, 46
  generation of, 12, 36, 111, 203
  kinetic of peroxidation, 54, 66, 122
  life time of, 8, 12, 13, 28, 33, 40, 62, 111, 112, 139, 205
  oxidation mechanism, 135, 159, 214
    1,2-addition, 116, 119
    1,4-addition, 116
    cycloaddition, 9, 117, 119, 227
    ene reaction, 9, 116, 121, 126
  potential energy curves, 7
  quenching of
    charge transfer quenching, 42
    mechanism of quenching, 37, 38, 39, 42, 45, 46, 214
    physical deactivation, 17, 24, 36, 56, 57, 67, 69, 80, 200, 215, 332
    quantum yield of formation of, 56
    quencher definition, 69
    quenching by
      amines, 42, 45, 56, 73, 78, 127
      azomethine dyes, 42
      bilirubin, 42, 45
      biliverdin, 57
      inorganic ions, 101, 130
      leucomalachite green, 59
      metal chelates, 33, 94, 237, 322
      ozone, 17
      phenols, 90, 214
      solvents, 41, 42, 66
      sulphides, 42, 45, 89, 127
      sulphoxides, 127
      tocopherol, 45, 56, 91
    rate constant of quenching, 30, 34, 38, 128
  reactions with polymers, 2, 211, 226
    humic acids, 294
    lignin, 302, 311
    polydienes, 235, 264
      polybutadiene, 172, 218
      polyisoprene, 282
      rubber EPDM, 282
    polyethylene, 215
    polystyrene, 216, 244, 254
    polysulphides, 288
    poly(vinyl chloride), 218
  stabilization against, 320
  β-value (half-value acceptor concentration), 28, 55, 56, 113, 122, 166, 220
Solvent-induced electronic relaxation, 55
Spin conversion requirements, 54
Spin-flip, 25
Spin forbidden process, 52
Spin–spin interaction, 25
Stabilizers for polyolefins, 316
State correlation diagrams, 9, 10
Stern–Volmer method, 37, 39, 102, 138, 144
Stilbene, 53
Sulphides, 144
Superoxide ion (O_2^-.), 136, 300

α-Terpinene, 116, 130
Tetracene, 57
1,1,4,4-Tetraphenylbutadiene, 220
Tetraphenylporphine, 159
Tocopherol, 56, 91
Trinitrobenzene, 246
Triplex complex, 179
Trypoflavin, 1
Tryptophan, 86, 189

UV-stabilizers, 230
UV-stabilizing mechanisms, 237

Vinyl halide structures, 159
Vitamin E, 56, 91

Weller function, 44
Woodward–Hoffman rules, 9, 142

Xanthone, 53

Young's technique, 140

Zero-point energies, 27
Zwitterion, 129